Canadian Chemical News /
L'Actualité Chimique canadienne

ADVANCES IN CATALYSIS

VOLUME 34

Advisory Board

ADVANCES IN CATALYSIS

VOLUME 34

Edited by

D. D. ELEY
The University
Nottingham, England

HERMAN PINES
Northwestern University
Evanston, Illinois

PAUL B. WEISZ
University of Pennsylvania
Philadelphia, Pennsylvania

1986

ACADEMIC PRESS, INC.
Harcourt Brace Jovanovich, Publishers
Orlando   San Diego   New York   Austin
Boston   London   Sydney   Tokyo   Toronto

ACADEMIC PRESS, INC.
Orlando, Florida 32887

United Kingdom Edition published by
ACADEMIC PRESS INC. (LONDON) LTD.
24–28 Oval Road, London NW1 7DX

LIBRARY OF CONGRESS CATALOG CARD NUMBER: 49-7755

ISBN 0–12–007834–1

PRINTED IN THE UNITED STATES OF AMERICA

86 87 88 89   9 8 7 6 5 4 3 2 1

Contents

Spillover of Sorbed Species

W. CURTIS CONNER, JR., G. M. PAJONK, AND S. J. TEICHNER

Mechanistic Aspects of Transition-Metal-Catalyzed Alcohol Carbonylations

THOMAS W. DEKLEVA AND DENIS FORSTER

Quantum-Chemical Cluster Models of Acid–Base Sites of Oxide Catalysts

G. M. ZHIDOMIROV AND V. B. KAZANSKY

Near-Edge X-Ray Absorption Spectroscopy in Catalysis

Jan C. J. Bart

Preface

There is something notable about this volume. In four chapters, it demonstrates the diversity of science that constitutes the field of catalysis as well as the cohesiveness of its parts. It covers *technique* of materials analysis, quantum chemical *theory,* a branch of catalytic *practice,* and the discussion of a major *phenomenon* of solids.

Each chapter is a comprehensive overview, but the emphasis is on the current horizon of knowledge. Each should serve not only the researcher, teacher, or practitioner of catalytic science alone, but those in chemistry and materials science with spillovers to solid-state physics and other "disciplinary" practices that deal with the transformation of solids or the fate of molecules that contact solids.

The state and capabilities of quantum chemical modeling of silicon dioxide framework structures (silica, aluminosilicates, zeolites) are discussed by Zhidomirov and Kazansky. Here are basic theoretical approaches to a class of catalytic compositions which contribute the probably most massive quantities of catalysts used by man's (petroleum and chemical) industries.

The action of transition metals in the carbonylations of alcohols is treated by Dekleva and Forster. This is a field of important chemistry, from the point of view of coordination chemistry as well as catalysis. It is also another example of cross-stimulation between basic science and technological utility.

Conner, Pajonk, and Teichner discuss the once mysterious phenomenon of molecular "spillover," or migration of hydrogen and other molecular species across boundaries of heterogeneous compositional regions of solids. Its implications extend to many aspects of solid-state science and practices.

The pursuit of catalytic science has itself helped to catalyze something, namely the development of sophisticated physical tools for probing the physical–chemical state of solid constituents at the molecular and atomic scale. Bart discusses near-edge X-ray absorption spectroscopy, still a rapidly moving activity. It also is an interesting example of an important symbiosis between seemingly remote scientific endeavors: between "big machine" and high-energy particle physics and the science of chemical catalysis and solid-state structures.

PAUL B. WEISZ

Georg-Maria Schwab, 1899–1984

Georg-Maria Schwab died on 23 December 1984, just 5 months after he had presided so successfully over the 8th International Congress on Catalysis in Berlin. Schwab had originally studied chemistry in that city, graduating Ph.D. in 1923 with a thesis directed by E. H. Riesenfeld on the Thermodynamic Properties of Ozone. As part of that work Schwab dissolved ozone in liquid hydrogen, isolating a blue–black complex, which must say something about the imperturbable nature of his character. His first catalyst work, with E. Pietsch on the decomposition of methane, led to his Habilitation thesis, taken at Würzburg on the catalytic decomposition of ammonia. In 1928/9, now in Munich, he collaborated again with Pietsch in working out the kinetic consequences of reaction along the line between two adjacent surface phases, the "Adlineation" theory. In 1929 he published with E. Cremer on the Compensation Effect, which he called the theta rule, theta referring to the catalyst preparation temperature assumed to fix the thermodynamic distribution of active centers. In 1931 appeared his comprehensive textbook, "Katalyse vom Standpunkt der Chemischen Kinetik," which appeared in English translation in 1937. In 1938 he published, with Elly Agallidis (Mrs. Schwab), his first paper on the parahydrogen conversion by organic radicals, which he continued on migrating to Greece in 1939. Here he was Head of the Inorganic, Physical and Catalytic Chemistry Department at the Nikolaos Kanellopoulos Institute in the Piraeus, at which he carried out his important studies of the formic acid dehydrogenation on a series of metal alloys. This he reported to the Faraday Society Discussion at Liverpool in 1950, a pioneer effort in which he successfully correlated activation energy changes on alloying with the Hume–Rothery electron concentration parameter. In 1950 he became Director of the Institute of Physical Chemistry at Munich University, and there followed a long series of investigations inspired by the ideas of electron transfer at catalyst surfaces, first on semiconducting oxides, and later on supported catalysts, both metals on oxides, and oxides on metals. In his very latest papers (1983/84) Schwab returned to the Compensation Effect, seeking an explanation in terms of an activation energy distributed among a number of square terms. Although catalysis was his main theme, Schwab also contributed to solid-state reactions, chromatography, infrared and Raman spectroscopy, and was much interested in the history of chemistry. Schwab was active on a number of committees and served as President of the Research Council of the Fraunhofer Foundation for Applied Science.

Among the honors received by Schwab were membership of the Bavarian, "Leopoldina," Heidelberg, and Vienna Academies, and an Honorary Professorship of the University of Caracas. He received Honorary Doctorates from the Universities of Berlin, Paris, Liège, and Hamburg, and the Liebig Medal of the German Chemical Society. He was an Honorary Member of the German Bunsen Society for Physical Chemistry. He received the St. George's Order of Greece, and was an officer of the Order of the Belgian Crown.

Schwab was a man of equable and humorous temperament, a rock-climber in earlier days and latter a skier, with a mountaineer's preference for deep snow over the piste. Professor and Mrs. Schwab had a party with old students each semester in the convivial surroundings of a famous Munich beer-cellar and the laboratory staff and students made an annual expedition to the mountains for skiing. During the time my wife and I were in Munich in 1963, Professor and Mrs. Schwab went to great lengths to make our stay there as enjoyable as possible, and I have no doubt other visiting Professors found the same hospitable reception. Professor Schwab is survived by his wife and by a son and two daughters who are married with families.

DANIEL D. ELEY

Georgii Konstantinovich Boreskov, 1907–1984

Catalysis science has suffered a severe bereavement—on August 12, 1984, a leading scientist in the field of catalysis, an eminent chemical engineer, an outstanding organizer of science and industry, the Director of the Institute of Catalysis of the Siberian Branch of the USSR Academy of Sciences, a great teacher, and a remarkable man, academician Georgii Konstantinovich Boreskov passed away.

G. K. Boreskov was born in Omsk on April 20, 1907. On graduating from the Odessa Chemical Institute in 1929 he became head of an industrial laboratory of catalysis. It was under his leadership that the first home catalysts for SO_2 oxidation and methods of calculating commercial reactors were developed. Later he summarized his experience in the book "Catalysis for Sulfuric Acid Production," which was published in 1954 and retains its value to the present day.

Boreskov emphasized the necessity of taking into account the effect of the reaction medium on the catalyst, particularly when describing kinetics of heterogeneous catalytic reactions. This problem was also the subject of his last report at the 8th International Congress on Catalysis in West Berlin shortly before his death.

People who knew Professor Boreskov at the beginning of his scientific career recall him as a man with an inquisitive mind, encyclopedic knowledge, diligence, and energy. He was also a gifted engineer and organizer. Georgii Konstantinovich took an active part in all steps of research from laboratory experiments and catalyst investigations to projection of large-scale reactors and optimization of processes.

Boreskov supported the idea of a chemical approach to catalysis, according to which the mechanism of catalytic action consists of an intermediate chemical interaction of the catalyst with a reactant.

The theoretical activity of Boreskov is intimately associated with all states in the formation and development of the modern science of catalysis. He treated the catalytic process by considering all related phenomena. It is from just such a concept that a new trend, mathematical modeling of catalytic (and chemical, in general) processes, arose. Now, with mathematical modeling of chemical reactions as the basis, a special branch of science, concerned with the theoretical basis of chemical technology, has been formed.

As a technologist, Boreskov showed a keen interest in new possibilities of catalysis, i.e., the expansion of the application of catalytic methods both in technology and energetics, as well as for new nontrivial solutions of chemical engineering problems.

Boreskov had a sharp sense of the new and always supported the appearance of novel trends in catalysis. He directed experiments on the application of catalysis for fuel combustion and participated in the development of new methods to carry out catalytic processes—performance of reactions in unsteady state conditions (a promising way of performing exothermal reactions with low adiabatic heat).

The creation of the Institute of Catalysis was Boreskov's life work. He was the director of the Institute from the time of its foundation in 1958. He formulated the main research problems of the Institute: the development of scientific bases for foreseeing catalytic action and scientific bases of catalyst preparation and mathematical modeling, combined with solutions of applied problems essential for modern industry. In a fairly short period of time the Institute has gained world-wide recognition. In 1980 it received an International "Gold Mercury" prize.

Boreskov was a tireless supporter of international cooperation in catalysis. He contributed to the Organization of International Congress on Catalysis by serving as president from 1972 to 1976 and on his initiative, regular bilateral seminars on catalysis—Soviet–Japanese (1971) and Soviet–French (1972)—were organized. Boreskov was the chief coordinator in the USSR in the Soviet–American Scientific cooperation on the problem of "Chemical Catalysis" in 1973–1980. He was a member of the editorial board of a number of authoritative journals, both foreign and Soviet.

The scientific activity of Boreskov has been highly valued by the Soviet and foreign scientific communities. He was a full member of the USSR Academy of Sciences and was elected an Honored Member of the New York Academy of Sciences, Foreign Member of the Academy of Sciences of the German Democratic Republic, and Honored Doctor of Poitiers University (France) and of Wroclaw Polytechnical Institute (Poland). The Government of Bulgaria decorated Boreskov with a first degree Kirill and Mefodii Order. The Soviet Government gave him the highest award of the USSR—the Hero of Socialist Labor.

Fond memories of Professor G. K. Boreskov will always remain in the hearts of those who knew this great scientist and this great man.

K. I. Zamaraev
Yu. I. Yermakov

ADVANCES IN CATALYSIS, VOLUME 34

Spillover of Sorbed Species

W. CURTIS CONNER, JR.

Chemical Engineering Department
Goessmann Laboratory
University of Massachusetts
Amherst, Massachusetts 01003

AND

G. M. PAJONK AND S. J. TEICHNER

Université Claude Bernard LYON 1
Laboratoire de Catalyse Appliquée et Cinétique Hétérogène (CNRS)
69622 Villeurbanne, France

I. Introduction

Solid catalyst surfaces are by their very nature heterogeneous. Steps, kinks, edges, and vacancies on supported metals, mixed oxides, and sulfides give rise to catalytically "active sites." This concept of active sites was used by Langmuir and Hinshelwood to express the kinetics of heterogeneous reactions. For supported metals in particular, the support was initially perceived only to disperse the active metal, increasing and preserving its effective surface area. More recently, it has been accepted that the support can influence the activity of the metal [e.g., strong metal–support interaction (SMSI)].

Langmuir–Hinshelwood kinetics assumes that the number of active sites initially present or induced by the reactants is fixed and that the reacting species adsorb and interact at the isolated sites. No mobility from a site across the surface is taken into account.

Limited mobility of sorbed species has long been understood. The exchange of species from one position to another, either on the surface or through the bulk, has been well established (1). More unique is the mobilization of a sorbed species from one phase onto another phase where it does not directly adsorb. This has been defined as "spillover."

The phenomenon of spillover was first noticed in the 1950s (2) when it was observed by Kuriacose that the decomposition of GeH_4 on a Ge film was

1

increased by contact with a Pt wire used to measure the conductivity. Taylor proposed that the wire provided a "porthole" for the recombination of H to H_2 (3) ("reverse spillover"). The first direct experimental evidence for spillover was presented by Khoobiar in 1964, who documented that the formation of tungsten bronzes (H_xWO_3) was possible at room temperature for a mechanical mixture of WO_3 with Pt/Al_2O_3 (4). Sinfelt and Lucchesi postulated that reaction intermediates (presumably H) had migrated from Pt/SiO_2 onto Al_2O_3 during ethylene hydrogenation (5). During the ensuing score of years over 400 papers involving spillover have been published. The First International Symposium on the Spillover of Adsorbed Species (FISSAS) was held in 1983 in Lyons (6).

Reviews of the publications have appeared in 1973 (7), 1980 (8), and recently from a historical perspective in 1983 (9). The focus of this article is to present our current understanding of the phenomenon and articulate the unknown aspects of spillover.

First, it is necessary to define spillover. During the recent international symposium a definition was proposed and affirmed by the congress (10).

Spillover involves the transport of an active species sorbed or formed on a first phase onto another phase that does not under the same condition sorb or form the species.
And offered as a comment: The result may be the reaction of this species on the second phase with other sorbing gases and/or reaction with, and/or activation of the second phase.

Mechanistically this can be generally visualized as a sequence of steps, viz:

$$S_{fluid} + A \rightleftharpoons SA \qquad \text{sorption} \qquad (1)$$

$$SA \rightleftharpoons S_aA \qquad \text{activation of sorbing species} \qquad (2)$$

$$S_aA + A' \rightleftharpoons A + S_aA' \qquad \text{exchange} \qquad (3)$$

$$S_aA' + \theta \rightleftharpoons A' + S_{sp}\theta \qquad \text{spillover to second phase} \qquad (4)$$

$$S_{sp}\theta + \theta' \rightleftharpoons \theta + S_{sp}\theta' \qquad \text{exchange or surface diffusion} \qquad (5)$$

$$S_{sp}\theta \rightleftharpoons S_{sp}\theta^* \qquad \text{creation of active surface}$$

$$\text{site on second phase} \qquad (6)$$

$$S_{sp}\theta^* \rightleftharpoons {}^* + P_{fluid} \qquad \text{desorption} \qquad (7)$$

where S_{fluid} is the species sorbing from fluid (gas or liquid), A, A' are the sites able to adsorb S, S_a is the activated sorbed species, S_{sp} is the spiltover species (usually equivalent to S_a), θ, θ' are the phases unable to sorb S directly, $S_{sp}\theta^*$, * are the possible active sites or new phases resulting from θ, and P_{fluid} is the possible product of the creation of the new active site or phase.

These seven steps initially simplify the discussion of the phenomena related to spillover. As the most common example, hydrogen is adsorbed from the gas phase onto a metal (Pt, Pd, Ni, etc.), where it dissociates to atomic hydrogen ($S_aA = HPt$, etc). Spillover occurs if the atomic hydrogen is then able to be transported onto the support (PtH $+ \theta \to$ Pt $+ H_{sp}\theta$, equivalent to Step 4).

We hope objectively to access what is known and what is not known about spillover. Our specific focus is on the phenomena and the catalytic consequences.

Spillover may result in a spectrum of changes in the nonmetallic phase onto which it occurs. In the weakest sense, the spiltover species (e.g., H_{sp}, for spiltover hydrogen) is transported across the surface of this phase as a two-dimensional gas. It may exchange with similar surface species (e.g., OH $+ D_{sp} \leftrightarrow$ OD $+ H_{sp}$). The spiltover species may react with the surface (e.g., $H_{sp} +$ M—O—M' $\to$ M—O—H $+$ M'), which can result in the creation of surface defects and/or active sites. Further, the bulk of the solid may be transformed into a different structure (as in the transformation of oxides to hydrogen–bronzes). In each of these cases, the second phase is no longer an inert. It is not serving to promote the inherent activity on the first phase. The second phase is participating directly in the transport, exchange, and reaction with the spiltover species. In some cases it is able to become catalytically active on its own and thereby to participate directly in subsequent catalysis.

The remainder of this article will be divided into five sections. First, we will discuss the types of phenomena associated with spillover. This will include enhanced adsorption, exchange with the surface, and the bulk transformation of the second phase. In the second section we will discuss the mechanism of spillover, the nature of the surface and spiltover species, and the kinetic and mechanistic implications. In the third section we will consider the spillover of species other than hydrogen. In the fourth section we will discuss the aspect of spillover with the most significant catalytic implications: surface activation of the second phase (e.g., the support of the metal) and the resulting catalysis. In the final section we will bring together the general conclusions of the earlier sections and discuss the implication of spillover to catalysis and other heterogeneous processes.

We recognize that the phenomenon of spillover is still controversial, and yet the potential implications are far reaching. Therefore, we will take this opportunity to advance some hypotheses concerning the phenomena associated with spillover and their consequences.

II. Phenomena Associated with Spillover

For the sake of clarity the most important consequences of spillover are summarized and discussed in the following four subsections. First, the enhanced adsorption by mono- and bimetallic supported catalysts is described. This deals mainly with abnormal hydrogen adsorption quantities and rates. The second subsection describes isotopic exchange with the surface. The third subsection relates bulk changes in materials exposed to spillover, and the last subsection reports on strong metal–support interaction (SMSI) and its relation to spillover. These last two phenomena occur with less refractory and more easily reducible inorganic oxides.

The very important problems of catalytic activity and reaction mechanism involving spillover will be discussed in Sections III and V.

A. ENHANCED ADSORPTION

Literature data are mainly centered around hydrogen adsorption on group VIII, VI, and I metals (mono- or bimetallic systems) supported on refractory inorganic oxides such as Al_2O_3, SiO_2, zeolites, or carbons. The studies involving SMSI have focused on TiO_2 as a support.

Enhanced adsorption of H_2 is generally reported to occur at atmospheric pressure (or lower pressures of H_2) at temperatures about 200°C. However, many cases are described for adsorption at ambient temperatures (and below).

Although this article is devoted to recent experimental and theoretical developments in the ever-growing field of research on spillover, it is of interest to recall here, very briefly, some important results published earlier in the literature (7, 8). These results are expressed as ratios H/M_t, where H stands for the number of H atomic species adsorbed per total number of metallic atoms, M_t, in supported catalysts, and are shown in Table I according to the review published by Sermon and Bond in 1973 (7). Figure 1 represents the three combination cases where spillover may occur; it distinguishes the primary spillover effect (case A) from the secondary spillover effects (cases B and C) (7, 11). In general, the ratio H/M_s, where M_s is the total number of *surface* metal atoms of the catalyst, rarely exceeds 2 (12) for the most commonly encountered metals. Large values of H/M_t indicated enhanced adsorption attributed to a spillover of adsorbed hydrogen according to one (or more than one) of the schemes depicted in Fig. 1. Values of H/Pt_t as high as 64 and 6.8 for Pt/Al_2O_3 and Pt/SiO_2 catalysts, respectively, have been quoted by Altham and Webb (13), while H/Pt_t ratios between 3 and 277 have recently been calculated by Sermon and Bond (11) for reverse spillover involving the two platinized bronzes H_xWO_3 and H_xMoO_3.

Now, turning to the kinetics of spillover-enhanced adsorption, it may be induced that if spillover occurs, the rate of adsorption increases. For instance, Fujimoto and Toyoshi (14) found increased rates of H_2 adsorption with active carbon as a support for cobalt (case A spillover) or added as a diluent to this metal (cases B and C). The extent of the increases depended on the

TABLE I

Maximum Values of n_H/n_M Measured for Some Supported Metals by Hydrogen Adsorption[a]

Metal	Support	Maximum[b] n_H/n_M	Temperature of adsorption (°C)	Hydrogen pressure (cm Hg)	Ref.[c]
Pd	SiO_2	$(4.0)^d$	−196	40	47
	Al_2O_3	5.0			49[e]
	Al_2O_3	3.2	25		50
Pt	SiO_2	1.35	20	0.1	5
	SiO_2	$(1.5)^d$	−196		33
	SiO_2	$(1.6)^d$	−196	10^{-3}–10^{-6}	18
	Al_2O_3	2.44	250	12	34
	Al_2O_3	2.19	250	12	24
	Al_2O_3	2	300		31
	Al_2O_3	1.5	200	0.9	37
	Al_2O_3	>1	200	0.9	39
	Al_2O_3	10.0			49[e]
	C	75	250	60	23
	Zeolite	1.4	21		10
	Zeolite	2	100		25
	Zeolite	2	200	5–20	42
		1.92	250	<25	14
Rh	Al_2O_3	1.39	25	0^f	54

[a] After Sermon and Bond (7).

[b] n_M is the total number of metal atoms; n_H is the number of chemisorbed H atoms.

[c] These references belong to the original paper (see Ref. 7 of the present chapter).

[d] The validity of characterization of supported metals by hydrogen adsorption at −196°C is questionable.

[e] Values obtained by D. V. Sokolskii and E. I. Gil'debrand, reviewed by Popova (49).

[f] Extrapolated to zero pressure.

contact between the metallic component and active carbon, in this case Co/C (noted CC). This is shown in Fig. 2, where TM stands for "treated mixture" [active carbon (AC) mixed with CC and bound with polyvinyl alcohol]. SM is the simple physical mixture of AC with CC. Moreover, Fujimoto and Toyoshi found that the amount of adsorption does not vary with the source of spiltover hydrogen (Ni, Cu, Fe), with the metal loading, or with use of pure or sulfided metals. The amount of hydrogen uptake at 400°C at equilibrium is the same. Similar H_2 adsorption is measured on pure AC. Thus, its extent

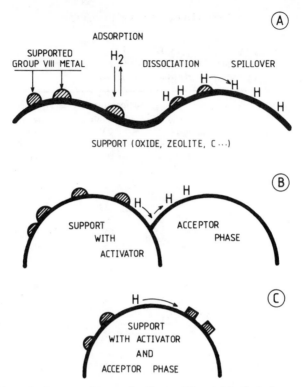

FIG. 1. Schematic diagrams of types of spillover: (A) metal (activator) on a support; (B) metal (activator) on a support with an admixed support (acceptor); (C) metal (activator) on a support with a reactant (acceptor).

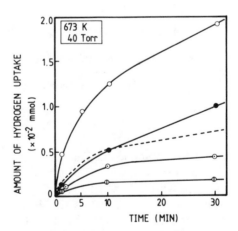

FIG. 2. Hydrogen uptake on carbon catalysts: ○, TM 0.4 g = AC 0.36 g + CC 0.04 g; ●, SM 0.4 g = AC 0.36 g + CC 0.04 g; ⊙, CC 0.04 g; ⊕, AC 0.36 g; ---, ⊙ + ⊕ (14). Reprinted with permission from North-Holland Publishing Company, Amsterdam.

depends only on the number of accepting sites on the carbon (14). Similar results were reported for 10% palladized carbon (15) and Cs, Mo, Ni, and Cu (pure or presulfided) on carbon catalysts (16).

For metal-on-carbon systems two aspects should be noted: (i) metals and sulfided metals can act as sources of spiltover hydrogen species, and (ii) as active carbon adsorbs hydrogen by itself at high temperatures [$T >$ 300°C (15)], the effect of spillover is to increase the rate of uptake but not the extent. With the inorganic oxides, spillover most often results in an increase in the amount of adsorption.

The kinetics of the chemisorption of spiltover hydrogen species have been fully detailed from a mathematical and physical point of view for carbon as well as for alumina by Robell et al. (17) and Kramer and Andre (18), respectively. Both groups of authors concluded that surface diffusion is the rate-determining step in the overall process of hydrogen spillover. This is shown in Fig. 3 for platinized carbon and Fig. 4 for platinized alumina.

In a more general way, Berzins et al. (19), in their recent paper discussing isothermal chemisorption upon oxide-supported platinum, attracted attention by carefully demonstrating that in most cases the gradient dg_a/dp (g_a is the amount of adsorbed gas, H_2, O_2, or CO; p is their corresponding partial pressure) never falls to zero with increasing p. They hypothesized that the inclined plateau (where $dg_a/dp \neq 0$) may be due to spillover. Indeed, such a case has been clearly reported by Anderson et al. (20) in their study of H_2 chemisorption on several Pt–Au-on-Aerosil catalysts at 20°C, as shown in Fig. 5 for Pt 67–Au 33 mole % and Pt 15–Au 85 mole % samples.

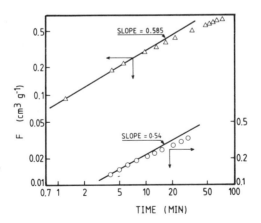

FIG. 3. Net adsorption of H_2 at 300°C on 0.2% Pt sample: $\triangle$, $P = 600$ Torr; $\bigcirc$, $P = 300$ Torr (17). Reprinted with permission from *Journal of Physical Chemistry*. Copyright (1964) American Chemical Society.

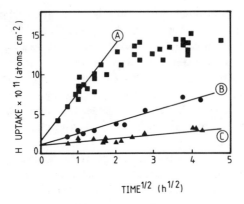

FIG. 4. Amount of spiltover hydrogen versus $t^{1/2}$ at 710 Torr: (A) 400°C; (B) 320°C; (C) 250°C (18).

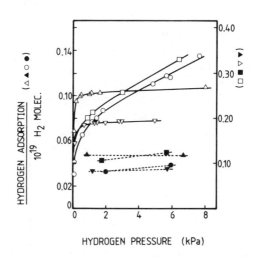

FIG. 5. Hydrogen adsorption isotherms at 293 K with platinum–gold/Aerosil catalysts: ▽, Pt 98, Au 2 mol %, 1.0 wt % metal; 0.516 g catalyst; △, Pt 90, Au 10 mol %, 0.9 wt % metal, 0.510 g catalyst; □, Pt 67, Au 33 mol %, 0.9 wt % metal, 0.500 g catalyst; ○, Pt 15, Au 85 mol %, 1.0 wt % metal, 0.450 g catalyst; "standard" pretreatment (cf. text). Filled symbols, amount of adsorbed hydrogen remaining after pumping at 293 K for 30 min, after equilibration at indicated pressure. Catalyst samples identified from corresponding symbols above. Within the limits of experimental accuracy, no adsorption could be detected on a Pt 0, Au 100 mol %, 1.0 wt % catalyst, using a 0.500 g sample (20).

Another typical hydrogen isotherm on Pt/Al$_2$O$_3$ [published by Ostermaier et al. (21)] exhibited such a plateau, the gradient of which was not zero. With their conditions [P(H$_2$) = 760 Torr, T = 25°C], they found a H/Pt$_s$ ratio as high as 5.6. They reported a very slow uptake of hydrogen for all their experiments, which agrees with the other studies of sorption rates.

Supported Rh also has attracted the attention of many authors. According to the results of Gajardo et al. (22), Rh/CeO$_2$ showed an irreversible chemisorbed hydrogen-to-Rh-atom ratio equal to 3 after hydrogen adsorption between 200 and 400°C. Two papers by Apple et al. (23, 24) depicted the spillover of H$_2$ on Rh/TiO$_2$ samples using NMR and TPD methods. The authors came to the conclusion that hydrogen spillover gives rise to an extra upfield NMR resonance line due to H species interacting with paramagnetic Ti^{3+} centers. They also demonstrated that this extra adsorption was suppressed by CO preadsorption (24) as well as by coadsorption. A recent investigation of hydrogen adsorption by Jiang et al. (25) with Ni, NiFe, or Pt on the same support, TiO$_2$, indicated a very slow equilibrium of hydrogen during adsorption–desorption. This slow uptake was explained as being due to hydrogen spillover from the metal to either Ti^{3+} cations (TiO$_x$) or to OH groups on TiO$_2$.

According to volumetric, IR, and ESCA results published by Van Meerbeck et al. (26), even traces of metals, on the order of tens of parts per million when contaminating pure silica gel, are able to provoke a very large and slow uptake of hydrogen at high temperature (1000°C). The efficiencies of the metals involved were in the sequence Ta > Pt > Ni > W. The IR measurements showed that a Si-H species was formed from H$_2$ at 1000°C. (see below). These authors calculated for Ta on SiO$_2$ that hydrogen adsorbed to an extent greater than 100 times the number of metal atoms present. The formation of Si-H groups had been observed previously by Morterra and Low (27) after surface methoxylation of SiO$_2$ by CH$_3$OH. Pyrolysis at 700°C opens the siloxane bridges; H$_2$ is then adsorbed at room temperature to form Si-H species.

So far, all the characteristics mentioned for monometallic-supported catalysts can be generalized for bimetallic systems. For instance, Pt-Ir on alumina, studied by Engels et al. (28), exhibited an enhancement of H$_2$ adsorption (more than three times the sum of the individual capacities) for 80% Pt-20% Ir (by weight). Sinfelt and Via (29) also observed a rather high ratio [H/(Pt + Ir) = 1.7] for H$_2$ chemisorption at room temperature on a series of Pt-Ir/alumina catalysts with low metal loading. They considered spillover as a possible explanation for the H/(Pt + Ir) ratio that they found.

Hydrogen spillover has also been recognized on many Pt-Re-on-alumina systems by Dowden et al. (30) and more recently by Barbier et al. (31). A maximum in the amount of spillover occurred with a metal–metal ratio of 40% Pt to 60% Re. In both studies the catalysts were treated by hydrogen at high temperatures, 480 and 500°C, respectively.

Iron on alumina as well as iron and iridium supported on the same carrier were suspected of exhibiting hydrogen spillover after treatment with hydrogen at 400 to 500°C. The TPD analysis is described by Paryjczak amd Zielinski (32,33). The high-temperature hydrogen desorption peak was attributed to the spiltover species.

Takeuchi *et al.* observed that the amount of tritium irreversibly adsorbed on Fe/Al_2O_3 and Fe–Cu (1:1)/Al_2O_3 was increased 10-fold compared to pure Fe and Fe–Cu (1:1) sheets (34). Adsorption of tritium was measured at 250°C with a mixture of H_2–T_2 at 15 Torr.

In a study of the adsorption of H_2 between −195 and 500°C on a physical mixture of Pt–Ru + Al_2O_3, or as Pt–Ru deposited on the same alumina, Engels *et al.* (35) detected a spillover effect for adsorptions conducted at temperatures higher than 300°C for each system.

Hydrogen spillover was also often detected with even more complicated catalysts such as Ru–Ni–La_2O_3 supported by active carbon, as reported by Inui *et al.* (36, 37). They measured the adsorption capacities with H_2 at room temperature to 320°C at a H_2 pressure of 0.04 atm. Once more the slow uptake was attributed to the spillover of hydrogen species from (Ni + Ru)/La_2O_3 to the active carbon.

It is generally accepted that spillover depends on at least two prerequisites: a source for the spilling species (a group VIII metal, for example) and an acceptor (an oxide or an active carbon, for instances). Two papers give evidence for hydrogen spillover occurring on group VIII–group IB metal alloys: Pt–Au on various supports (20) and unsupported Ni–Cu alloys (38). In both studies a spillover effect was deduced by comparing the adsorption measurements to the surface compositions of the alloys. It was assumed that the group IB metals were able to chemisorb hydrogen atoms but not dihydrogen.

Factors influencing the spillover of hydrogen measured as the enhanced adsorption (extent and rate) are relatively well recognized, but some differences concerning the relative influence can be noticed among the literature data. The following is a list of these factors: the range of the temperatures of hydrogen adsorption (below room temperature or significantly above); the possible necessity of a cocatalyst such as water or other proton acceptors (39); the amount and the percentage dispersion of the source of hydrogen spillover (11, 13); the nature of the contact between the source and the acceptor (see Fig. 1); the specific areas of the coupled source–acceptor (40, 41); the possibility of migration through the gas phase and not exclusively on the surface; the role of the partial pressures of H_2; the strength of the bond between spiltover hydrogen and acceptor sites; the chemical nature of the acceptor and source (11); the presence of chlorine and sulfur ions (14, 16, 42–44); the duration of chemisorption (25); the effects of O_2 (45) and CO (24); and the nature of the spiltover hydrogen (11).

Among these numerous factors some are detrimental to hydrogen spillover, especially O_2 and CO, or limited contact between the source and the acceptor (cases B and C in Fig. 1). When carbon is the acceptor, an acidic or alkaline pretreatment does not affect subsequent spillover experiments. Added water is either an accelerating promoter or without influence, though it may be formed from O_2 in the system and in this case it lowers the rate of spillover (46).

Reverse spillover or back-spillover is observed to proceed by surface migration of the spiltover species from the accepting sites to the metal, where it desorbs as H_2 molecules or reacts with another hydrogen acceptor such as O_2, pentene, ethylene, etc. Reverse or back-spillover (primary as well as secondary) is hindered by H_2O (11), whereas secondary spillover is promoted by H_2O (case B in Fig. 1). Hydrogen spillover depends on the acceptor surface; it is thought to be easier on silica than on alumina (45) for hydrogen–molybdenum–bronze preparation.

Increases of metal contents above 0.8% do not influence the spillover process. When the metal percentages are less than 0.0008%, no spillover occurs even as primary spillover. For Pt, spillover starts to become significant for loadings of 0.03% (11).

B. EXCHANGE WITH THE SURFACE

Hydrogen spillover can be interestingly investigated with the help of isotopic molecules such as D_2 and T_2. Exchange reactions between the surface and the gas phase, and equilibration reactions involving hydrogen and deuterium, are easily monitored by IR spectroscopy, chromatography, and mass spectroscopy.

Carter et al. (47), using IR spectroscopy, were among the first to study the increased rate of exchange of OH–OD surface hydroxyl groups on alumina containing Pt compared to pure alumina. Altham and Webb used radiotracers and found similar results for H_2-T_2 exchange on Pt–alumina and Pt–silica (13) compared to the pure oxides. Both studies indicated a decrease in the enhancement of the exchange rate when the proportion of Pt increased to about 1%. This last finding was attributed by both groups to the influence of the chlorine content, which increased with the content of Pt and consequently diminished the number of exchangeable OH groups of the supports. They also suggested that reverse spillover increased as the dispersion and the percentage of the precious metal was increased.

Dowden *et al.* (*30*) have reported that on bimetallic Pt–Re/Al$_2$O$_3$ catalysts, spiltover deuterium exchanges with the alumina OH groups. They used differential hydrogen analysis and mass spectrometry to determine the amount and rate of H$_2$ exchange in the gas phase. The highest rate of exchange was observed for 40% Pt–Re 60%/Al$_2$O$_3$ between −13 and −3°C. Pure Re and catalysts with percentages of Re above 60% (Re:Pt > 60:40) were very poor catalysts and required far higher exchange temperatures. They proposed that spillover from Pt was blocked by Re: with excess amounts of Re, all Pt ensembles or isolated particles could be covered by Re. This assumption is in agreement with H$_2$ chemisorption data (see subsection II, D).

Cavanagh and Yates (*48*) used transmission IR spectroscopy to study OH–OD exchange on Rh–Al$_2$O$_3$. They suggested that the alumina hydroxyls reacted with D$_2$ by a rapid spillover from Rh to alumina, followed by a slow migration on alumina and a rapid exchange of the spiltover D with the OH's. They confirmed the role of CO as a poison as claimed by Apple and Dybowski (*24*).

The role of water during the surface exchange was emphasized by Ambs and Mitchel (*49*), who studied the OH–OD exchange with a chromatographic technique based upon the retention times of pulses of HD. The process was studied in a continuous stream of D$_2$ at 150°C with an argon tracer. Water was proposed both to ease the spillover of the D species from Pt to alumina and also to contribute OH groups as bridges and sites for additional spillover. However, the presence of water may be detrimental to the exchange via spillover, as shown below. No gaseous exchange between D$_2$ and aluminas OH's took place in the absence of Pt (no retention of the HD pulse) for their experimental conditions.

Similar studies were described by Scott and Phillips (*50*) for Ni/SiO$_2$ above 120°C. They complemented their experiments with IR spectrometry and reported only a fairly limited exchange by D$_2$ between 350 and 500°C. Further, they found that a physical mixture of their catalyst with pure WO$_3$ in flowing H$_2$ saturated with water did not induce bronze formation (no color change) below 300°C. Above this temperature a grey–blue color developed but "only in the region adjacent to the Ni/SiO$_2$." This observation is consistent with other experiments involving Ni on alumina aerogel and WO$_3$ (*51*), and demonstrates two points: the presence of water is not always beneficial in spillover, and the nature of the metal, Pt or Ni, is of prime importance (Section V).

Spiltover hydrogen was found by Van Meerbeck *et al.* (*26*) to reduce the surface of silica gel containing only traces of metals such as Ni, Pd, Pt, W, and Ta. When their silica samples were heated to about 1000°C under an H$_2$ atmosphere, they detected an IR bond at 2300 cm^{-1} that was attributed to a

Si–H stretching mode. This Si–H band, however, did not appear in the absence of the minute traces of metals. The reaction was controlled by diffusion whose rate was independent of the nature of the metal. The extent of such diffusion, however, was strongly related to the nature of the metal. Van Meerbeck *et al.* inferred that this Si–H surface species is the same as the species described by Morterra and Low with their "reactive silica" (*27*). Tiller *et al.* (*52*) also obtained stable surface Si–H bonds by exposing silica pellets to a H_2 plasma, provided the silica sample was dehydrated before exposure.

All these authors have deduced or assumed that surface migration is rate controlling. Indeed, diffusion over a distance from the source of spillover was clearly shown by IR and FTIR techniques on $Pt–Al_2O_3/SiO_2$ at 90°C and Pt/SiO_2 wafers at 240°C using OH–OD exchange (*53, 54*). In each series of experiments, a concentration gradient for the OD band appearance was found: a Pt spot was either on the edge or on the center of the pellet. The role of water in exchange was carefully studied by Bianchi *et al.* (*53*), who showed how the results may depend on traces of oxygen (or oxygenated compounds) accidentally introduced into the cell. In all cases no gradient in the exchange was found if either water or O_2 were present; this was due to the D_2O–OH exchange on the overall surface of SiO_2 and did not depend on the position of the Pt spot. NMR investigations of OH–OD exchange on $Pt–SiO_2$ have also shown a slow exchange at 25°C with a time scale of several hours. Spiltover hydrogen, assumed to be atomic (and not protonic), also contributed significantly to SiOH proton relaxation processes according to Sheng and Gay (*55*).

In contrast to these studies, Finlayson-Pitts (*56*) observed by IR either OH-to-OD surface exchange or new OD creation without exchange from D atoms generated by microwave discharge. The reaction depended on the nature of the silica tested: nonporous Cab-O-Sil giving exchange and Vycor porous glass creating new ODs. His silica samples were carefully evacuated at 650°C previous to the microwave discharge experiments. It must be noted that the Cab-O-Sil sample was very pure silica, while the porous Vycor glass contained a small percentage of boria and alumina. However, Finlayson-Pitts did not find any evidence of Si–H or Si–D band formation in either case. Therefore it may be concluded that the interaction between a hydrogen atomic beam is not necessarily identical to that which occurs with hydrogen spillover from trace metals. It should be noted, however, that his pretreatment temperature (650°C) was lower than the pretreatment temperature (1000°C) used by Van Meerbeck *et al.* (*26*) in their observations of Si–H formation.

Similar studies have been extended to HNaY zeolites using IR and mass spectroscopy (*57–59*). Deuteration of the surface of the HNaY zeolite at room temperature was accelerated by a Pt on NaY zeolites; the results agreed with the earlier studies published by Dalla-Betta and Boudart (*60*), although the interpretation of their data was not the same. The rate of exchange did not depend on the nature of materials such as oxides (HNaY, SiO_2, ZnO, Fe_2O_3) connecting the HNaY wafer and the platinum catalysts, provided the materials contained surface O^{2-} ions. Moreover, the acidity of the various OH groups of the zeolite did not affect the kinetics of the isotopic exchange reaction.

FTIR diffuse reflectance spectroscopy was recently applied by Kazanski *et al.* (*61*) to monitor the interactions of molecular hydrogen with silica, alumina, magnesia, and zeolites at low temperatures ($\sim -195°C$). Molecular hydrogen absorption spectra were found for all oxides examined. Adsorption bands attributed to hydrogen interacting with oxygen vacancies were found for all samples except silica. The site density of this type of hydrogen was estimated to be $\sim 10^{13}$ cm^{-2}; which is similar to the concentration of H_{sp} measured by hydrogen adsorption isotherms at higher temperatures ($T > \sim 20°C$; see Section V). In addition, the aluminas exhibited Al-H or Al-D hydride absorption bands (as well as OH and OD bands), although similar bands have never been detected with spillover systems except on silica gel pretreated at very high temperatures [1000°C (*26*)]. The location of the molecular hydrogen, seen in the IR bond, was very sensitive to the presence of OH group, $Al^+ + O^-$ (acid–base) pairs, etc. It seems that this technique would be useful in the investigation of the various spillover phenomena.

C. INDUCTION OF BULK CHANGES

The lowering of the temperature of reduction by H_2 of numerous metal oxides by addition of a transition metal is well documented and may be attributed to spillover (*7, 8, 62*). Among all the metal dopants added, Pt and Pd are seen to be the most efficient and have been the most often studied. It should be noted that spillover does not only give rise to the reduction of solid or adsorbed phases: oxidations of inorganic compounds such as UF_4 (as well as carbon) can occur when Pt is present and the sample is exposed to oxygen (*63–65*).

Both types of reactions involving the spillover of either H_2 or O_2 have been termed topochemical heterogeneous catalysis (*62*). Besides the catalyzed reduction of metal oxides either to metals or suboxides, the formation of new and specific reduced oxides, such as the well-known hydrogen bronzes of W, Mo, and V, have attracted considerable attention (*7, 66–68*). In many cases the reduction of the corresponding oxides by spillover of H_2 led to reduced compounds not otherwise obtainable (*69*).

A large number of metal oxides has been reduced by spillover from Pt or Pd on silica: Co_3O_4, V_2O_5, UO_3, Fe_2O_3 (63, 64, 69) and MoO_3, WO_3, Re_2O_7, CrO_4 (69) on the one hand, and Ni_3O_4, MnO_2 (70) and NiO, CuO, Cu_2O, ZnO, CdO, SnO_2 (69) on the other. These studies have been performed by differential thermal and thermogravimetric analysis. The first group of oxides could easily be catalytically reduced by small amounts of Pt metals. Each step leads to a decrease in the oxidation state of a metal one at a time, as in the hydrogen bronze formation. All metals in this group were in their highest oxidation valency state (this is not the case for the second group of metals). It should be noted that in general these catalyzed reductions proceeded without evidence of an induction period. However, for H_2 reduction without spillover, the well-known sigmoidal kinetic curves are often found. Spillover has been invoked to explain the reduction. Spiltover H atoms are known to react with nonrefractory metal oxides very rapidly even at low temperatures, as was shown recently by Che et al. (71). A general stoichiometric equation was proposed by Bond (69) to depict the first step:

$$M^{x+} + O^{2-} + H \longrightarrow M^{(x-1)+} + OH^- \tag{8}$$

The second group of metal oxides was found to be reduced in the presence of Pt metals without either a large increase in the rate or a decrease in reduction temperatures. These compounds contained relatively stable low-valency cations. It was assumed that these oxides are more or less readily reduced to the metallic state in a one-step process. Further, the metal may then be able to activate H_2. The reduction would therefore be autocatalytic. This is often found for nonspillover reductions.

For both series of metal oxides the apparent activation energies of the catalyzed (spillover) reduction reactions were found to be similar to those of the noncatalyzed reductions. Generally, the effect of the Pt was to increase the available reactive hydrogen and/or to increase the rate of the nucleation (pre-exponential factors). Thus, this "catalysis" increases the availability of H but does not ("classically") decrease the activation energy.

A recent study of the reduction of CuO by temperature-programmed reduction (TPR) (72) in the presence of group VIII and IB metals was carried out by Gentry et al. (73). Pd and Pt were extensively investigated in the CuO reduction. In these studies, both Pd and Pt metals were effective in substantially reducing the starting reduction temperatures (as shown by the TPR profiles). The effects were not identical. Pd gave Pd–Cu alloys and promoted type-A spillover (Fig. 1). With Pt–Cu alloys a less efficient spillover process occurred, that is, type-B spillover (see Fig. 1) or interparticle spillover. The presence of group IB metals did not affect the TPR profile. This is consistent with the metal's inability to dissociatively chemisorb molecular hydrogen.

The reduction of Ag_2O by hydrogen was carried on in a differential scanning calorimeter by Szabo and Konkoly-Thege (74), who reported two phenomena: (i) in the presence of Pd the rate of reduction of Ag_2O supported on Al_2O_3 was dramatically increased compared to pure Ag_2O and Ag_2O/Al_2O_3; and (ii) finely dispersed Ag_2O (and partially reduced) on Al_2O_3 was proposed to dissociatively adsorb H_2 and promote its spillover; however, this did not take place on pure silver oxide because of the rapid sintering of Ag atoms.

In a similar way, the TPR investigation of the $Pt-Re/Al_2O_3$ catalyst described by Mieville (75) showed that rhenium oxide cosupported with Pt on Al_2O_3 can be reduced by H_2 spillover from Pt. This results in the nucleation of Re metal centers, which in turn catalyze the further reduction of Re oxides. This hypothesis contrasts with the proposal by Isaacs and Petersen (76), who calculated that there were too few spiltover species to account for their results. They thus discarded spillover as an explanation.

Praliaud and Martin (77) proposed the formation of Ni–Si and Ni–Cr alloys on silica and chromia supports, respectively, under H_2 at sufficiently high temperatures. They suggested that hydrogen spilt over from Ni to the Cr_2O_3 carrier and partially reduced it to Cr^0, which was then alloyed with Ni as indicated by magnetic measurements. The same technique in conjunction with IR spectroscopy and volumetric adsorption of H_2 was applied to partially reduced Ni-on-alumina and Ni-on-zeolite catalysts by Dalmon *et al.* (78). These supported Ni systems contained Ni^0 and Ni^+. H_2 was found to be activated only when the couple Ni^0/Ni^+ was present according to

$$H_2 \longrightarrow 2H \quad (\text{on } Ni^0) \tag{9}$$

and

$$2H + 2Ni^+ \longrightarrow 2Ni^0 + 2H^+ \tag{10}$$

A spillover of protons was inferred from the IR data. The protons migrated onto the basic sites of the support. The origin of the Ni^+ species could be either a disproportionation of Ni^0 and Ni^{2+} or reverse spillover according to Eq. (10) during evacuation.

Interesting, too, is the short paper by Grange *et al.* (79) which refers to the enhanced reduction of MoO_3 by hydrogen in the presence of tungsten carbide as the source of spiltover hydrogen. Again, water is beneficial during the reduction. A very sharp decrease of the induction period depended on the intimacy of the mixture of WC and MoO_3. Thus WC is comparable to the other Group VIII metals.

Systems other than metal oxides have not often been studied. It is worthwhile to mention that $CoCl_2$ was reduced by H_2 spillover from Pt, Pd, and Ni supported on alumina or on silica–alumina (80). The authors noted a strong influence of support acidity on the chloride reduction by H_2 spillover (chloride trapping).

An intercalated hydrogen sulfur compound, $H_{0.05}WS_2$, has been induced by the addition of Ni and Co to WS_2 (81). Hydrogen spillover across intermetallics like $SmMg_3$ treated with aromatics was detected by Imamura and Tsuchiya (82). Hydrogen adsorption and also absorption on $SmMg_3$ was enhanced [and observed also on Mg_2Ni (83)]. The decomposition of the intermetallic into samarium hydride (SmH_2) and metallic magnesium was seen by XRD. As in the earlier studies of Vannice and Neikam (84, 85), electron donor–acceptor complexes between the aromatic and the metal(s) were formed. This formation was promoted by the spillover of monatomic hydrogen.

Special mention must be given to the numerous studies concerning hydrogen bronze formation induced by spillover. In particular, tungsten and molybdenum oxide hydrogen bronzes have been thoroughly studied since Khoobiar's experiments (4).

Extensive work on hydrogen bronzes of WO_3 and MoO_3 (produced by Pd/SiO_2, Pt/SiO_2, and Pt/MoO_3 or Pt/WO_3) has been published by Sermon and Bond (11, 66) and Bond and Tripathi (45). They clearly identified $H_{0.46}WO_3$ and $H_{1.63}MoO_3$ bronzes by a number of techniques including XRD, IR, ESR, TGA, and DTA. They noted that the formation of $H_{1.6}MoO_3$ (monoclinic) violated the generally observed rule predicting maximum values for $x \simeq 1$ (for H_xMO_3). The thermal stabilities as well as the chemical reactivities of these bronzes were discussed in relation to their stoichiometry. $H_{1.6}MoO_3$ and $H_{0.5}WO_3$ are stable from $0°C$ (temperature of formation) up to $90°C$, but it is easier to reduce H_xMO_3 (e.g., to Mo_2O_5) than H_xWO_3. Retrieval of inserted hydrogen by acceptors (such as O_2) is more difficult with the molybdenum bronze than with the tungsten bronze.

A new hexagonal form of $H_{0.4}WO_3$ was recently described by Gerand and Figlarz (67). It was obtained by spillover of hydrogen supplied by Pt. The authors found little variation of the W–O matrix structure during the insertion of spiltover H. This new variety of hexagonal bronze of tungsten, however, seems less stable than the monoclinic bronze described above.

The kinetics and mechanism of the formation of molybdenum bronzes was studied in detail and published in a series of three papers by Erre et al. (86–88), who studied the (100) face of MoO_3. Pt was added and the sample was exposed to H_2 under very low partial pressures (10^{-8}–10^{-5} Torr), forming monoclinic $H_{1.6}MoO_3$. The kinetics unexpectedly exhibited three steps: (i) activation of H_2 by Pt with an increasing rate with time on stream; (ii) a kinetic plateau where the rate remained constant; and finally (iii) a decrease in the rate of bronze formation. The salient results were that once step (i) starts, Pt is not required for the insertion of hydrogen into the lattice, which is now able to adsorb and dissociate molecular hydrogen directly from the gas phase. A similar situation was found for δ or amorphous aluminas. However, this occurred at much higher temperature (430°C); spiltover H was reformed in the absence of any metal catalyst. It was found that H produced on Al_2O_3 by spillover from a metal activator or directly from the gas phase was responsible for the induction period detected during ethylene hydrogenation (*89, 90*) (see Section V). Erre *et al.* were able to duplicate all their findings on MoO_3 using H atoms provided by a hot W wire (*88*).

The vanadium bronze $H_xV_2O_5$ was prepared by H_2 spillover by Tinet *et al.* (*68, 91*). Its formation, together with the formation of tungsten and molybdenum bronzes, was investigated by microcalorimetry. The maximum value of x for the vanadium bronzes was between 3 and 3.8. The bronze was amorphous to X-ray and neutron diffraction; it was a semiconductor (instead of the metal-like conduction of molybdenum and tungsten bronzes). Hydrogen self-diffusion coefficients in molybdenum and tungsten bronzes, measured by NMR spectroscopy, appeared to be 7 to 10 orders of magnitude greater than the diffusion coefficients for spiltover H. It was concluded that the rate-determining step in the formation of these bronzes is the penetration of the spiltover species below the surface of the initial corresponding oxides (*92*).

D. METAL–SUPPORT INTERACTIONS

Metal–support interactions have been recently reviewed by Bond (*93*), who drew special attention to catalysts that gave evidence for strong metal–support interactions (SMSI). This condition was first observed in 1978 by Tauster *et al.* (*94*) for Pt on titania catalysts. The catalysts seemed to lose their capacity for H_2 and CO chemisorption but nevertheless retained and enhanced their activity for only two types of reaction: methanation and Fischer–Tropsch synthesis. Since then a considerable number of papers devoted to SMSI studies have been published all over the world.

As will be seen below, in each of the mechanisms proposed by the various authors to account for SMSI (chiefly on M/TiO_2 systems), hydrogen spillover plays an important and often major role. Hydrogen spillover can be considered as a necessary condition for creation of the SMSI state. The principal metals studied were Pt, Ni, Rh, Ru, Pd, Ir, Au, and Ag. In addition to TiO_2, other supports were found to exhibit the same effects; these include V_2O_3, Ta_2O_5, Nb_2O_5 (*95, 96*), TiO, and Ti_2O_3 (*97*).

Common to each proposed SMSI mechanism is the reduction of the support by hydrogen. The reductions occur near 500°C in the presence of a metal able to chemisorb and dissociate molecular hydrogen.

A topographic explanation of the results of hydrogen spillover on Pt/TiO_2 catalysts has recently been provided by Baker et al. (*98*) from high-resolution electron microscopy studies. In their earlier studies these authors developed a model for the SMSI state (reduction at $T > 500°C$) (99). The metal dissociates H_2 into atoms that spillover onto the TiO_2 and reduce it to Ti_4O_7. Subsequently, the Pt particles assume a "pill-box" structure on the Ti_4O_7; they suggest that this is indicative of the SMSI state. To see if the dissociation (and spillover) of hydrogen is a necessary requirement of the metal to produce the SMSI state, Ag/TiO_2 systems were examined because Ag does not dissociate H_2. After the reduction by H_2 at 550°C in this system, the metal particles appeared to be globular in outline, which is normal. The electron diffraction simultaneously showed that TiO_2 remains as an unreduced rutile. A subsequent introduction of Pt onto these specimens, followed by a further reduction cycle, showed that Ag is now in the form of a pill-box structure and that TiO_2 is reduced to Ti_4O_7. Therefore, in the presence of Pt, silver exhibits characteristics attributable to SMSI-state bonding on Ti_4O_7. Consequently, the function of Pt is to provide a source of H atoms which spillover to titania and are responsible for its conversion to Ti_4O_7.

The reduction of TiO_2 is not only observed after high-temperature reduction (HTR) ($T > 500°C$) of the titania-supported catalyst (the SMSI state) but is also observed after low-temperature ($\sim 300°C$) treatment (not the SMSI state). The ESR studies by Huizinga and Prins (100) of Pt/TiO_2 reduced at 300°C gave evidence for the formation of Ti^{3+} with hydrogen present. If the sample is evacuated, the signal disappears; however, it reappears if H_2 is reintroduced at room temperature. The formation of Ti^{3+} is not reversible if the reduction is carried out at 500°C. These observations have been explained by hydrogen atom spillover from Pt and the formation of Ti^{3+} and OH^- species during the low-temperature reduction (LTR), viz:

$$1/2H_2 \xrightarrow{\text{Pt}} H_{sp} \quad [H_{sp} \equiv \text{spiltover } H_2] \qquad (11)$$

$$Ti^{4+} + O^{2-} + H_{sp} \longrightarrow Ti^{3+} + OH^- \qquad (12)$$

During the high-temperature reduction the hydroxylated TiO_2 surface adjacent to Pt is dehydrated. A Ti_4O_7 layer is formed that inhibits the reversibility. Conesa et al. (101) observed the same results for Rh and Pt/TiO_2.

An interesting explanation for Pt/TiO_2 catalysts in the SMSI state, which tends to suppress the chemisorption of H_2, has been advanced by Chen and White (97). If TiO_2 is reduced by H_2 at 875°C before deposition of Pt, the SMSI state is also observed. If this reduced TiO_2 is reoxidized before the deposition of Pt, the catalyst irreversibly adsorbs H_2 at room temperature, since any Pt/TiO_2 reduced at low temperatures and the SMSI state is not achieved. It may, therefore, be concluded that the nature of Ti ions should be correlated with the SMSI state. Further, Pt deposited on Ti_2O_3 or on TiO does not adsorb H_2 (as in the SMSI state). High-temperature reduction of the

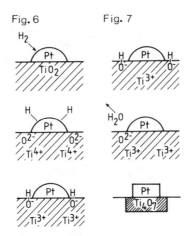

Fig. 6   Fig. 7

FIG. 6. Formation of Ti^{3+} ions at low reduction temperatures. Reprinted with permission from *Journal of Physical Chemistry*. Copyright (1981) American Chemical Society.

FIG. 7. Dehydration of TiO_2 surface at high reduction temperatures (*100*). Reprinted with permission from *Journal of Physical Chemistry*. Copyright (1981) American Chemical Society.

support leads to the SMSI state, which seems to be stabilized by the dehydroxylation, as depicted in Figs. 6 and 7.

These two oxides, Ti_2O_3 and TiO, are highly conducting *n*-semiconductors, but their surface, as disclosed by XPS spectroscopy, contains only Ti^{4+} ions. The explanation provided by the authors involves conduction-band electrons tunneling through the surface TiO_2 layer. These are finally trapped by Pt particles, a well-documented process (*102*). For *all* previous catalysts containing Pt, the activity in H_2–D_2 exchange is very high (at 25, 100, or 200°C) and is independent of the presence or absence of the SMSI state. It follows that Pt particles, with or without trapped electrons, are always able to dissociate H_2, but the particles of Pt enriched in electrons (SMSI effect) do not bond strongly with hydrogen. The following diagram represents the situation:

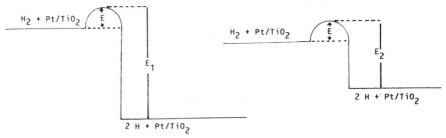

not SMSI state   SMSI state

The activation energy E of the dissociation of H_2 is the same in both cases, although E_1 and E_2, the heats of adsorption of H, differ, with $E_1 > E_2$. In other words, the SMSI state is observed with a support of high electronic conductivity. In this state the work function of the support is smaller than that of the supported metal. If the TiO_2 support is not initially a conductor, it becomes one by reduction with H_2 in the presence of Pt at high temperature. This reduction may not be limited to the surface. For low reduction temperatures, surface OH^- groups and Ti^{3+} ions are formed by hydrogen spillover, as shown previously (100). But this formation of Ti^{3+} and OH^- is not the source of SMSI, since the electrons are not transferred into the conductivity bands to be trapped by Pt.

Herrmann et al. (103, 104) arrived at similar conclusions concerning electron trapping by electrical conductivity measurements of Pt/TiO_2 under H_2 in both temperature ranges (LTR and HTR). Conductivity measurements showed that UV illumination did not affect the spillover process (104, 105)

Using temperature-programmed oxidation (TPO) and reduction (TPR), Kunimori et al. (106, 107) again found that the titania was reduced by spiltover hydrogen. Even alumina, in the case of Pt/Al_2O_3, was reduced by H_2 at 500°C (or lower temperatures). They suggested that sulfur promoted the H spillover. The degrees of reduction of TiO_2 and Al_2O_3 are shown in Table II. For Pt/Al_2O_3 they propose the formation of a Pt-Al alloy as suggested in the work of Den Otter and Dautzenberg (108). In the studies by Beck et al. (44, 109), two states of hydrogen adsorption on TiO_2 are identified: one is located near Pt atoms, while the other is spatially separated from the first. A pill-box model was also proposed by Foger (110) for Ir on TiO_2 in the SMSI state obtained under HTR (above 400°C) in hydrogen. He

TABLE II

Information on the SMSI State Estimated from the Consumption
Measurements of Pt/Al_2O_3 and Pt/TiO_2 Catalysts[a]

Catalyst	Degree of reduction	$N_O{}^b$	Local structure[c]
5 Pt/Al_2O_3	Al_2O_{3-x} ($x = 0.03$)	$\sim 4 \times 10^{13}$	$(Pt-Al_2O_2)_n$ or $(Pt-Al_{2/3})_n$, etc.
1 Pt/TiO_2	TiO_{2-x} ($x = 0.01$)	$\sim 1 \times 10^{14}$	$Pt_n(Ti_4O_7)_{2n}$, etc.

[a] After Kuhimori and Uchijima (106).

[b] Number of O atoms (per cm^2) eliminated from the oxide surface.

[c] The number of Pt atoms of each crystallite in the normal state is given as n.

observed that the strength of the SMSI was correlated with the Ir particle's dimensions: the strength increased as the particles sintered.

Migration of a titanium suboxide (with TiO_x due to reduction of TiO_2 by hydrogen spillover) onto Pt particles has also been advanced by several authors (*25, 111, 112*) to account for SMSI states. Resasco and Haller (*112*) have shown that TiO_2 is reduced either by a beam of H atoms generated by a microwave discharge or by chemical means (Rh in this case). Exposure to H atoms (by either means) is a sufficient condition (if not a necessary one) to promote SMSI. Jiang *et al.* (*25*) advocate the migration of hydrogen from metal particles to adjacent regions of the titania support to explain their results for Ni, Pt, and NiFe on titania. They argue that the subsequently reduced oxide TiO_x induces electronegativity in Pt and therefore suppresses its ability to chemisorb H_2 as well as CO.

Burch and Flambard (*113*) have recently studied the H_2 chemisorption capacities and CO/H_2 activities of Ni on titania catalysts. They attributed the enhancement of the catalytic activities for the CO/H_2 reaction (after activation in H_2 at 450°C) to an interfacial metal–support interaction (IFMSI). This interaction is between large particles of Ni and reduced titanium ions; the Ti^{3+} is promoted by hydrogen spillover from Ni to the support, as pictured in Fig. 8. The IFMSI state differs from the SMSI state since hydrogen still chemisorbs in a normal way; however, if the activation temperature is raised to 650°C, both the CO/H_2 activity and the hydrogen chemisorption are suppressed. They define this condition as a "total SMSI state." Between the temperature limits, they assumed a progressive transition from IFMSI to SMSI. Such an intermediate continuous sequence had been

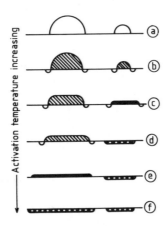

FIG. 8. Interactions in the Ni/titania system as the activation temperature is increased: ⌒, NiO; ⌒⌒, bulk Ni metal; ▬ surface Ni (≤ 1 nm) in the SMSI state; ▭, Ni in a partially ionized subsurface state; ▽, oxygen anion vacancy (*113*).

suggested by Conesa et $al.$ (101), who studied the appearance of Ti^{3+} species by ESR under mild reduction conditions by H_2 (200°C) and suggested Eqs. (13)–(15) to account for the transformation:

$$1/2H_2 + Pt_s \longrightarrow Pt_s \cdots H_{ads} \qquad (13)$$

$$H_{ads} + \overset{O}{\underset{\|}{-Ti^{4+}}} \longrightarrow \overset{O-H}{\underset{|}{-Ti^{3+}}} \qquad (14)$$

$$H_{ads} \longrightarrow H^+ + e^-; \quad e^- + Ti^{4+} \longrightarrow Ti^{3+} \qquad (15)$$

Here H_{ads} is the precursor of the spillover species and is the reactant which creates the reduced Ti^{3+} ion. This contrasts with the proposal by Apple et $al.$ (23).

Therefore, at least on titania, transition metals promote the spillover of hydrogen to the support; this is a necessary step in the reduction of the support (and hence modification of the global solid's catalytic properties). In other words, hydrogen spillover is a prerequisite in each of these recently recognized metal–support interactions (SMSI and IFMSI). Evidently these very specific metal–support interactions are, from the point of view of the spillover phenomena, merely the reduction of more or less easily reducible metal oxides, as mentioned in the preceding subsection.

III. Mechanisms and Kinetics

Studies of spillover have concentrated on one aspect or another of the phenomenon. Many have involved characterizing the species adsorbing and spilling over. Another focus has involved transformation occurring on the surface or throughout the solid. The specific nature of the active site created on the "support" surface and of the catalysis proceeding on these sites has also been investigated. Underlying these mechanistic studies is the rate by which the processes occur and the extent of their influence.

The discussion of mechanism and kinetics will be initially drawn along these lines:

(a) mechanism of sorption and spillover,
(b) nature of the solid-state interaction,
(c) the kinetics and extent of spillover.

Since over 90% of the studies have involved the spillover of hydrogen, this will be our focus. However, we will discuss the evidence for the spillover of other species in the fourth section of this article. We realize that there is considerable overlap between the parts, but the division will facilitate the discussion.

A. Mechanism of Sorption and Spillover

The nature of spiltover hydrogen is the subject of considerable controversy. Common to each proposal mechanism is the dissociative sorption of hydrogen onto a metal or other sorbing surface (or other oxides capable of sorbing H_2: ZnO, Cr_2O_3, Co_3O_4). The controversy involves the interpretation of the mechanism of the initial adsorption (or of the subsequent spillover) as homolytic or heterolytic. This can result in three possible species being created and subsequently transported on the sorbing surface: H^+ or H^-, $H\cdot$, and $H—$. The atomic species may be an ionic, a radical, or a bonded species.

Several studies have attempted to generalize hydrogen spillover as one of the three types of species or to reject spillover based on studies disproving one of the assumed species types. We firmly believe that spillover is more general and may occur by a variety of mechanisms involving, under certain circumstances, any of the species.

Sinfelt and Lucchesi first proposed that activated hydrogen migrated from Pt onto the SiO_2 or Al_2O_3 support, where it reacted with ethylene from the gas phase (5). The discussion implies that the migrating species is atomic hydrogen (not necessarily ionic). No proposals were offered to differentiate between a radical or a bound hydrogen.

Khoobiar utilized electron conductivity measurements to suggest that protonic hydrogen is formed on Pt and diffuses across Al_2O_3 onto the separate WO_3 particles (4).

Vannice and Neikam used ESR to measure the enhanced decrease in anthracene and perylene radicals by the presence of Pt supported on base-exchanged zeolites (84). The hydrogen free radicals, $(H\cdot)$ formed and spilled over from the Pt are able to react with the organic radicals sorbed onto the zeolite. In a subsequent (85) study, Niekam and Vannice found that the rate and extent of hydrogen uptake corresponded to the decrease in the ESR-measured perylene radicals. The perylene was suggested as a quantitative sink for the radicals $(H\cdot)$ created and diffusing from Pt. They also suggested that the hydrogen radicals spilling over eventually react with oxygen sorbed onto the zeolite surface.

Studies of the effect of other sorbed species give considerable insight into the nature of spillover. Boudart *et al.* studied the influence of sorbed water on the rate of spillover (114). The ratio of WO_3 conversion at room temperature to the hydrogen bronzes and the sorption of hydrogen from the gas phase were dramatically increased by the presence of water. Levy and Boudart used a spectrum of alcohols and acids as coadsorbents, and found that the increase in the rate of reduction of WO_3 was related to the proton affinity of the coadsorbent (39, 115). The proton, produced on the Pt, is proposed to be solvated by the coadsorbent and diffuse via the layer of coadsorbent to the WO_3 surface, where the tungsten bronze is readily formed.

Jiehan et al. interpreted the "blue shift" in the IR band of CO with H_2 sorption as indicative of proton (not atomic H) spillover from Pt onto TiO_2 (containing CO) (116). The studies were reinforced with parallel studies of the conductivities of the samples.

Gonzales and Gadgil created atomic hydrogen species in the gas phase from a tungsten filament (117). These species were then sorbed onto a variety of oxide surfaces at low temperatures. The rate of adsorption was different depending on the nature of the "support" surface with a constant rate of hydrogen atom production from the gas phase. This implies that the sorption process was activated. The rate of adsorption of atomic hydrogen and the amount adsorbed were small compared to the rate of hydrogenation found to occur on the surfaces. This underscores the conclusion that the dominant effect is activation of the surface and not reaction of the hydrogen being spilled over.

Che has confirmed this latter interpretation with similar experiments (71, 118). H atoms produced in a microwave discharge were found to activate SiO_2 to catalyze ethylene hydrogenation. The proposal is that hydrogen radicals create F center defects that are catalytically active in hydrogenation. Nogier et al. used a hydrogen plasma to induce hydrogenation activity on commercial (Davison and Degussa) silicas (119). The hydrogenation activity was found to be an order of magnitude less than for silica activated by hydrogen spiltover from Pt; without the plasma treatment there was no activity. In similar studies Minachev et al. concluded that only the external exposed surface is able to interact with gaseous atomic species (120).

The above studies are not necessarily in conflict with more conventional methods of spillover activation of an oxide in the presence of a metal. Each of these studies involved attempts to activate oxides at low temperatures. However, the results of these studies and more conventional spillover experiments confirm that the process is activated and may be enhanced by treatment at higher temperatures. Yet each study suggests that atomic hydrogen (as a radical or an ion) is involved in the creation of active sites on various oxides.

Studies by Steinberg et al. (58, 121) involved a variety of surfaces between the source of spillover and a reacting surface (OH–OD exchange). They found that n- and p-semiconductors and insulators equally promoted the transport from the source to the reacting surface, although the rate of transport was much less on stainless steel. They concluded that the species spilling over was uncharged and that its transport did not depend on the semiconductor properties of the oxide. However, the oxide (or hydroxide) surface was involved.

Turning full circle, Duprez and Miloudi studied the activation of Rh/TiO_2 by magnetic measurements. They proposed that hydrogen radicals created on the metal were able to spill over as hydrogen protons, while electrons diffused in parallel within the conduction band of the TiO_2 support (*122*).

The studies seem to contradict each other in the specific proposals concerning the nature of the spiltover species. This conflict is resolved, however, by suggesting that hydrogen spillover does not occur via a single entity. Indeed, atomic hydrogen may spill over in a charged, bound, or radical state. If a cocatalyst is present (such as alcohol, acid, or water), protonic spillover can dominate the transport. The subsequent activation of the support to create catalytic activity involves a slower process with a finite activation energy. Even above 400°C considerable time is required to fully activate SiO_2 or Al_2O_3 (or other refractory oxides). Lower temperature adsorption processes may involve similar species of atomic hydrogen, although the activation of the support may not be complete. For processes with smaller activation energies, such as sorption or bronze formation, kinetic control is shifted to the formation, spillover, or transport of the activated atomic hydrogen.

To further complicate the picture, there is strong evidence that more than a single form of hydrogen can exist as a spiltover species under similar conditions. The evidence is that the possibility of multiple states of spiltover hydrogen species varies depending on the support; similarly, these various species can be involved in the reaction on and activation of the surface in different ways. This multiplicity will therefore be discussed in more detail in the following two sections, even though the evidence implies that studies focusing on proving or disproving a single form of hydrogen on the support surface may not negate the existence of spillover.

B. Nature of the Solid-State Interaction

There is a broad spectrum of solid changes that are induced by spillover. In the extreme, spillover facilitates a bulk crystallographic change in the solid. This transformation may involve the incorporation of the spiltover species into the solid lattice or the reduction of the lattice by reaction with the spiltover species. The effect may be confined to the surface. Other adsorbed species may facilitate the transport or be removed by the spiltover species. In the mildest sense, the surface may only exchange with the spilling-over species.

As we have seen, isotopic exchange of the hydroxyls on oxide surfaces by spiltover deuterium has been known since 1965 (*47*). For exchange to occur over a distance from the source of atomic hydrogen (e.g., Pt), a mobile hydrogen must be present on the surface. The picture is, however, somewhat clouded. If either water or oxygen is present, the exchange may occur via $H_2O-HDO-D_2O$ intermediates (53).

More profound transformation can occur. A methoxylated surface of silica is transformed by hydrogen spilled over from Pt to a hydroxylated surface with the production of methane (123, 124). The sample was prepared as an aerogel with a surface area of 900 m^2/g. These samples were physically mixed with Pt/Al_2O_3. The rate of demethoxylation was found to be first order in the methoxyl concentration and three-halves order in hydrogen. Compression of the admixed solids increases the rate. At 400°C with continuous spillover of H_2 provided by Pt/Al_2O_3 under dynamic conditions, the surface of a previously methoxylated silica aerogel could be specifically changed into a fully hydroxylated one, releasing CH_4. One investigation (123) was conducted in such a manner as to determine if surface migration of the spiltover H species was the rate-determining step. The results observed by IR and kinetic methods led to the conclusion that for these experimental conditions (high temperature), the surface reaction between the methoxyl groups and the spilling species, and not the diffusion, was the slowest step. The most activated step was found to be the migration of the spiltover H species across the Pt/Al_2O_3 catalyst to the silica interface.

The intermediate states in this two-hydrogen atom reaction are not known, although Teichner et al., basing their view on the overall kinetics, have suggested that H_3 is involved in the reaction; H_3 has been observed spectroscopically in other studies (125). Alternatively, because of the low concentration of spiltover hydrogen on the surface, a two-dimensional-three-center $(2H_{sp} + OMe)$ reaction seems improbable. It may be that the reaction occurs stepwise by the association of spiltover hydrogen with a methoxyl as a first step, i.e.:

$$H_{sp} + OCH_3 \longrightarrow HOCH_3 \qquad (16)$$

followed by

$$HOCH_3 + H_{sp} \longrightarrow OH + CH_4 \qquad (17)$$

The extension of this concept to other surfaces would mean that spiltover hydrogen may be able to associate reversibly with surface hydroxyls.

The incorporation of spiltover hydrogen into the lattices of bronze-forming oxides was recognized early in the study of spillover (4). A variety of studies focused on the nature of the changes that occur in the solid, and both molybdenum and tungsten bronzes have been used as indicators for spillover. The extent of hydrogen incorporation into MoO_3 was measured as $H_{1.6}MoO_3$, and $H_{0.4}WO_3$ for tungsten trioxide. It has been suggested that

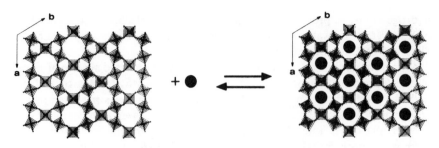

FIG. 9.   Reduction of hexagonal WO_3 considered as an insertion reaction in which the W–O framework retains its structure (*67*).

reduction occurs directly to the four-valent metallic species (8). X-ray analysis shows the starting monoclinic tungsten oxide converts to a simple cubic structure, while IR analysis gives no evidence of hydroxyl formation. For an initial hexagonal tungsten trioxide structure, the maximum extent of hydrogen incorporation is $H_{0.4}WO_3$. The minor shifts in X-ray structure and the stoichiometry led Gerand and Figlarz to propose that the atomic spiltover hydrogen resides in the hexagonal holes of the host lattice (see Fig. 9) (*67*).

The results for bronzes indicate that spiltover hydrogen can become an integral part of oxide lattices. It can change the crystallographic structure and influence the oxidation state of the metal without lattice oxygen removal or hydroxide formation. The hydrogen can come from metal impregnated onto the trioxide or from a mechanical mixture, although the ease of hydrogen incorporation depends on the nature of the bronze that is formed. Indeed, a variety of other oxides and mixed metal oxides are able to incorporate the atomic hydrogen within the lattice (*126*).

The host oxide lattice, moreover, is able to be reduced by spiltover hydrogen, producing water. Spillover induces lower temperatures of reduction for vanadium, uranium, chromium, cobalt, cadmium, and tin oxides (*127*), among others. The reduction may involve bulk transformation or it may be confined to the surface. The most studied example of this phenomena involves TiO_2 and the resultant changes in sorption capabilities of the surface (SMSI), as discussed above. SMSI seems to be an extreme example of the change in chemisorptive properties with reduction and subsequent occultation of the supported metal.

A variety of surfaces have been shown to react with spiltover hydrogen. The above discussion focused on oxide surfaces where polar association is possible. Boudart *et al.* found that spillover occurs from Pt to carbon at modest temperatures (*114*). Indeed, it has been suggested that a hydrocarbon bridge assists in spillover from Pt onto the supports (*46, 85*). The spiltover hydrogen is able to react with the carbon lattice to form methane (*128*). The most important extension of this concept involves the removal of carbon from Pt, reforming catalysts (*129*). It has been proposed that both oxygen and hydrogen spill over and remove surface coke (*130*). This implies that under certain circumstances catalysts can both contribute to their own destruction (by forming coke precursors) and help reverse the destructive process. By enhancing the spillover there is the potential of increasing the lifetime or decreasing the regeneration requirements for reforming catalysts. It has been suggested that interparticle transport of spiltover hydrogen is inhibited between carbon surfaces (*40*).

Spillover has also been shown to induce the reduction of sulfides, specifically silver sulfide (*131, 132*). In the studies by Fleisch and Abermann (*132*), electron microscopy was used to follow the decrease of the area of a silver sulfide island as a function of time (see Fig. 10). Baker investigated these phenomena in relation to the spillover-induced reduction of titania (giving rise to SMSI) (*98*). The phenomena surrounding spillover are not distinct but reflect diverse extents of interaction.

In Section V below we show that spillover can induce catalytic activity on the support. The nature of the active site created on the support may result from the surface reduction, or the adsorbed hydrogen may be a center and site for reaction (*123*). On the other extreme, spiltover hydrogen has been shown to inhibit ortho-para conversion over sapphire and ruby surfaces

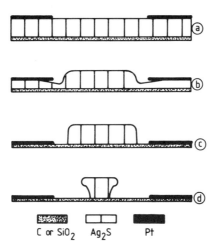

C or SiO_2 Ag_2S Pt

FIG. 10. Schematic cross section through the specimen at characteristic moments of the reduction: (a) unreduced sample; (b) start of the reduction of the sample; (c) start of the spillover measurement; (d) widening of spillover gap after further reduction of the Ag_2S in the Pt shadow area (*131*).

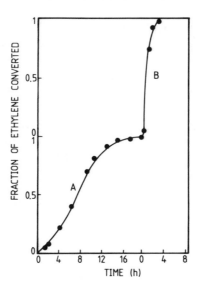

FIG. 11.   Hydrogenation on SiO_2 at 200°C of two consecutive doses (A and B) of ethylene.

(*133*). Similarly, H_{sp} is found to inhibit catalytic reaction on spillover-activated surfaces; this gives rise to an induction period for hydrogenation (shown in Fig. 11) (*134*).

This brings the discussion of the changes in the solid full circle. Spiltover hydrogen can exchange with the surface. It may react with and replace methoxyls with hydroxyls. It may be incorporated into the bulk with a change in the bulk crystal structure. Bulk reduction may occur. The species spilling over may react only with the surface, with coke, or with other sorbed species. In addition, spillover may promote or inhibit reaction on the surface.

The picture that emerges is that spillover provides an indirect source of reactive hydrogen. Reaction and activity modification of the bulk support and of other surface species may occur.

When first proposed, the concept of spillover was not easily accepted. It has been presumed that spiltover hydrogen is present as a single species. Many recent studies have discussed the unique nature of the interface between the activating metal and the support. It is not clear that this is or is not a distinct energetic state. There is very recent evidence that there may be more than one form of adsorbed–spiltover hydrogen. These states are distinguishable from hydroxyls.

Recent studies by Maret *et al.* have investigated the process of spillover activation of amorphous alumina (*135, 136*). If the sample was cooled in H_2 from 430 to 180°C with the source of spillover present (supported Pt), no induction period was found for ethylene hydrogenation. This implies that there is no spiltover hydrogen on the surface. However, if the sample was reheated to 430°C in H_2 for 10 h and cooled back to 180°C without contact between the sample and the source of spillover (Pt), an induction period was found, implying that under these conditions spiltover hydrogen was present. Several conclusions are consistent with these facts. Spiltover hydrogen sorbed on the active sites on alumina is able to desorb during the cooling if Pt is present; otherwise, "reverse spillover" is not easy. An activated alumina surface is able to adsorb hydrogen at 430°C without the help of Pt and restore the spiltover-type hydrogen species; however, the recombination of hydrogen on and desorption from the alumina sites is not easy without Pt. More hydrogen is adsorbed on Al_2O_3 at high temperature without Pt than at low temperature with Pt; this implies a second state of adsorption for any net exothermic processes (compared to H_2 gas; see Section V). This is probably due to the creation of active sites by hydrogen spillover at high temperature.

Additional evidence of the ability of active sites to sorb hydrogen directly comes from recent studies of H_2–D_2 exchange (137). A silica surface that has been activated by H_2 spillover is able to promote H_2–D_2 exchange without the metal present.

Further evidence of the multiple nature of spiltover hydrogen comes from the pioneering studies of Beck and White (44, 109, 138). Hydrogen and deuterium were sorbed separately on Pt/TiO_2 at 227 and 27°C, respectively; the sorptions and evacuation between the sorptions were performed in rapid sequences and the sample was cooled to -133°C and subsequently programmed with an increasing temperature to over 480°C. Separate D_2 (at 77°C) and H_2 (at 327°C) peaks were observed (Fig. 12). An insignificant

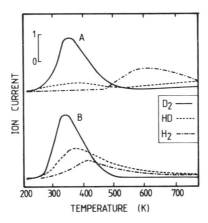

FIG. 12. TPD spectra following sequential dosing of H_2 followed by D_2, each for 100 s at 0.077 Torr: (A) H_2 dosed at 500 K, D_2 dosed at 300 K; (B) H_2 and D_2 dosed at 300 K (109). Reprinted with permission from *Journal of Physical Chemistry*. Copyright (1984) American Chemical Society.

amount of HD was produced, and when Pt was not present, little H_2 adsorption–desorption was found. When both species were sequentially sorbed at 27°C, considerable HD was found and no peak at 327°C was evident.

The authors interpreted the separate peaks and lack of HD as being due to the spatial separation of H_{sp} and D_{sp} on the surface. One of the states was located in the vicinity of Pt and assumed to be H on reduced titania TiO_x, where $1 < x < 2$ (Ti^{3+} ions), the other one being situated in the bulk.

We find this explanation not totally satisfactory. Unless subsequent spillover displaces previous spillover, diffusion will involve a monotonic gradient from the source. Subsequent spillover should not displace but intermix with previously sorbed species. Since it is generally accepted that the spiltover species is atomic, it is difficult to accept that little HD is formed and the desorption peaks occur with a 250°C difference in temperature.These studies seem to give credence to the hypothesis that multiple states of spiltover hydrogen (or deuterium) exist on the surface. Further studies are needed to clarify the nature of the sorbed states, their energetics, and their number on the support surface.

C. THE KINETICS AND EXTENT OF SPILLOVER

It follows from the above discussion that there is a sequence of steps involved in the process of spillover, though few studies have isolated them. A variety of assumptions have been made to simplify the analysis. Further, some of the phenomena associated with spillover, particularly the induction of support activity, have not been fully recognized.

Now, concerning the densities of accepting sites for hydrogen spillover, much data have been published dealing essentially with alumina, silica, and carbon(s). Thus Bianchi et al. (139) measured by volumetric adsorption a concentration of $\sim 10^{12}$ sites cm^{-2} for alumina aerogel at temperatures ranging between 300°C and ambient. Kramer and Andre (18), using TPD of hydrogen on Pt/Al_2O_3 samples, reported values of $\sim 2 \times 10^{12}$ sites cm^{-2} at 400°C and 710 Torr H_2. The same site density ($\sim 10^{12}$ sites cm^{-2}) was also found, this time on silica, by Bianchi et al. (124) using volumetric H_2 chemisorption and confirmed by Lacroix et al. (140), who titrated the spiltover hydrogen with ethylene as the quantity corresponding to the end of the induction period (see below). Sermon and Bond (41) measured the same value by titration of the spiltover hydrogen on silica by pentene-1. The same order of value, that is, 1.5×10^{12} sites (or H) cm^{-2} of SiO_2, was deduced from NMR experiments carried on a Pt–SiO_2 catalyst at room temperature by Sheng and Gay (55).

Hence, at least on alumina and silica, it seems that the density of sites accepting hydrogen spillover is in the range of 10^{12} cm^{-2}—assuming that one accepting site corresponds to one monoatomic hydrogen spiltover species. This figure is extremely low and represents only a fractional surface coverage of about 0.1 %.

The situation is strikingly different on carbon, where concentrations of 10^{13} to 10^{16} sites cm^{-2} have been recently reported by many authors (14, 46). These are in good agreement with results published earlier by Robell et al. (17) and Boudart et al. (40). It appears that carbon is by several orders of magnitude a better acceptor than alumina or silica.

For alumina, the values of the site densities of the accepting centers, 10^{12} cm^{-2}, were fully confirmed in experiments where a physical rather than chemical means of supplying atomic hydrogen species was employed (Pt on alumina, for instance). Gadgil and Gonzalez (117) generated the atoms in the gas phase on a hot tungsten filament and let them adsorb on pure alumina powders at 0°C. They calculated a value of 10^{11} atoms of H cm^{-2} based on desorption data at 250°C for the same alumina aerogel described by Gardes et al. (141). Kramer and Andre (18) reported a value of the same order at 400°C in their work with partially dissociated H_2 obtained through a high-frequency discharge.

The capacity for spiltover hydrogen varies from surface to surface, although there is less discrepancy on similar supports. For oxides, in general, the coverage is less than 1 % of the surface and is, therefore, far less than the number of surface hydroxyls (25).

Quantification of rate constants for this multistep process hinges on the assumed rate-controlling step. Depending on the steps that are assumed to dictate the rate, reaction rates or diffusion constants are calculated from the net kinetics of reaction or sorption. Various studies have assumed that either of two steps are rate controlling: either the surface diffusion or the actual spillover from the source. All analyses have assumed a first-order dependence of the concentrations of atomic hydrogen for each step in the sequence.

During the early studies of spillover, the diffusion of hydrogen away from the source was assumed to be rate controlling. This was based on the pioneering studies of Kramer and Andre (18) for Pt/Al$_2$O$_3$ as well as those of Fleisch and Abermann (132) and Schwabe and Bechtold (131) involving the reduction of Ag$_2$S by spiltover hydrogen. A dependence of the initial rate of reduction on the square of the distance from the islands of Pt to the receding Ag$_2$S surface led the authors to conclude that diffusion was rate controlling. The agreement is expressed in terms of classical diffusion theory, where the concentration is dependent on the square of the distance from the source. However, the production of H_2S would involve two atomic hydrogens. Is the rate, therefore, first order in hydrogen? For a very dilute (generally $<1\%$ of the surface) two-dimensional system, are homogeneous analogs appropriate?

On other surfaces both the concentration and strength of bonding of the spiltover species will vary. The rate-controlling step may shift, and since the activation energies of the individual steps are not equal, the rate control can shift with temperature. For example, the diffusion coefficient may behave as $T^{3/2} \rightarrow T^{1/2}$ for two-dimensional diffusion as a surface gas. If a "jump-like" diffusion occurs from point to point, the dependence may contain an appropriate activation energy.

Nevertheless, based on the assumption that diffusion was the rate-controlling step in the process, the effective diffusion coefficients and activation energies that were found are shown in Table III.

There is considerable variation between the estimated diffusion coefficients and activation energies for similar systems. The techniques used to measure the rate were temperature-programmed desorption (TPD) (*18*), sorption studies (*66, 85*), and electron microscopy (*41, 131, 132*). NMR has also been used to attempt to estimate hydrogen surface diffusion. In 1972, Mestdagh *et al.* (*142*) estimated the diffusion coefficient as 10^{-15} cm^2 s^{-1} at 25°C on Y zeolites. More recently, pulse NMR has been used involving both the temperature dependence of the longitudinal relaxation time and pulsed-field gradients (*92, 143*). For tungsten bronzes at room temperature, diffusion coefficients of hydrogen have been estimated as 7×10^{-6} cm^2 s^{-1} and 7.9×10^{-6} cm^2 s^{-1}, respectively.

Infrared spectroscopy was used to measure the concentration gradient of exchange deuterium from a single point source on silica (*54*). This concentration as a function of time and distance at 200°C is shown in Fig. 13. The diffusion coefficient can be estimated from these studies as 10^{-5} cm^2 s^{-1}. Most importantly, if the diffusion coefficient is this high, then diffusion will not be the rate-controlling step in the sequence.

TABLE III

Estimated Surface Diffusion Parameters

Surface	Temp. (°C)	D_{eff} (cm^2/s)	Distance (Å)	E_a (kcal/mole)	Ref.
Al$_2$O$_3$	400	10^{-15}	2000	28.5	*18*
WO$_3$	50	10^{-16}			*66*
MoO$_3$	50	10^{-13}			*66*
C or SiO$_2$	100	10^{-14}	2000	15.5	*132*
C	200	10^{-12}	1000	21	*131*
C	119	10^{-17}		39.2	*41*
Ce/Y-Zeo.	20	10^{-10}			*85*

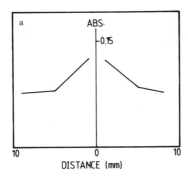 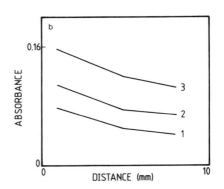

FIG. 13. (a) Infrared absorbance of the OD band at various positions across the sample, 40 min after the beginning of the experiment. (b) Concentration of spiltover deuterium as the function of position for various times; curves 1, 2, and 3 are the absorbance profiles after 10, 40, and 80 min (54).

By measuring the changes of the electroconductivity of a film of ZnO exposed to hydrogen spilt over from Pt/SiO_2 or Pt/Al_2O_3. Lobashina et al. (144) produced evidence for the surface migration of spilling hydrogen atoms across a SiO_2 surface. At 100°C the rate of diffusion was 2×10^7 particles s^{-1}, with an estimated diffusion coefficient on quartz of 1.3×10^{-7} cm^2 s^{-1}. This agrees fairly well with corresponding values published elsewhere for comparable metal oxides (as discussed above).

Several recent studies have involve the kinetics of exchange (59) and demethylation of a silica aerogel surface (123) with a variation of contact between the source of spillover and the reacting surface. Each concluded that the surface migration is the easy step. In agreement with the latter studies (cited above), Dmitriev et al. estimated the diffusion coefficients at room temperature as 10^{-10} cm^2 s^{-1} and at 200°C as 10^{-6} cm^2 s^{-1} (59). Separately, they estimated the OH group diffusion coefficients as 10^{-16} cm^2 s^{-1} at 20°C to 10^{-11} (at 200°C), which agrees with the estimates of the earlier studies. Depending on the dominant mechanism of transport and the temperature, the diffusion may vary by over 10 orders of magnitude.

The variation in rate-determining step with coadsorbed species was noted by Bond in his recent review (9). Bond cited the extensive evidence that spillover (passage from the metal to the support) can be rate determining. For systems where H_2O (or other sorbed species) promotes spillover, it may do so by increasing the interfacial transport, a potentially slow step in the process.

Our interpretation of these diverse studies is that for most systems the spillover from metal to support is the rate-controlling step, although this may not always be the case. Our intuition is that spiltover hydrogen has only a weak bond with the support compared to its bond with the metal. The spillover step must be either endothermic or close to energetically neutral—at least it should not be very exothermic. Unless the activation energy from position to position on the support is high, the spillover will be rate determining. If the activation energy from point to point were great, desorption would be favored during the transition. In the experiments involving atomic hydrogen produced by discharge, the ability of this atomic hydrogen to induce activity was reduced compared to hydrogen spillover at higher temperatures on dispersed metallic systems (*25, 71, 117, 118*). The energetic coordinate consistent with these results is shown in Fig. 14.

Considerable speculation has concerned the mechanism of spillover and the species involved. As discussed above, the mechanism depends on the specific system (notably coadsorbents) studied. Similarly, the dominant rate-controlling step is usually the spillover from metal to support; however, coadsorbents and the nature of the metal–support interface can affect the relative dominance of this step.

To access the potential influence of spillover on catalysis and interfacial transport, more qualitative studies are required. Further, it is, for instance, necessary to isolate the individual steps in the phenomena and account for the reaction kinetics of the process. As an example, what is the difference between inter- and intraparticle transport on the support?

Further cautions should be discussed. Whereas transport of hydrogen may occur at temperatures well below 400°C, the induction of catalytic activity on the support by spiltover hydrogen is an activated process and requires considerable time (up to 12 h of treatment at 430°C in hydrogen). Comparison of catalytically active surfaces must be done with similar temperatures and times of spillover pretreatment. To further complicate the analysis, there is evidence that an activated support (e.g., Al_2O_3) may be able to dissociate hydrogen. The process may, therefore, be autocatalytic: that is, the support first activated by spillover may be able to adsorb, dissociate, spill over, and consequently activate more support surface (*137*).

The relative significance of spillover can only be put in perspective if we isolate the specific steps in the phenomena and quantify the relative rate of the steps. *A priori* assumptions as to the controlling steps should be considered with caution.

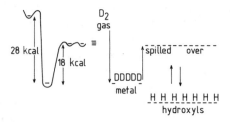

FIG. 14.   Energy diagram of the surface species (*54*).

IV. Spillover of Species Other Than Hydrogen

Considerable experimental evidence has been collected for the spillover of hydrogen species. Hydrogen is not, however, the only species which can migrate from an activator to the support (see below). Although the amount of data concerning the spillover of other species is much smaller, it seems at the present time that the spillover of oxygen, carbon monoxide, isocyanate, and various organic radicals is well established, with a marked abundance of data concerning oxygen spillover.

A. SPILLOVER OF OXYGEN AND OF CARBON MONOXIDE

The oxidation of carbon is the simplest reaction involving oxygen spillover. For the graphite basal plane there is no need for a metallic activator. Indeed, oxygen is first chemisorbed on the basal carbon and then migrates (spills over) to carbon atoms exposed at the steps between layers on the graphite surface, where it reacts with the edge carbons. The reaction also proceeds by a direct collision of O_2 molecules with these carbons (*145*). A more pertinent example of oxygen spillover in the catalytic oxidation of graphite at 650°C (20% O_2 in Ar) is given in the work of Yang and Wing (*146*), where graphite was admixed with carbides (WC, TaC, and Mo_2C) or oxides (Cr_2O_3, WO_3, Ta_2O_5, and MoO_3). The transition-metal carbides exhibit electronic structures similar to Pt and are able to dissociate O-O and H-H bonds. It is therefore not surprising that these carbides behave as dissociation centers of O_2, providing O atoms which diffuse across the basal plane of graphite and react with carbon atoms at edges of the monolayer pits. These pits (TEM) are hexagonal and similar to those created by *atomic* oxygen; however, for the reaction carried out in O_2 in the absence of a carbide catalyst, the pits are circular. In the presence of oxide catalysts, deformed circular pits are found. Therefore, the mechanism advanced for the oxides is not the same for the metal carbides. This mechanism takes into account the fact that the reaction temperature is near or above the Tammann temperature of the oxides. Molecular species or clusters of the oxides migrate on the basal plane of graphite and are subsequently trapped on the edges of the etch pits. Gasification of the edge-carbon is thus catalyzed by direct contact with the trapped oxide. In the case of carbide catalysts the spillover of oxygen on the surface of graphite seems to be well established; in the case of oxide catalysts, it is rather the spillover of the oxide which finally promotes the reaction.

This explanation concerning the role of oxide catalysts in the gasification does not hold for temperatures below the Tammann temperature. In the kinetic studies of gasification of coal without an activator, it is frequently assumed that the reaction is not catalytic but is induced by direct gas–solid interaction. On the other hand, the gasification of carbon from lignite, in air, between 252 and 452°C, is accelerated by the presence of $CaO-CaO_2$ (*147*). The dissociative chemisorption of O_2 occurs on CaO (and also probably on carbon), and the extra oxygen (in CaO_2) spills over onto carbon, where the gasification occurs (a redox mechanism). Now, active carbon may be completely oxidized at low temperatures ($\sim 500°C$) by composite catalysts (*148*). This carbon (1230 m^2 g^{-1}) is simultaneously a support of the catalyst-activator $Fe-La_2O_3-Pt$ and a reactant. Pt is proposed to adsorb oxygen and the particles of $Fe-La_2O_3$ allow its spillover to carbon. For a single-component catalyst the sequence of the catalytic activity in the gasification of the active carbon is Cu > Co > Fe > Ni > Pt. It may again be possible that an activator redox mechanism prevails in the spillover of oxygen onto carbon.

In the presence of H_2O between 652 and 852°C, Pt is more efficient than other metals or oxides (*149*) in the gasification of Graphon, Sterling F-T, and Norit-A carbons. Since the reaction products are $CO + H_2$ (and only CH_4 if the gasification is carried out in H_2), the spiltover species may be the products of dissociation of H_2O, probably H, O, and OH.

Hydrogen and oxygen spillover from Pt/Al_2O_3 catalyst were advocated by Parera *et al.* (*130*) in the elimination of coke during naphtha reforming. Indeed, coking pure Al_2O_3 for short times by passing naphtha or methyl-cyclohexane produces a coke which can be *partially* eliminated by direct hydrogenation and *completely* eliminated if hydrogenation is performed in the presence of admixed Pt/Al_2O_3 catalyst. Similarly, in the presence of this catalyst the coke is eliminated by oxidation (as CO_2) at lower temperatures than without Pt/Al_2O_3. As in the case of hydrogen, oxygen spillover from the catalyst is implicated in this gasification.

The oxidation of reactants other than carbon also has been found in spillover catalysis. Bond *et al.* (*150*) reported one of the first examples of the spillover of species other than hydrogen. Pd, pure or supported on SiO_2, and pure SnO_2 are well-known catalysts of the oxidation of CO. For SnO_2 as for other oxides the redox mechanism is assumed and the partial kinetic dependencies (orders) for the reaction are 0.2 for O_2 and 1 for CO, whereas for Pd they are 1 for O_2 and -1 for CO. Now for Pd admixed with SnO_2 the partial orders are completely different: 0.5 for O_2 and 0 for CO. A synergetic effect of the mixture $Pd-SnO_2$ is observed in the conversion of CO at 150°C with similar contact times. For Pd/SiO_2 the CO conversion is 0.03 % and for SnO_2 it is 0.23 %. When the two components are mixed together (without grinding) the conversion increases to 0.33 %, but when they are mixed *and* ground the conversion peaks to 3.75 %. The role of each activator was unraveled through a gravimetric study of reduction by CO and reoxidation of SnO_2. The initial rate of reduction increases in proportion to the Pd content in the mixture (up to 4.5%). Similarly, the rate of reoxidation increases proportionally to this content (also up to 4.5%). This behavior shows that CO is chemisorbed on Pd and then spills over to SnO_2, producing a faster reduction than in the absence of Pd. The concentration of reducing species on the surface of SnO_2 is therefore higher in the presence of Pd than in its absence, where CO can only come from the gas phase. Similarly, oxygen is chemisorbed on Pd, and once dissociated, spills over to reduced SnO_2.

In the catalytic oxidation of CO in the presence of pure SnO_2, the reduction of the catalyst by CO (redox mechanism) is the rate-limiting step (order 1 for CO). If in the mixture $Pd-SnO_2$ the spillover of CO from Pd to SnO_2 is faster than the spillover of O atoms, this creates a higher concentration of CO chemisorbed on SnO_2. The reoxidation of the catalyst becomes the rate-limiting step and the orders are now 0 for CO and 0.5 for O_2, as noted above.

In this example CO and oxygen are the species which simultaneously spill over from Pd to SnO_2. In a recent paper Inui *et al.* (*37*) propose a simultaneous spillover of H_2 and CO for Ni + Ru + La_2O_3 supported on SiO_2, methanation catalysts, by dynamic adsorption studies. For this three-component catalyst the adsorption of H_2 and CO is increased with respect to a single component Ni/SiO_2 catalyst. This behavior is explained by the role of Ru as the transport agent of reducing gases to Ni, which is the acceptor. The La_2O_3 is proposed only to increase the dispersion of Ni.

Oxygen spillover was also advocated for reactions other than simple oxidations. Tascon *et al.* (*151*) studied the transformation of formamides to nitriles at 275°C on BiPMo catalysts prepared by a physcial mixture of powders of MoO_3 and $BiPO_4$:

$$\underset{\text{formamide}}{\overset{\displaystyle O \atop \displaystyle \|}{R-NH-C}} \overset{-H_2O}{\xrightarrow{\hspace{1cm}}} \underset{\text{isonitrile}}{(R-N{=}C{:})} \xrightarrow{\hspace{1cm}} \underset{\text{nitrile}}{R-C{\equiv}N} \qquad (18)$$

The activity and the selectivity of this catalyst remain stable with time on stream *in the presence* of O_2, which is not a reactant. The synergetic effect of the two components depends, as we would expect, on the quality of the contacts between the two solids. The explanation concerning the effect of oxygen is based on the assumption that MoO_3 is the active phase. But MoO_3 may be slightly reduced by the organic reactant, giving rise to the formation of Mo^{5+} species, which are known as acid centers and are active in dehydration. The centers which are too much reduced lose their activity. The role of $BiPO_4$ is to activate (or to reactivate) the first phase. This activation involves the adsorption and dissociation of O_2 by $BiPO_4$ which spills over onto MoO_3 in order to maintain this last phase in the optimum conditions of activity and selectivity (reduced slightly). This emphasizes the requirement of a small amount of O_2 in the stream of reactants (formamide). Only one phase is catalytically active (MoO_3 containing Mo^{5+}), but the second phase and a gas which is not a reactant (O_2) promote the activity of the first phase by spillover.

Ru and Rh (on alumina) are known to be the active components of catalysts for the simultaneous removal of NO_x, CO, and hydrocarbons in automotive exhaust gas. The products of this catalysis are N_2, CO_2, and a small amount of NH_3. Rh may be used in both oxidative and reducing atmospheres, whereas Ru shows low activity in an oxidative atmosphere. If Rh is alloyed with Sn, the temperature of the simultaneous reduction of nitric oxide and oxidation of carbon monoxide is lowered (152). Chemisorption of oxygen (by dissociation of NO) on Rh decreases the rate of decomposition of NO. But Sn is an oxygen scavenger since it is more easily oxidized than Rh. The authors assume that oxygen formed from NO on Rh migrates onto Sn by spillover and is removed by CO (as CO_2), which maintains the activity of Rh. The bond of Ru with oxygen is stronger than that of Rh. Also, the spillover of oxygen from Ru to Sn (in the Ru–Sn alloy) proceeds to a much smaller extent (153). The cleaning effect of Sn on oxygen adsorbed on Ru and Rh in the oxidation of CO by NO or by O_2 was recently reported by Masai *et al.* (154). The affinity of oxygen for the three metals decreases in the sequence Sn > Ru > Rh. The spillover of oxygen from Rh or Ru to Sn maintains "clean" Rh or Ru sites for the adsorption of NO or CO. This mobility of oxygen from Rh or Ru toward Sn increases the chance of recombination of oxygen atoms, which allows the desorption of O_2. The possibilities of the oxidation of Sn into SnO_2 [which is a catalyst for the CO oxidation (150)] and of a subsequent redox mechanism were not hypothesised by these authors (154).

Finally, oxygen spillover has also been advocated in typically inorganic reactions. In an earlier work by Batley and Ekström (155), the phenomenon of spillover was not mentioned. Instead, the authors referred to a topochemical heterogeneous catalysis for the reaction of UF_4 with O_2. This, however, can be understood through the spillover of oxygen. The reaction

$$2UF_{4(s)} + O_{2(g)} \longrightarrow UO_2F_{2(s)} + UF_{6(g)} \tag{19}$$

is accelerated if the solid reactant (UF_4) is physically admixed with a supported metal like Pt, Ir, Os, Pd, Rh, Au, or Ag. The solid UF_4 may also be directly used as a support of these metals. In this case the rate of the oxidation increases as the distance between the reactant and the catalyst decreases. In all cases the effect of the catalyst is to increase the preexponential factor of the Arrhenius equation, but the activation energy of the reaction remains the same for all catalysts. The variation of the catalytic activity is therefore explained by the differences in entropy of adsorption of O_2 on these various metals. The rate of the reaction is controlled by the rate of formation of active species on the metal surface, or more probably by the concentration of adsorbed oxygen and the rate of oxygen spillover from the metal to UF_4. Since the activation energy of oxidation does not change for all metal catalysts employed, it is probable that the bond strength between the metal and adsorbed oxygen is not involved.

More recently, Chadwick and Christie reported evidence for oxygen spillover in the oxidation of lead monolayers on copper (*156*). Polycrystalline copper or single crystal (111) or (210) surfaces were covered by less than a monolayer of lead. O_2 reacts with Pb to give PbO. If a complete monolayer of Pb was formed on Cu, the rate of oxidation of Pb decreased. Now H_2S reacts with Cu but not with Pb. If H_2S was adsorbed at saturation on Cu containing less than a monolayer of Pb (which does not adsorb H_2S), the oxidation of Pb again decreased. These results were explained by the spillover of oxygen from Cu to Pb (and its oxidation into PbO). Oxygen was therefore adsorbed on a free surface of Cu, not covered either by Pb or by H_2S, and then spilled over to Pb.

B. Spillover of Isocyanate Species

In connection with the problem of the automotive exhaust gas catalysis (NO + CO), Solymosi *et al.* (*157*) found that for supported Pt the support affected the formation and the stability of surface isocyanate (–NCO) adsorbed on Pt. The stability of isocyanate on Pt increased in the sequence $Pt/TiO_2 < Pt/Al_2O_3 < Pt/MgO < Pt/SiO_2$. But in the case, for instance, of Pt/Al_2O_3, the number of isocyanate species exceeded the number of Pt surface atoms. The authors concluded that isocyanate migrated from Pt to Al_2O_3. On the contrary, the lack of stability of isocyanate on Pt/TiO_2 tends to show that either isocyanate did not migrate onto TiO_2 or, if it did migrate, it reacts with the oxygen ions of TiO_2. At around 250°C during the catalytic exhaust-gas reaction (CO + NO), the isocyanate was formed on Pt according to the sequence

$$Pt\text{–}NO + Pt \longrightarrow Pt\text{–}N + Pt\text{–}O \qquad (20)$$

$$Pt\text{–}N + CO \longrightarrow Pt\text{–}NCO \qquad (21)$$

The spillover of isocyanate would be represented by

$$Pt\text{–}NCO + support \longrightarrow support\text{–}NCO + Pt \qquad (22)$$

This research was carried out mainly on silica-supported metals using IR spectroscopy. The asymmetric stretching vibration of Pt–NCO gave a band at 2180 cm^{-1}, whereas that of Si–NCO was shifted to 2310 cm^{-1}.

If the reaction NO + CO is carried out at 200 to 400°C with Pt/SiO$_2$ catalyst separated by 8 to 10 mm from a wafer of pure SiO$_2$, the band at 2310 cm^{-1} which is characteristic of Si–NCO is not detected on the silica wafer. The spillover of –NCO from Pt to SiO$_2$ therefore cannot proceed through the gas phase (*158*). But isocyanate on pure silica may also be formed by the reaction with isocyanic acid (HNCO) between 100 and 400°C. Now on Pt/SiO$_2$, –NCO is formed on Pt even at −83°C (band at 2180 cm^{-1}), which shows a dissociative adsorption of isocyanic acid on Pt:

$$HCNO + 2Pt \longrightarrow Pt–H + Pt–NCO \tag{23}$$

This species may spill over from Pt to SiO$_2$ by a mere heating in vacuum of the previous system. Indeed, the band at 2310 cm^{-1} (–NCO on SiO$_2$) started to appear at room temperature and was maximum at 200°C, whereas the band at 2180 cm^{-1} (–NCO on Pt) decreased simultaneously. This research is one example of the use of IR spectroscopy to follow the spillover of adsorbed species. In the same work (*158*) it was reported that chemisorbed oxygen increased not only the stability of –NCO on Pt but also the extent of spillover of –NCO onto SiO$_2$. Other metals like Rh or Pd supported on SiO$_2$ also favor the formation of metal–NCO by interaction with HNCO at −83°C. The isocyanate species spills over onto SiO$_2$ at 100 to 150°C but simultaneously it is partially decomposed into CO, which remains adsorbed on the metal (*159*). It is of interest to note that Si–NCO, formed on Rh/SiO$_2$ catalyst by the interaction of NO + CO, was very stable, as shown by Hecker and Bell (*160*). It is not decomposed at around 330°C by 100% of H$_2$, 2% of O$_2$, 2% of NO, 3% of CO, or even 3% of H$_2$O. In the presence of air at 25°C Si–NCO disappears after 40 h.

A second argument supporting the proposed spillover of –NCO from Rh to SiO$_2$ is based on the following observations. During the interaction of NO + CO at 210 to 242°C, the species Rh–NCO was formed almost instantaneously (band at 2170 cm^{-1}), whereas the species Si–NCO was formed after a much longer period (165 min for increase of the band at 2300 cm^{-1}). When the reactants (NO + CO) were exhausted, the species Rh–NCO disappeared, whereas the species Si–NCO remained stable. The mechanism (*160*) is very much similar to that reported previously for Pt:

$$Rh–N + CO \longrightarrow Rh–NCO \tag{24}$$

$$Rh–NCO + SiO_2 \longrightarrow SiO_2(–NCO) + Rh \tag{25}$$

with simultaneous decomposition of Rh–NCO:

$$Rh–NCO \longrightarrow \tfrac{1}{2}N_2 + CO + Rh \quad (or\ Rh–CO) \tag{26}$$

C. Spillover of Organic Species

Examples of the spillover of organic species are rather scarce; direct evidence is rarely provided. One of the first reports by Webb and MacNab (*161*) concerned the reaction of 1-butene with D_2 over Rh/SiO_2 catalysts. Indeed, hydrogenation and exchange of 1-butene with D_2 occurred on the metal, but isomerization was correlated with the presence of silica. This example may have a more general application to dual-function catalysts, and therefore deserves a more detailed analysis. It now becomes apparent that the origin of the exchange activity (with D_2) seems to involve the terminal hydroxy (or OD) groups on the surface of SiO_2 or Al_2O_3 rather than the acidic properties of the oxide. We have seen that hydrogen can migrate or spill over from the metal to the support and vice versa. A support with a metal on it can easily be deuterated by D_2 (–OD groups) at moderate temperatures. The question now is whether a hydrocarbon can also migrate from the metal to the support. Since in the absence of Rh, silica itself shows no isomerization activity, the reversible adsorption of 1-butene on pure silica does not seem to occur to any appreciable extent. In the presence of Rh this isomerization proceeds easily. Therefore Rh must activate the surface of SiO_2 by either (i) creating sites on the silica surface (in the presence of H_2 or D_2) capable of adsorbing 1-butene from the gas phase or (ii) adsorbing 1-butene, which then migrates from the metal to the support before undergoing isomerization. The second hypothesis is preferred by Webb and MacNab (*161*); it will be shown, however, in the next section that the first hypothesis is supported by several other studies, that is, studies involving the reactions of hydrocarbons occurring on SiO_2 or Al_2O_3 after exposure of these oxides to spillover at high temperatures. Subsequently the metal activator was removed. The ability of the support directly to adsorb a hydrocarbon helps us to understand the exchange capacity of the hydrocarbon on the surface via the $-OH \leftrightarrow OD$ groups induced by the spillover of deuterium from the metal.

An infrared study of the surface interaction between H_2 and CO_2 over Rh on various supports led Solymosi *et al.* (*162*) to conclude that spillover is *not* involved in the formation of formate species. Indeed, in the case of Rh supported on MgO, TiO_2, or Al_2O_3 (but not SiO_2), the presence of H_2 strongly increases the adsorption of CO_2 at 100°C. The presence of formate ion on the support was identified. Two routes for its formation are possible: (i) it is formed on Rh by

$$Rh + H_{(ads)} + CO_{2(g)} \longrightarrow Rh-O-C\overset{\displaystyle O}{\underset{\displaystyle H}{\big\|}} \qquad (27)$$

and the formate ion migrates (spills over) onto the support (other than SiO_2); (ii) hydrogen is activated on Rh and spills over onto the support, where it reacts with adsorbed CO_2 to give the formate ion.

For Rh/SiO_2 the presence of H_2 did not increase the CO_2 adsorption, which tends to favor the second hypothesis. Indeed, CO_2 is only very weakly adsorbed on SiO_2, and HCOOH is adsorbed on silica in a molecular and not in an ionic form. Since the IR spectra provides no evidence for the presence of the formate group on Rh with either Al_2O_3, MgO, or TiO_2 supports, the second proposal is preferred by the authors.

Different conclusions concerning the methoxy species and also the formate species have been reached by Palazov *et al.* (*163*), who used IR spectroscopy to study the interaction between CO and H_2 on supported Pd. The adsorption of methanol on pure Al_2O_3 (but not on SiO_2) at 35°C produced methoxy species:

$$\begin{array}{c} CH_3 \\ | \\ O \\ | \\ -O-Al-O \end{array} \qquad \text{(bands at 1480, 2850, and 2960 cm}^{-1}\text{)}$$

whereas at 210°C and above formate species are formed:

$$\begin{array}{c} HC-O \\ | \quad | \\ O-Al \end{array} \qquad \text{(bands at 1390 and 1590 cm}^{-1}\text{)}$$

The same IR bands are produced on Pd/Al_2O_3 (but not on Pd/SiO_2) by the interaction of $CO + H_2$. For high H_2/CO ratios methoxy species predominate, whereas for low H_2/CO ratios (~ 3) the formate species is found. The reactants CO and H_2 can only react on Pd, but the species produced may migrate onto Al_2O_3, which is a "trap" for those formed on the Pd. Indeed the methoxy species is unstable at room temperature on the surface of the metal, and in the presence of an excess of $H_{2(g)}$ it migrates onto the alumina. If the gas phase is evacuated, the methoxy species migrate (by reverse spillover) onto Pd, where they decompose, leaving adsorbed CO. For $H_2/CO \approx 3$, the formate species is formed at 122 to 222°C on Pd and spills over onto Al_2O_3, where it is strongly adsorbed. By increasing the temperature the IR bands of formate decrease, and methane is detected. This product can only be formed on Pd by reverse spillover of formate species. It appears therefore that for the synthesis of CH_3OH from $CO + H_2$, the intermediate species is the methoxy group, whereas for methanation the intermediate species is the formate ion. These studies therefore favor the spillover of methoxy and formate species from Pd to Al_2O_3 as the interaction of CO and H_2 occurs *on* palladium (IR evidence of interaction between preadsorbed CO and H_2) where the surface species (methoxy or formate) are produced.

A slightly different approach to the interaction of $CO + H_2$ in the methanation on Pd/TiO_2 catalysts was suggested by Bracey and Burch (164). Pd/TiO_2 catalysts, either in the SMSI state (reduced at 500°C) or not (reduced at 200°C), always exhibited higher activity for the methanation than Pd/SiO_2. It was concluded that this high activity was not due to SMSI but to new active sites at the interface between the metal and the support (TiO_2). The support acts directly in the catalytic process by adsorbing or assisting in the adsorption of CO. The active site in the case of titania is assumed to be an exposed Ti^{3+} cation (reduction of Ti^{4+} by hydrogen spillover) on the support, adjacent to a normal metal particle. This exposed cation is then able to hold the CH_xO species (formate or methoxy). In the author's opinion, however, whether CH_xO species are formed first on one phase (either on the metal or on the support) and are transferred to the other phase is an open question (164).

A somewhat different model of the active site for $CO + H_2$ reaction on Pt/TiO_2 catalysts has been proposed by Vannice and Sudhaka (165). A reverse procedure for the preparation of such catalysts was used. Platinum powder ($0.5\ m^2\ g^{-1}$) was impregnated by titanium n-nonylate $[(C_9H_{19}O)_4Ti]$ and the compound was then converted, by oxidation, into TiO_x. Either a hypothetical monolayer or 10 monolayers were formed on Pt powder. The oxide was finally reduced at 200°C (below the temperature needed to produce the SMSI state) or at 500°C (SMSI state temperature). Even after reduction at 200°C, the activity of these catalysts in methanation was much higher than that of Pt powder. The authors assume that TiO_x is partially reduced by hydrogen spillover from Pt, and Ti^{3+} ions and in concert anionic vacancies are created. These new sites in the neighborhood of Pt are involved in the chemisorption of CO, which adsorbs on Pt with its oxygen in the vicinity of anionic vacancy on TiO_x. This interaction assists in the dissociation of CO, which is the rate-limiting step for methanation on Pt or Pd catalysts.

V. Spillover Activation of the Support and Induction of the Catalytic Activity

After the studies of Khoobiar et al. (4) showing that hydrogen spilt over from Pt can convert WO_3 into a hydrogen bronze at room temperature, a trend developed to consider this hydrogen only as a reactant for various hydrogenations. In particular, it has been presumed only to increase the amount of available hydrogen due to spillover from a supported metal admixed or in contact with a support.

A. INTERACTION BETWEEN SPILTOVER HYDROGEN,
THE SUPPORT, AND THE REACTANTS

The first attempt at taking advantage of spiltover hydrogen for a hydrogenation reaction seems to have been that of Sinfelt and Lucchesi (5), who
observed that the rate of hydrogenation of ethylene was increased for a
Pt/SiO_2 catalyst which was physically admixed with several times its volume
of pure Al_2O_3. Similarly, Khoobiar *et al.* (4) studied the conversion at 425 to
480°C of cyclohexane into benzene (naphthene dehydrogenation) in the
presence of $Pt(0.5\%)/Al_2O_3$ catalyst admixed with pure Al_2O_3 in the
proportions of 1:80 to 1:5000 (by volume). Only a fraction of the observed
conversion could be ascribed to the "catalytically active" material
(Pt/Al_2O_3). Since the inert packing in the reactor (Al_2O_3) had been shown to
be completely inactive by itself, the reaction was proposed to start on the
surface of Pt/Al_2O_3 and then to continue on inert pellets of Al_2O_3, Khoobiar
postulated the hydrogen atoms spill over from Pt to the Al_2O_3 diluent
and dehydrogenation of cyclohexane to benzene occurs according to the
mechanism

$$n\text{H}_{sp} + m \bigcirc \; \rightleftharpoons \; \bigcirc \; \rightleftharpoons \; \bigcirc \; \rightleftharpoons \; \bigcirc + (n + 6m)\text{H}_{sp} \quad (28)$$

This mechanism was later shown to occur by reverse spillover (11, 41) and a
similar mechanism may be inferred for the dehydrogenation of *n*-heptane as
discussed in Section V,D,4 and 5.

The migration of spiltover hydrogen onto the admixed support has been
demonstrated more recently by deuterium exchange experiments (see Section
II). Oxides other than alumina or silica are also effective in receiving the
spiltover hydrogen. It has been shown by Kosaki *et al.* (166) that on
Pt/Al_2O_3 catalyst the reduction by H_2 of nitric oxide at 300°C produced
NH_3, which subsequently reacted with NO to give N_2 and N_2O. If the
catalyst was physically admixed with oxides like V_2O_5, MoO_3, WO_3, or
SnO_2, the selectivity for NH_3 production decreased. The first step is the
formation of ammonia, which, however, decreases as the availability of
spiltover hydrogen (from Pt) decreases. This is because the admixed oxide is
being reduced by the spiltover hydrogen. More recently, Antonnucci *et al.*
(167) studied the hydrogenation of benzene over Pt/Al_2O_3 catalyst diluted
with pure Al_2O_3 in proportions varying from 1:3 to 1:100 (by volume). The
activity per gram of Pt increased with the dilution up to a ratio 1:50. At this
dilution the activity was the same as that of the catalyst diluted in a ratio
1:100. Since the activation energy of the reaction (between 60 and 120°C) is
the same for all dilutions (9.2 kcal mol^{-1}), the slowest step does not seem to
be the spillover of hydrogen on alumina, whose activation energy, as
determined by Kramer and Andre (18), is of the order of 28 to 30 kcal mol^{-1}.

The rate-determining step would be the reaction between benzene, adsorbed or gaseous, and the spiltover hydrogen. The mechanism of the reaction on pure or diluted catalyst would be the same, with the same activation energy. It is known that benzene adsorbed on Pt (partial order 0 for C_6H_6) interacts with atomically chemisorbed hydrogen. This suggests a similar mechanism on the diluent involving the spiltover hydrogen. The rate of hydrogenation would be the sum of the rates on the Pt/Al_2O_3 catalyst and on the Al_2O_3 diluent.

Lau and Sermon (168) showed that ethylene may be hydrogenated at 200°C on $SiO_2-Al_2O_3$ (13% or 25% Al_2O_3), MoO_3, $\gamma-Al_2O_3$, but not on pure SiO_2, if hydrogen initially flows over a bed of $Pt(1.7\%)/Al_2O_3$ catalyst. The reactant ethylene was diluted with N_2 ($N_2:C_2H_4 = 1000:1$) and then passed over the oxide without exposure to the Pt. An interesting point, which will be developed in the next subsection, is that the whole content of the reactor (the two superimposed beds) had been, prior to the reaction, reduced by H_2 at 200 or 300°C. If the bed of Pt/Al_2O_3 catalyst was removed, ethylene was still hydrogenated (with a smaller rate) on the remaining bed of oxide ($SiO_2-Al_2O_3$ or MoO_3 but not on pure Al_2O_3). This behavior shows that the oxide in the second bed may become catalytically active for ethylene hydrogenation after activation by hydrogen spillover. Tests on pure alumina are negative and also favor this interpretation (see below).

B. Reverse Spillover on Carbon

In the preceding subsection it was pointed out that hydrogen spilt over from the metal onto the support may be used by an accepting reactant mainly for hydrogenation. It is appropriate to mention at this point that reactions releasing hydrogen may occur on the support. For those that generate atomic hydrogen, the hydrogen atoms may migrate to the metal (reverse spillover), where they recombine and desorb as molecules. Aspects of reverse spillover have been discussed above. A typical example of the reactions favored by reverse spillover will now be briefly considered.

The dehydrogenation of isopentane, giving isopentenes, or cyclohexane, given benzene, may be performed on active carbon between 380 and 450°C. These reactions are accelerated if a transition metal (up to 5%) is deposited on the active carbon. An increasing rate is also observed without the deposition of the metal but with a hydrogen acceptor, like C_2H_4 or NO, in the gas phase. The accelerating process is certainly related to the desorption of adsorbed hydrogen generated on the surface, as Fujimoto et al. have shown (169). These authors also observed that the adsorption of hydrogen on carbon is accelerated by the presence of a deposited metal. The maximum amount of H_2 adsorbed does not depend on the nature of the metal but only on the nature of the active carbon. This behavior tends to show again that the metal is only a porthole (or gate) for the adsorption (dissociative) of hydrogen; the maximum amount of spiltover hydrogen depends on the nature of the acceptor and not of the activator. The authors observed a correlation between the initial rates of adsorption of hydrogen at 400°C (the initial rate of hydrogen spillover) and the rates of dehydrogenation of cyclohexane at the same temperature on active carbons containing a supported metal. Similarly, TPD experiments showed an increase of the rate of desorption of H_2 in the presence of a deposited metal (*14*). This behavior suggests that the rate of the reverse spillover of hydrogen controls the rate of dehydrogenation. Without the metal activator the rate-determining step would be the migration of generated hydrogen on active carbon and finally its desorption from a particular carbon porthole.

C. Spiltover Hydrogen Fixed by the Support (Bronzes)

The formation of WO_3 and MoO_3 bronzes with the spiltover hydrogen has already been reported in the preceeding sections. At the present stage it is appropriate to emphasize that the bronzes formed in this way may become catalytically active. In the earlier work of Marcq *et al.* (*170*) the activity for the hydrogenation of ethylene was not attributed to the MoO_3 bronze but to the Pt deposited (0.05 to 2%) on the MoO_3 support that facilitates bronze formation (with H_2 at 60°C). Ethylene can then be hydrogenated, in the absence of gaseous hydrogen. Hydrogen comes from the bronze by reverse spillover to Pt, where the hydrogenation takes place. Gaseous oxygen is a better acceptor (on Pt) of this hydrogen than ethylene. In both cases the rate-limiting step is the migration of hydrogen toward Pt, which may involve the ease of release of hydrogen by the bronze and its reverse spillover to Pt.

In a more recent work, Marcq *et al.* (*171*) prepared, by the same method, hydrogen bronzes of V_2O_5 of composition $H_{3.3}V_2O_5$. The initially crystalline V_2O_5 leads to an amorphous bronze. Hydrogen included in the bronze may be partially recovered by reaction with ethylene at 100°C or by desorption at 120°C. The titration by H_2 of an exhausted bronze was used to determine the initial composition. In these studies the Pt was still present on the oxide support and any activity of the bronze could not be studied independently in the presence of the gaseous H_2. The bronze serves mainly as a reservoir of hydrogen in the absence of hydrogen gas. Very recently, however, it has been definitely shown by Benali *et al.* (*172*) that at least one composition of hydrogen bronze of MoO_3 is a catalyst for the hydrogenation of ethylene by molecular hydrogen *in the absence of Pt*. When Pt is deposited

(0.1 %) on MoO_3, as in the earlier work, four bronzes H_xMoO_3 ($x = 0.34$, 0.93, 1.68, and 2) can be formed in the presence of H_2 at temperatures between -10 and $60°C$. However, the bronze $H_{1.6}MoO_3$ can also be obtained with H_2 at $160°C$ without direct deposition of Pt on the surface of MoO_3. Pt/Al_2O_3 is contacted with the MoO_3 and exposed to H_2. The Pt/Al_2O_3 is then removed from the reactor by a windlass device which will be described below. Hydrogenation of ethylene on that bronze in the absence of gaseous H_2 (and of Pt) starts at about $80°C$ and proceeds via a progressive transformation of $H_{1.6}MoO_3$ into $H_{0.9}MoO_3$ (structurally different). In this experiment the bronze is again a reservoir of hydrogen but is not the catalyst. It is, however, interesting to note that a Pt porthole is no longer necessary to induce the reverse spillover of hydrogen toward ethylene. The initial rate of hydrogenation of ethylene at $160°C$ is comparable to that observed by Marcq et al. (170) on $Pt(2\%)/H_{1.6}MoO_3$. The bronze $H_{1.6}MoO_3$, without Pt, seems to have very efficient sites (other than Pt) for the reverse spillover of hydrogen and its capture by ethylene. The apparent activation energy of hydrogenation of ethylene by hydrogen from the bronze $H_{1.6}MoO_3$ is 13 kcal mol^{-1}, which is comparable to the value of activation energy of diffusion of protons in this bronze (173).

In the presence of gaseous H_2, $H_{1.6}MoO_3$ bronze (without Pt) catalyzes the hydrogenation of ethylene, between 120 and 160°C, without structural modifications of this bronze. In the presence of an excess of H_2 the rate of hydrogenation is constant with time (zero order) almost up to a complete exhaustion of ethylene in a batch reactor. Many doses of this reactant can be hydrogenated in this manner. The bronze is functioning as a catalyst. The activation energy in this process is 9 kcal mol^{-1}. The rate of hydrogenation of ethylene, which is constant with time, increases proportionally to the initial partial pressure of ethylene (first order in ethylene) to a maximum where the reaction order becomes zero in ethylene (site saturation). This behavior shows that hydrogenation sites (not involving Pt) are created by the interaction between the bronze and the reactants. This is the first study to demonstrate that a bronze can catalyze ethylene hydrogenation ($H_2 + C_2H_4$) without the presence of the metallic activator (Pt).

D. ACTIVATION OF THE SUPPORT BY HYDROGEN SPILLOVER

It has been shown in the previous sections that the addition of small amounts of a transition metal to various metal oxides lowers the temperature required for their reduction by H_2. This phenomenon has been attributed to hydrogen spillover. It follows that a partial reduction of the host oxide can induce or modify the catalytic activity of the host material.

A new explanation for catalyst synergy between two solid phases of a catalyst has recently been advanced by Delmon (*174*) in studies of hydrodesulfurization (HDS) with mixtures of MoS_2 and Co_9S_8. Both the activity and the selectivity of the HDS reaction increased if the contact between the admixed phases was improved. Spiltover hydrogen from the Co_9S_8 partially reduces the MoS_2; modest reduction creates hydrogenation sites and further reduction creates HDS sites.

Hydrogen atoms (or ions) can migrate from a metal such as Pt or Pd, or even from a nonmetal (*174, 175*), to another substance, in contact with the first. In the case of a metal-supported catalyst, a carrier such as alumina is the first substance which accepts the spiltover hydrogen (primary spillover). However, this migration may extend further to a second hydrogen acceptor in mechanical contact with the catalyst, like tungsten trioxide (*4*) or molybdenum trioxide (*170, 171*), which form bronzes. Now, the spiltover hydrogen can also be accepted by an organic reactant to be hydrogenated. By admixing supported metal catalyst with pure support (Al_2O_3), enhanced activity has been observed, in comparison with a nondiluted catalyst. A Pt(0.05%)/SiO_2 catalyst, diluted with alumina (1:9) was seven times more active in the hydrogenation of ethylene than the undiluted catalyst (*5*). Similarly, the activity in the hydrogenation of benzene was increased by dilution (1:200) with further alumina of a Pd (2.2%)/Al_2O_3 catalyst (*176*). The enhancement of the activity was explained by the reaction of the hydrocarbon, adsorbed on the diluent surface, with the spiltover hydrogen. It is possible, however, that the diluent surface may become a catalyst for the direct $H_2 + C_2H_4$ reaction without involving spiltover hydrogen.

These results were questioned, however, by some authors (*84, 177*) who were unable to reproduce these experiments. If the alumina diluent can act as a scavenger for any catalyst poison, an enhancement of the catalytic activity may be observed; but this effect would not be associated with hydrogen spillover (*177*). It was therefore uncertain whether the increase of the activity of supported metal, after dilution with the support, was due to hydrogen spillover or to some other phenomenon. This problem was difficult to solve because the effect of the known catalyst (e.g., the metal) in a reaction may be a major one, and the added contribution of the admixed carrier would escape observation. It has also been suggested (*41*), in the case of the hydrogenation of *n*-pentene on Pd/SiO_2 catalyst, that this hydrogenation occurs on Pd. However, hydrogen from the support may, by reverse spillover, participate in the hydrogenation. To yield unambiguous results a reactor was imagined by Gardes *et al.* (*178*) in which the catalyst (like Pt/Al_2O_3) was first in contact with the carrier (acceptor) and exposed to H_2. The catalyst may be withdrawn and isolated from the acceptor after hydrogen spillover. The interaction between the hydrocarbon to be hydrogenated and hydrogen, molecular and Spiltover, is therefore observed in the absence of the supported metal, with only the carrier present in the reactor. The first results obtained with this "reactor with an elevator" involved Ni/Al_2O_3 activator and various aluminas as acceptors of the spiltover hydrogen (*139, 141, 178–80*). It appeared that this system (catalyst and acceptors) is rather a complicated one in the hydrogenation of ethylene, whereas the behavior of Pt/Al_2O_3 catalyst and SiO_2 acceptor is easier to understand. Indeed, SiO_2 activated by hydrogen spillover seems unable to reform the spiltover hydrogen in the absence of the activator, whereas alumina is able to reform this type of hydrogen (see below). Hence, the $Pt/Al_2O_3 + SiO_2$ system will be described first. The reactions of ethylene, benzene, and hexadienes, and finally *n*-heptane, are catalyzed by silica activated by hydrogen spillover, after the Pt/Al_2O_3 has been removed.

1. *Hydrogenation of Ethylene and Acetylene on Silica*

The principle of the "reactor with an elevator" in Pyrex glass (*139, 178*) is given schematically in Fig. 15. The Pt/Al_2O_3 catalyst (20–30 mg) is placed at the bottom of a pyrex nonporous pan (A) and is covered by 1 g of silica held in a pyrex holder (B) with a porous glass bottom (position L). The lifting device (C) allows the removal of the catalyst (with some silica) in the pan, which is suspended by an inert wire and glass rod and is raised to position H. The stopcock (D) isolates the catalyst in the pan from silica in (B) during the catalytic run.

More recently, a system using greaseless valves and a magnetic lifting mechanism was used (*135*), but no differences were found in the behavior of solids activated in either type of reactor. The activation procedure consists in the evacuation (10^{-5} Torr) of the system at the position L at 430°C for 8 h;

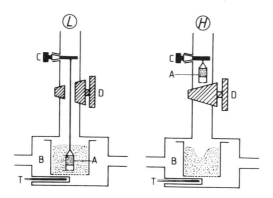

FIG. 15.   Scheme of the reactor with the "elevator." For details see the text.

then 760 Torr of H_2 are introduced in the reactor for 12 h at 430°C, which is the spillover activation temperature. The temperature is afterwards decreased to the desired temperature of the catalytic reaction (100–200°C) and the catalyst in the pan (A) is lifted and isolated by the stopcock (D). A second reactant, like ethylene, may then be introduced under a partial pressure P (usually 10–50 cm^3 of the organic reactant as compared with 1000 cm^3 of H_2 introduced initially into the reactor). The temperature of the activation by hydrogen spillover (430°C) allows a demethoxylation of silica (or alumina) aerogel (with the formation of methane, see above) used as the acceptor (*123*).

Figure 11, curve A, shows the results of the hydrogenation of the first dose (50 cm^3) of ethylene into ethane at 200°C. The induction period has to be noted; for the second dose (50 cm^3, curve B) and all subsequent doses the reaction is fast and does not exhibit the induction period (*124, 134*).

This retardation effect may be due to spiltover hydrogen on the silica, which is exhausted after the first dose run. Indeed, if prior to the reaction with ethylene the fresh activated silica is first evacuated at 200°C (0.5 h), the subsequent introduction of $H_2 + C_2H_4$ mixture at 200°C gives the type B curve of Fig. 11, without the induction period. It is shown below that the spiltover hydrogen can also be scavenged by benzene. The subsequent hydrogenation of ethylene again gives a curve of type B of Fig. 11. The amount of the spiltover hydrogen is easy to determine volumetrically. SiO$_2$ does not adsorb H_2 directly and the adsorption on Pt is known. By difference, it is calculated that activated SiO$_2$ contains 1.4 cm^3 of H_2 per gram of SiO$_2$, which would correspond to about 10^{12} atoms per cm^2 of silica. It thus appears that the catalytic activity in the hydrogenation of ethylene, which has a very reproducible character for many successive doses (50 cm^3) of ethylene, is not related to the amount of spiltover hydrogen, which would give only 1.4 cm^3 of ethane. Hydrogenation chiefly proceeds with H_2 from the gas phase, whereas the spiltover hydrogen is rather a retardant (first run). The activation of silica by hydrogen spillover therefore creates permanent hydrogenating sites which are insensitive to oxygen, even at 430°C, since the activity is again very similar to that given by curve B of Fig. 11 after exposure to O_2 (*134*). This reaction is also not affected by ammonia (0.1 cm^3) introduced into the reaction mixture ($H_2 + C_2H_4$) during the run. However, ammonia is a poison of other catalytic reactions perfomed on the activated silica (see below).

The activated silica is unable to reform the spiltover hydrogen when heated in H_2 at 430°C (12 h) without the Pt/Al$_2$O$_3$ catalyst. Indeed, the hydrogenation of ethylene at 200°C shows that the activity pattern is very close to that of Fig. 11, curve B (*134*).

Various tests, including neutron activation analysis, showed that traces of Pt from the catalyst did not migrate on SiO$_2$ during activation and therefore silica may be converted by hydrogen spillover activation into a very unusual hydrogenation catalyst, active at about 150°C (E_a = 12 kcal mole^{-1}) and not poisoned by O_2 (or even H_2O). The spiltover hydrogen which is required for the creation of active sites on SiO_2 (no catalytic activity is recorded if in the activation procedure Pt/Al_2O_3 catalyst is omitted or if in its presence He is used instead of H_2) inhibits hydrogenation (first run). Other silicas (e.g., Degussa fumed silica) behave in the same manner.

The kinetic studies of the hydrogenation of ethylene indicate (181) that the interaction of both reactants can be represented by Langmuir–Hinshelwood kinetics. They are strongly adsorbed on distinct sites (zero order with respect to each reactant for a stoichiometric feed). However, with hydrogen in large excess, a competition in adsorption is observed, and ethylene is displaced from its sites by hydrogen (positive order with respect to C_2H_4 and negative with respect to hydrogen).

The interaction between benzene or cyclohexadienes and activated silica forms acetylene as an intermediate. Its catalytic hydrogenation will be examined first. Figure 16 shows the conversion of the first dose (curves A) of acetylene (50 cm^3) at 200°C (181). Ethylene and ethane are formed simultaneously.

· The reaction is slow (lower abscissa) due to the presence of spiltover hydrogen. This is no longer the case for the second dose (curves B, upper

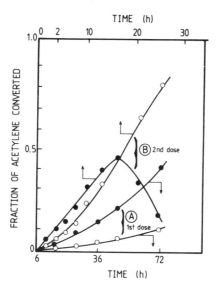

FIG. 16.   Hydrogenation of acetylene on SiO_2 at 200°C: (A) first dose; (B) second dose. ●, Ethylene; ○, ethane.

abscissa) or for the hydrogenation performed after the evacuation (by desorption, following the activation) of the spiltover hydrogen. This behavior shows again that the spiltover hydrogen is strongly adsorbed on silica and is a less active reactant than the molecular hydrogen which is also adsorbed, as pointed out previously. The conversion of acetylene into ethylene and ethane is a sequential reaction, and a maximum (curve B) in the conversion into ethylene is found. The hydrogenation of acetylene is insensitive to ammonia or to an oxygen pretreatment of silica, as previously shown for ethylene. Although 1,4-cyclohexadiene does not affect the hydrogenation of ethylene into ethane, it completely inhibits the conversion of acetylene into ethylene. The 1,3-cyclohexadiene isomer has no effect. This behavior shows that silica activated by hydrogen spillover acquires at least two types of hydrogenating sites, those poisoned by 1,4-cyclohexadiene in the conversion of acetylene into ethylene and those unpoisoned by this cycloolefin in the conversion of ethylene into ethane.

2. *Reactions of Benzene and Cyclohexadienes on Silica*

The spiltover hydrogen can simply be added to benzene (forming cyclohexane and cyclohexene) in a noncatalytic reaction which exhausts entirely this hydrogen species, as shown below. Also, in order to have a clearcut picture, the reaction of benzene is carried out after evacuation of silica, which has been activated. The evacuation desorbs the spiltover hydrogen. Figure 17 shows the conversion at $170°C$ of benzene (8 cm^3) with hydrogen (1000 cm^3) into ethane and initially into acetylene (*182*).

This behavior shows that the first gas phase product of the reaction of benzene with H_2 is acetylene, which is then normally hydrogenated into ethane. Second, third, and subsequent doses of benzene (curve B) or a much higher initial dose of benzene (40 cm^3) present the same kinetic characteristics. The net result is therefore a catalytic hydrogenolysis of benzene into ethane at $170°C$ on silica activated by hydrogen spillover. Such a hydrogenolysis of benzene is not observed on conventional metal catalysts at $170°C$. The reoxidation of silica (curve C) inhibits the reaction of benzene, whereas there is no effect on the subsequent hydrogenation of ethylene (curve D). Injection of a small amount of ammonia also poisons the reaction of benzene. Since the reaction of benzene starts by its cracking into acetylene, the inhibiting effect of oxygen or ammonia involves this first step of the reaction. In conditions preventing the hydrogenation of acetylene into ethane, that is, in the absence of molecular hydrogen, only acetylene is formed over activated silica with a kinetic order zero from a mixture of benzene (50 cm^3) and He at $170°C$. This cracking of benzene is poisoned by ammonia and inhibited by oxygen pretreatment. Therefore a third type of site (for cracking) seems to be created

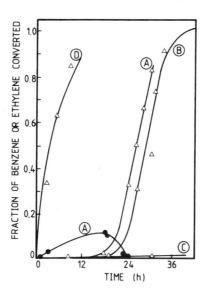

FIG. 17.   Hydrogenolysis of benzene at 170°C and hydrogenation of ethylene at 200°C on SiO₂: (A) cracking and hydrogenolysis of benzene after evacuation of the spiltover hydrogen; (B) hydrogenolysis of benzene after run (A); (C) hydrogenolysis of benzene after interaction between SiO₂ and O₂ at 430°C; (D) hydrogenation of ethylene at 200°C. ●, Acetylene; △, ethane.

on silica activated by hydrogen spillover and may be of an acidic nature. The energetic balance for cracking of benzene at 170°C is quite unfavorable, and some other concerted reactions are probably involved. This very unusual reaction requires further detailed study. It is of interest to point out that silica activated by hydrogen atoms from a plasma, instead of spiltover hydrogen species from Pt/Al₂O₃, behaves exactly in the same way in the reactions with ethylene or benzene (*119*).

The behavior of 1,3-cyclohexadiene in the presence of hydrogen at 170°C is very similar to that of benzene (Fig. 17), with a transient formation of acetylene. Similarly, in the presence of He, 1,3-cyclohexadiene is cracked into acetylene, and this reaction can be repeated for many successive doses of the reactant. Now, the behavior of the isomer, 1,4-cyclohexadiene, is different because in the presence of hydrogen as well as He this reactant is only cracked into acetylene (*182, 183*). The explanation of this different evolution has been provided before: 1,4-cyclohexadiene is indeed a poison for the hydrogenation of acetylene and the stepwise reaction of hydrogenolysis of 1,4-cyclohexadiene (in the presence of H₂) stops with the production of acetylene. Finally, it should be mentioned that cyclohexene, either in H₂ or in He, is not catalytically transformed on silica activated by hydrogen spillover.

It may be hydrogenated by the spiltover hydrogen in a noncatalytic reaction (see below).

3. Reactions of the Spiltover Hydrogen

The spiltover hydrogen (1.4 cm^3 H$_2$ per gram of SiO$_2$ and 1.5 cm^3 H$_2$ per gram of Al$_2$O$_3$) causes an induction period (inhibiting effect) in the catalytic hydrogenation on SiO$_2$ (or on Al$_2$O$_3$) of the first dose of ethylene. But it is exhausted by this dose or by a preliminary evacuation of the activated silica (or alumina). The amount of 10^{12} hydrogen atoms per cm^2 of SiO$_2$ is in a good agreement with that obtained by titration of hydrogen adsorbed on a Pt/SiO$_2$ catalyst by its reaction with 1-pentene (*41*). It also agrees with the results of programmed thermal desorption from a Pt/Al$_2$O$_3$ catalyst (2 × 10^{12} h per cm^2 of Al$_2$O$_3$) (*18*). The behavior of the spiltover hydrogen toward the hydrogenation of benzene on activated silica at 170°C is shown in Fig. 18, for silica not evacuated after its activation.

It was observed that, in addition to ethane (curve A), as previously, small amounts of cyclohexane (curve B) and cyclohexene (curve C) were produced (*140*). This type of hydrogenation was limited. It corresponds to 1.5 cm^3 of hydrogen, an amount which is very similar to the amount of the spiltover hydrogen. Moreover, the reaction, which does not seem to be catalytic, very rapidly reached a maximum (curves B and C). This reaction was not observed

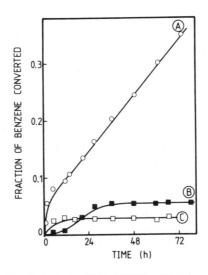

FIG. 18. Hydrogenation of benzene on SiO$_2$ at 170°C: (A) fraction of benzene converted into ethane; (B) fraction of benzene converted into cyclohexane; (C) fraction of benzene converted into cyclohexene. ■, Cyclohexane; □, cyclohexene; ○, ethane.

if prior to the reaction of benzene the spiltover hydrogen was evacuated (Fig. 17). In the presence of a small amount of ammonia the hydrogenolysis of benzene into ethane was inhibited (see above). Simultaneously, the addition of the spiltover hydrogen to benzene forming cyclohexane was also inhibited, but not the addition forming cyclohexene. Either the spiltover hydrogen which converts cyclohexene into cyclohexane reacted with NH_3 or the sites involved in the olefin reaction were blocked by NH_3. Other sites or another form of the spiltover hydrogen must still have been present since the reaction of benzene to form cyclohexene was insensitive to NH_3. These experiments, compared with those of Fig. 17, also show that the spiltover hydrogen is not required for the catalytic hydrogenolysis of benzene into ethane.

In this same way as for benzene, both cyclohexadienes are converted into cyclohexene and cyclohexane by addition of the spiltover hydrogen, as is shown in Fig. 19. Again the amount of cyclohexene and of cyclohexane is similar to the amount of spiltover hydrogen on silica. Finally, the only reaction observed between H_2 and cyclohexene involves spiltover hydrogen, and its addition to the reactant forms cyclohexane. This addition, which is inhibited by NH_3, exhausts the spiltover hydrogen; subsequently, the hydrogenation of ethylene does not show any induction period.

Table IV summarizes the results concerning the titration of the spiltover hydrogen.

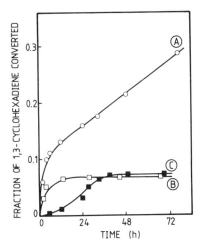

FIG. 19. Hydrogenation and hydrogenolysis of 1,3-cyclohexadiene on SiO_2 at 170°C: (A) fraction of 1,3-cyclohexadiene converted into ethane; (B) fraction of 1,3-cyclohexadiene converted into cyclohexene; (C) fraction of 1,3-cyclohexadiene converted into cyclohexane. ■, Cyclohexane; □, cyclohexene; ○, ethane.

TABLE IV
Titration of the Spiltover Hydrogen by Direct Adsorption and by Addition Reactions

Titration of H_{sp} by	Temp. (°C)	Volume (cm^3 STP g^{-1} SiO$_2$) of H_{sp} adsorbed
1. Volumetry	200	1.4
2. Induction period in $C_2H_4 + H_2$	200	~ 1
3. $C_6H_{10} + H_2 \longrightarrow C_6H_{12}$	170	0.58
4. $C_6H_6 + 2H_2 \xrightarrow{NH_3} C_6H_{10}$	170	0.91
5. $C_6H_6 \begin{cases} +2H_2 \longrightarrow C_6H_{10} \\ +3H_2 \longrightarrow C_6H_{12} \end{cases}$	170	1.51
6. 1,3-$C_6H_8 \begin{cases} +2H_2 \longrightarrow C_6H_{12} \\ +H_2 \longrightarrow C_6H_{10} \end{cases}$	170	1.54
7. 1,4-$C_6H_8 \begin{cases} +2H_2 \longrightarrow C_6H_{12} \\ +H_2 \longrightarrow C_6H_{10} \end{cases}$	170	1.49

These results help discriminate between the types of spiltover hydrogen or the types of sites adsorbing the spiltover hydrogen. Indeed, the amount of the spiltover hydrogen adsorbed on all sites is on the order of 1.5 cm^3 per gram of SiO$_2$ (Table IV, lines 1, 5, 6, and 7). The amount sorbed on sites poisoned by NH$_3$ is of the order of 0.6 cm^3 per gram of SiO$_2$ (line 3). Only this type of spiltover hydrogen may be added to cyclohexene to form cyclohexane. But spiltover hydrogen fixed on other sites or of a different nature (0.9 cm^3 per gram of SiO$_2$) can react with benzene (even in the presence of NH$_3$) to convert it into cyclohexene (line 4). However, NH$_3$ inhibits any further reaction. It may be speculated that a protonic form of the spiltover hydrogen is inhibited (as NH$_4^+$) by NH$_3$, whereas the nonprotonic forms are insensitive to this poison. Alternatively, acid sites (also active in the cracking of benzene or cyclohexadienes) are poisoned by NH$_3$, whereas other sites (also active in hydrogenation) are insensitive to NH$_3$.

4. Reactions of n-Heptane over Silica

n-Heptane is a good test reactant because it may either dehydrogenate, isomerize, or dehydrocyclize, depending on the nature of the catalytic sites.

If neither spiltover hydrogen nor molecular hydrogen are present after activation of silica and evacuation, a mixture of *n*-heptane (10 cm^3) and He is converted at 270°C into a decreasing series of products: methane, heptane and heptadienes, ethane, acetylene, toluene, and benzene (*183, 184*). Heptene and heptadienes can only result from the dehydrogenation of *n*-heptane over activated silica. Heptatriene is not detected, but toluene may result from the dehydrocyclization of this olefin. The transformation of toluene into benzene and methane, and the reaction of benzene or toluene to give ethane, is similar to the benzene reaction described above. An insufficient amount of hydrogen released by the dehydrogenation of n-heptane accounts for the formation of acetylene instead of ethane. A second run gives exactly the same results and suggests that the reaction is catalytic on the activated silica. With either NH_3 present or O_2 pretreatment of activated silica (at 270°C), the reaction of n-heptane in He does not occur.

In the presence of molecular hydrogen the reaction at 270°C is very much limited (20% conversion only) and the same products, except acetylene, are detected. However, the presence of the spiltover hydrogen is also required. No reaction in H_2 is observed if silica is evacuated after activation. The reaction in the presence of molecular and spiltover hydrogen is very probably not catalytic but results from the interaction of the spiltover hydrogen with n-heptane. The amount of n-heptane converted (2 cm^3) correlates well with the amount of the spiltover hydrogen present after activation (1.5 cm^3). The stoichiometry of the interaction between the spiltover hydrogen and n-heptane is not known. Assuming that one spiltover hydrogen atom reacts with one molecule of n-heptane, 3 cm^3 of n-heptane would be converted. Once the spiltover hydrogen is exhausted by the first mixture, a second run shows no conversion. These results indicate that if spiltover hydrogen is present the dehydrogenation of n-heptane on activated silica may proceed to some extent even in the presence of molecular H_2. But the reaction is not catalytic, whereas if molecular hydrogen is not present, it is catalytic and does not require the spiltover hydrogen. This problem will be examined again below in connection with the differing activites on aluminas.

5. Reactions over Activated Alumina

The behavior of amorphous alumina aerogel (135) or δ-alumina (Degussa P 110) in the hydrogenation of ethylene after hydrogen spillover activation was very similar to that of silica. The corresponding curves on δ-alumina at 145°C (89) may almost superimpose on the curves of Fig. 11 for silica at 200°C. The spiltover hydrogen was exhausted by the first dose of ethylene or by the evacuation after activation. Again ammonia or oxygen at 430°C did not detrimentally influence the catalytic activity. The main difference with silica was that the spiltover hydrogen was reformed by heating the activated alumina (after the first run, where the spiltover hydrogen was exhausted) in H_2 at 430°C without the Pt/Al$_2$O$_3$ catalyst. The subsequent hydrogenation of ethylene (at 145°C) showed again an induction period which was attributed to the presence of the spiltover hydrogen in the first run. Finally, a very small amount of NO (0.5 cm^3) introduced during the reaction did not affect the kinetics of ethylene hydrogenation on δ-Al$_2$O$_3$ activated by Pt/Al$_2$O$_3$ at 430°C. This point is of some importance as it differentiates this alumina from those activated at 300°C in the presence of Ni/Al$_2$O$_3$ catalyst. If the Pt/Al$_2$O$_3$ catalyst was used at 300°C in the activation of δ-Al$_2$O$_3$, no catalytic activity was recorded in the hydrogenation of ethylene. Conversely, if Ni/Al$_2$O$_3$ activator was used at 430°C, no activity was recorded (*185*). This emphasizes an unusual difference between alumina activated at higher temperatures by Pt/Al$_2$O$_3$ and at lower temperatures by Ni/Al$_2$O$_3$. Ni/Al$_2$O$_3$ catalyst was used in the early experiments of activation at 300°C in H$_2$ of methoxylated or demethoxylated alumina aerogels (*139, 141, 178–80*). Under these conditions demethoxylated alumina was active for the hydrogenation of ethylene even at 25°C; the comparable activity of methoxylated alumina is only observed at 110°C. This observation emphasizes that the initial surface of an oxide dictates the conditions of activation by hydrogen spillover.

Another significant difference was found for the lower temperature Ni/Al$_2$O$_3$ activation. After activation of δ-Al$_2$O$_3$ or amorphous alumina aerogel by Ni/Al$_2$O$_3$ at 300°C, these aluminas were only active for ethylene hydrogenation while H$_{sp}$ was present (*139*). If the spiltover hydrogen left after the activation (1.3 cm^3 per gram of Al$_2$O$_3$ aerogel at 110°C) was evacuated, only a very small activity with molecular H$_2$ was found. This shows that a permanent catalytic activity of alumina was not achieved after this activation and that the spiltover hydrogen may have initiated the reaction. But again almost the total amount (45 cm^3) of the ethylene dose (50 cm^3) was converted. Therefore the reaction was not a simple addition of the spiltover hydrogen (1.3 cm^3) to ethylene. Since NO, which may be a radical scavenger, inhibits the reaction, a scheme involving H· radicals was proposed:

$$H_{2(g)} \longrightarrow 2H^{\cdot}_{(ads\,on\,Al_2O_3)} \qquad \text{initiation} \qquad (29)$$

$$C_2H_{4(g,\,or\,ads)} + H^{\cdot}_{(ads)} \longrightarrow C_2H_{5(ads)} \qquad \text{propagation} \qquad (30)$$

$$C_2H_{5(ads)} + H_{2(g)} \longrightarrow C_2H_{6(g)} + H^{\cdot}_{(ads)}$$

$$C_2H_{5(ads)} + H^{\cdot}_{(ads)} \longrightarrow C_2H_{6(g)} \qquad \text{termination} \qquad (31)$$

In summary, if the activation of aluminas is performed at 430°C by a Pt/Al$_2$O$_3$ activator, a permanent catalytic activity is induced. The spiltover hydrogen has rather an inhibiting effect, as shown by the induction period. The reaction is much faster once this hydrogen is exhausted by the first run or by evacuation. If the activation of aluminas is performed at 300°C by a Ni/Al$_2$O$_3$ activator (which does not induce activation at 430°C), the spiltover hydrogen is required for the reaction and may be radical in nature since it is inhibited by NO.

Another interesting reaction on amorphous alumina activated by Pt/Al$_2$O$_3$ activator (at 430°C) was the isomerization of many doses of methylcyclopropane at 25°C, in H_2 or in He. The products were exclusively cis-2-butene and 1-butene; no trans-2-butene was detected (185). These catalytic properties were permanent and did not require the spiltover hydrogen.

Activated δ-alumina behaved differently from silica in the reaction with n-heptane examined previously. In the absence of molecular and spiltover hydrogen (in He), alumina exhibited (at 270°C) only a very limited catalytic activity for dehydrocyclization and hydrocracking (184). In the presence of molecular and spiltover hydrogen, in contrast to silica, the reaction proceeded easily, giving heptene and heptadienes. The reaction was catalytic, since many doses of n-heptane could be converted. Now, the spiltover-type hydrogen can be reformed on activated alumina (at 270°C) from molecular hydrogen, in the absence of the Pt/Al_2O_3 activator (89, 135). Since NO poisoned the conversion of n-heptane on alumina in the presence of H_2, it is again assumed that the spiltover hydrogen species are required for the conversion of n-heptane on alumina. The poisoning effect of NH_3 could equally be related to the scavenging of the protonic spiltover species (H^+ or H_3^+) (184). The difference in the behavior toward n-heptane of activated silica and activated alumina may be summed up by proposing that hydrogen spillover activation of silica creates a bifunctional catalyst with hydrogenating and acidic centers, whereas the activation of alumina gives mainly a monofunctional catalyst where acidic sites are weak or limited in number.

Finally, δ-alumina activated by Pt/Al_2O_3 (at 430°C) exhibited a completely different pattern from silica, activated in the same manner, for the conversion of benzene, cyclohexadienes (1,3- and 1,4-), and cyclohexene (181). The reaction occurred only in the presence of molecular hydrogen (at 160°C), giving cyclohexane in all four cases. No cracking was observed in the presence of He (contrary to silica). Ammonia and oxygen at 430°C did not affect the activity. This behavior is therefore similar to the activity for ethylene hydrogenation, which was also insensitive to NH_3 or O_2. The (acid) sites which were developed by the activation of silica do not seem to be induced on alumina. The activation by hydrogen spillover is, therefore, specific to the nature of the oxide to be activated. This conclusion is also reinforced by the behavior of magnesia activated by hydrogen spillover (89, 186) which, together with the behavior of silica and alumina, is summarized in the Tables V, VI, and VII (90).

Magnesia may also be activated by molecular H_2 at 430°C, without the need of a Pt/Al_2O_3 activator. However, the sites active for hydrogenation of ethylene, unlike those on alumina or silica, were destroyed by oxidation at 430°C. Spiltover hydrogen (but not molecular H_2) activates MgO already at 200°C and the active sites are insensitive to O_2 treatment but are blocked by NH_3. The active sites are therefore not the same as those created on SiO_2 and

TABLE V

Hydrogenation of Ethylene on Oxides Activated by Hydrogen Spillover with Pt/Al$_2$O$_3$

Oxide	Reaction temperature (°C)	Conversion of 50 cm^3 of C$_2$H$_4$	Induction period (first dose)[a]	Remarks
SiO$_2$	170–230	~100% in a few hours	Yes	Insensitive to O$_2$ at 430°C or to NH$_3$ at 200°C
Al$_2$O$_3$	110–150	~100% in a few hours	Yes	Insensitive to O$_2$ at 430°C or to NH$_3$ at 150°C
MgO	50	~100% in a few hours	No	Activation by splitover H at 200°C. Insensitive to O$_2$ at 430°C; poisoned by NH$_3$ at 50°C
	50–130	~100% in a few hours	Yes	Activation by splitover H at 430°C. Insensitive to O$_2$ at 430°C; poisoned by NH$_3$ at 130°C
	130	~100% in a few hours	Yes	Activation by H$_2$ at 430°C without Pt/Al$_2$O$_3$. Inhibited by O$_2$ at 430°C; insensitive to NH$_3$ at 130°C; inactive at T = 50°C

[a] Not recorded if the spiltover hydrogen is evacuated before the first dose is introduced.

Al$_2$O$_3$. The poisoning effects of NH$_3$ and O$_2$ on the reactions with the cyclohexadienes are also different on activated MgO from those observed on SiO$_2$ or Al$_3$O$_3$ (Table VII).

E. ACTIVATION OF THE SUPPORT BY OXYGEN SPILLOVER

It is well known that γ- or δ-Al$_2$O$_3$ can be activated for hydrogen adsorption and ethylene hydrogenation either by evacuation or by contact with O$_2$ at temperatures from 500 to 800°C (*136, 187*). The rate of hydrogenation of ethylene on δ-Al$_2$O$_3$ activated by thermal treatment in O$_2$ or in vacuum was of the same order of magnitude at 500°C (*136*) as that observed at 180°C only on δ-Al$_2$O$_3$ activated by hydrogen spillover (*89*). Spillover therefore produces a more active and efficient catalyst. According to Hindin and Weller (187), the activation of alumina at high temperature creates a surface strain (high-energy sites for catalytic reactions like the hydrogenation of ethylene) as a result of the dehydration. It was assumed by Hilaire (*188*) that active sites created by heating δ-Al$_2$O$_3$ in dry O$_2$ at high temperatures could be excess surface O^{2-} or O$^-$ anions, on which H$_2$ may be dissociatively

TABLE VI

Conversion of Benzene on Oxides Activated by Hydrogen Spillover with Pt/Al$_2$O$_3$

Oxide	Reaction temperature (°C)	Conversion in the presence of H$_2$	Conversion in the presence of He	Remarks
SiO$_2$	170	~100% ethane in a few hours	~100% acetylene in a few hours	Inhibition by NH$_3$ and by O$_2$ (430°C) pretreatment; C$_6$H$_6$ + H$_{sp}$ → cyclohexene + cyclohexane[a]
Al$_2$O$_3$	160	~100% cyclohexane in a few hours	No conversion	Insensitive to NH$_3$ and to O$_2$ (430°C) pretreatment
MgO	170	8% cyclohexane in a few hours	No conversion	No conversion of C$_6$H$_6$ on MgO activated by H$_2$ at 430°C without Pt/Al$_2$O$_3$ (unlike for C$_2$H$_4$)

[a] Reaction of addition.

TABLE VII

Conversion of 1,3- and 1,4-Cyclohexadienes (CHD) on Oxides Activated by Hydrogen Spillover with Pt/Al$_2$O$_3$

Oxide	Reactant	Reaction temperature (°C)	Conversion in the presence of H$_2$	Conversion in the presence of He	Remarks
SiO$_2$	1,3-CHD	170	~100% ethane in a few hours	~100% acetylene in a few hours	Inhibition by NH$_3$ and by O$_2$ (430°C). CHD + H$_{sp}$ → C$_6$H$_{10}$ + C$_6$H$_{12}$[a]
	1,4-CHD	170	~100% acetylene in a few hours	~100% acetylene in a few hours	Inhibition by NH$_3$ and by O$_2$ (430°C). CHD + H$_{sp}$ → C$_6$H$_{10}$ + C$_6$H$_{12}$[a]
Al$_2$O$_3$	1,3-CHD	160	~100% cyclohexane in a few hours	No conversion	Insensitive to NH$_3$ and to O$_2$ (430°C) pretreatment
	1,4-CHD	160	~100% cyclohexane in a few hours	No conversion	
MgO	1,3-CHD	170	~100% cyclohexane in a few hours	No conversion	Activation by H$_{sp}$ at 200°C; inhibition by NH$_3$
	1,4-CHD	170	~100% cyclohexane in a few hours	No conversion	Insensitive to O$_2$ (430°C) pretreatment
	1,3-CHD	170	~100% cyclohexane in a few hours	No conversion	Activation by H$_{sp}$ at 430°C; inhibition by NH$_3$
	1,4-CHD	170	~100% cyclohexane in a few hours	No conversion	Insensitive to O$_2$ (430°C) pretreatment; C$_6$H$_{10}$ as a transient
	1,3-CHD	170	~100% cyclohexane in a few hours	No conversion	Activation by H$_2$ at 430°C without Pt/Al$_2$O$_3$
	1,4-CHD	170	~100% cyclohexane in a few hours	No conversion	Inhibition by O$_2$ (430°C) pretreatment but insensitive to NH$_3$; C$_6$H$_{10}$ as a transient

[a] Reaction of addition.

adsorbed as OH groups. But activation by spiltover oxygen from Pt/AL_2O_3 catalyst (see Section IV,A) is more efficient than activation of alumina by molecular oxygen. Indeed, δ-Al_2O_3 activated by oxygen spillover at 430°C exhibited activity for the hydrogenation of ethylene at 180°C (without an induction period) that was an order of magnitude greater than an identical sample exhibited at 400°C after activation at 500°C without the Pt/Al_2O_3 activator (*136*).

Similar differences were also evident for the activation of amorphous alumina aerogel. This Al_2O_3 cannot be activated for hydrogenation of ethylene by a mere heating at 430°C in vacuum, under He (with or without Pt/Al_2O_3 activator) or in H_2 (without Pt/Al_2O_3) (*135, 136*). Heating in O_2 at 430°C without Pt/Al_2O_3 catalyst developed activity for ethylene hydrogenation at 180°C with $t_{1/2} = 40$ h. However, the activation by oxygen spillover from Pt/Al_2O_3 catalyst improved the activity since $t_{1/2} = 10$ h (*135, 136*). A better response to the activation by O_2 (with oxygen spillover or without) of the amorphous alumina as compared to δ-Al_2O_3 could be due to the more random surface of the amorphous structure. It is more difficult to create surface defects on the crystalline Al_2O_3. An argument that supports the formation of active sites by O_2 spillover activation of the type O^{2-} or O^- is the insensitivity of these sites to NH_3. This implies that hydrogenating centers are not acidic (Al^{3+} or protonic) in nature (*135, 136*). Now, if after O_2 spillover activation, the alumina aerogel was exposed to molecular H_2 at 430°C and then evacuated prior to the catalytic test, the introduction of small doses of NH_3 severely decreased the rate of hydrogenation. After H_2 treatment at 430°C the catalytic sites were converted from surface O^{2-} or O^- ions into more "conventional" sites, as for a hydrogen-activated catalyst. The response to NH_3 was as discussed above. A second argument in favor of O^{2-} or O^- sites is found in the severe poisoning by NO. This poisoning was not found for H_2-spillover-activated alumina (*180*). NO is an electron donor as is NH_3 (which has not effect). But NO may also react with the sites created by O_2 spillover by transforming them into NO_2^{-2} or NO_2^- species that may be inactive for the hydrogenation. Electron acceptor sites (Al^{3+}) are also likely to be inhibited by NO, but since NH_3 is without effect, they are probably not developed by O_2 spillover activation. Finally, it must be recalled that hydrogen spillover activation of alumina aerogel is, by far, the most efficient way to activate this oxide for the hydrogenation of ethylene at 180°C since $t_{1/2}$ is on the order of 0.5 h (*135*).

To sum up this section, it appears that some new and unforeseen catalytic properties may be induced on refractory oxides by hydrogen spillover, and to a lesser extent by oxygen spillover. If the activator, like Pt/Al_2O_3 catalyst, is not separated after the spillover activation from the activated oxide, these new properties could be masked by the catalytic activity of Pt/Al_2O_3.

VI. Conclusions and Implications

There are several areas where the consequences of spillover may be significant. Initially it is necessary to underscore that these effects will depend on the conditions and nature of the surface: the phenomena are not the same for different surfaces or under different conditions.

In this section we will consider the myriad studies discussed above. Loosely following our approach above, we will discuss

a. The Phenomena: What may be involved?
b. The Mechanism: What is the contribution to reaction mechanisms?
c. The Kinetics: What is the effect on catalysis?

Most of our insights in these discussions come from studies of hydrogen spillover; although the spillover of other species is known to occur and results in similar effects, we are somewhat limited in our perspective because these species have been less studied.

A. SPILLOVER—THE PHENOMENON

The above discussion demonstrates the multifaceted nature of spillover. The interphase transport of an activated species onto a surface (and sometimes into the bulk) where it is unable to be formed without the activator can induce a variety of changes on, and reactions with, the surface. All the reactions of atomic hydrogen are found to be induced by spillover: exchange, bronze formation, reduction, demethoxylation, and catalytic activation. An activated species is able to gain indirect access to the nonsorbing surface.

The effects are dependent on the receiving surface and the temperature of the system. Most oxides are able to exchange their hydroxyls via spillover of deuterium. The temperature must be sufficient for the deuterium to spill over onto the oxide surface. Coadsorbents (such as water) can facilitate the transport (Eq. 4) or the transport can be facilitated by an intermediate solid phase (such as a carbonlike or hydrocarbon phase). This implies that the method of catalyst preparation can influence the rate of spillover and the subsequent exchanges. Therefore, for example, not all types of Pt/SiO_2 will behave the same. This also means that the phenomena associated with Pt/SiO_2 may differ from Pt/Al_2O_3, Pd/SiO_2, $Ni/SiO_2-Al_2O_3$, etc. The interface between the phases influences the kinetics and also the nature of the effects of spillover.

Various phenomena associated with spillover occur after the spillover process; spillover is a necessary but not sufficient process to induce the effects that have been found. These effects can be independent of each other. As an example, the exchange of hydroxyls with deuterium on silica occurs readily at $\sim 100°C$ but the surface is not catalytically activated at this temperature ($T > 400°C$ is required). As above, this relative contribution depends on the systems being investigated. Atomic hydrogen generated by a microwave discharge induces the same catalytic properties in silica as the spiltover hydrogen (119).

Even for a single system, the conditions may be such that the history of the sample can have an effect. Spillover involves a sequence of steps and each of these steps may be influenced by the pretreatment of the sample. As an example, surface transport and surface activation may depend on hydroxyl content (via $H_{sp} + OH' \rightarrow OH + H'_{sp}$ and $H_{sp} + MOH \rightarrow M + H_2O$ mechanisms, respectively). Dehydroxylation of an oxide surface may, therefore, reduce the rate and extent of H_{sp} transport or subsequent surface activation (137).

The influence of spillover species on an acceptor phase can be in the extreme either subtle or profound. Many of the phenomena associated with hydrogen spillover are as subtle as the influences of type-2 hydrogen on the activity of ZnO (189) or as significant as bulk reduction, bronze formation, or catalytic activation. The effects may be similar to the exposure of a surface to a hydrogen plasma.

As discussed earlier (Section V,D,5), different forms of alumina (amorphous aerogel or crystalline δ) have been found to differ in the behavior after activation by hydrogen spillover. For instance, δ-Al_2O_3 (like silica) exhibits an induction period in the hydrogenation of the first dose of ethylene at $\sim 180°C$ which has been attributed to the presence of the spiltover hydrogen (89). The amorphous alumina aerogel does not exhibit this induction period and therefore it does not retain the spiltover hydrogen (or it retains only a very small amount) after activation by Pt/Al_2O_3 catalyst (136). When these aluminas, amorphous and δ, are activated by hydrogen spillover and the spiltover species is removed, either by reaction with the first dose of ethylene or by evacuation, they reform this type of spiltover hydrogen by exposure to gaseous molecular hydrogen at $430°C$ without the Pt/Al_2O_3 activator; however, silica doesn't seem to reform this type of hydrogen under the same conditions. This behavior could be explained by the existence of one (δ-alumina) or more (amorphous alumina) types of hydrogen species (or sites created by activation). This situation is schematized in Fig. 20 and corresponds to isobaric conditions (e.g., 1 atm H_2). Curve (A) shows the hydrogen adsorption isobar on δ-Al_2O_3 (and silica) after activation by hydrogen spillover at $430°C$ and cooling to the reaction temperature ($\sim 180°C$). The induction period which was found for ethylene hydrogenation is due to the spiltover hydrogen, which remains adsorbed at $180°C$. For amorphous

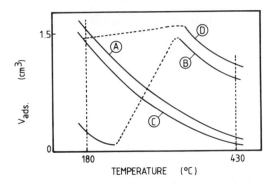

FIG. 20. Isobaric representation of spiltover hydrogen adsorption on activated aluminas, amorphous or δ (and silica): (A) δ-alumina (and silica), with Pt/Al_2O_3; (B) amorphous alumina with Pt/Al_2O_3; (C) δ-alumina without Pt/Al_2O_3; (D) amorphous alumina without Pt/Al_2O_3.

alumina, activated in the same way, the spiltover hydrogen is not detected at $\sim 180°C$ (no induction period in the hydrogenation of the first dose of ethylene). This may be explained, as above, by an adsorption isobar (B) of spiltover hydrogen which shows a transition between two states of (or sites for) adsorbed H_{sp}. When the aluminas (amorphous and δ) are first activated by hydrogen spillover and the spiltover species is then removed (by reaction with ethylene or evacuation), these aluminas both exhibit an induction period after exposure to H_2 at 430°C without the activator. This can be explained by isobar (C) of Fig. 20 for δ-alumina (one type of species or sites) and by isobar (D) for amorphous alumina (two types of species or sites).

The effects of temperature on the phenomena associated with spillover differ from study to study. We can, however, generally depict an energy coordinate for the various states. The relative energies will depend on the specific system. The hypothetical relationships are depicted for a metal (M) and a support (such as an oxide) in Fig. 21. The circled numbers refer to the mechanistic steps proposed in the introduction [Eq. (1)–(6)]. The dashed lines indicate mechanistic steps involving the sites formed on the support, which does not directly sorb H_2 (the high activation energy indicated by the dotted line). H_2 is readily adsorbed and activated on the metal (steps 1 and 2). Spillover (step 4) is activated but may be facilitated if a coadsorbent (such as H_2O) is present and provides an alternate path. Diffusion across the support (step 5) is relatively easy. It is shown as a uniform sequence of steps, although there is evidence that more than one path may exist (i.e., with or without association with hydroxyls, or intra- and interparticle paths). The support can become activated if the thermal energy is sufficient (step 6). This creates sites that *then* accept H_{sp} from the support or directly from the gas phase (for amorphous or δ-alumina, for example), but the direct adsorption may be

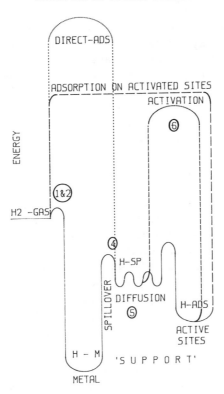

FIG. 21. Possible energy coordinate for the processes associated with spillover from a metal (M) onto an oxide support. The circled numbers refer to the equations in the introduction: (1 and 2), activated adsorption; (4), spillover; (5), surface diffusion on the oxide; (6), activation of the oxide support. The active sites are created by spiltover hydrogen at high temperatures. See text for more details.

difficult (a higher energy barrier, as an example, for silica). The relative energies of the states and the transitions will dictate the stability of the various species. As an example, it is evident that the spiltover hydrogen adsorbed on the sites created on δ-alumina differs in energy from the spiltover hydrogen on the corresponding sites of amorphous alumina. This gives rise to the differences in behavior during activation or exposure to hydrogen gas at high temperatures.

B. MECHANISMS OF SPILLOVER AND THE SUBSEQUENT CATALYSIS

The species being spilt over is not always unique. It can be ionic or radical-like; again, the mechanism depends on the surface of the acceptor, on the source of spillover, and on the interface. Further, parallel mechanistic paths may exist that give rise to the same net effect. As an example, bronze formation can occur by hydrogen spillover directly to the oxide, it can occur by spillover assisted by coadsorbents, and it can (autocatalytically) occur by spillover from sites created on the metal oxide. Further, the relative contribution of the parallel paths may change in the course of the process.

Each of the intermediate steps in the mechanistic sequence involved in spillover can differ. As an example, surface transport may occur as a two-dimensional gas, as species associated with an activated site, or as two-dimensional exchange (via hydroxyls). The literature has suggested a variety of mechanisms for spillover and mechanisms induced by spillover. The studies have not assumed that all of the mechanisms are possible and have not focused on discriminating between the alternative possibilities. This more open (albeit more disconcerting) approach is suggested to understand the phenomena of spillover. Further, novel experimental techniques are needed to isolate the catalytic effects occurring on the source of spillover (usually the metal) from the effects induced by spillover on the acceptor surface, and to access the relative contribution of each. Both the numbers of species involved and the relative rates need to be quantified.

C. The Effect on Heterogeneous Kinetics

There are several kinetic implications of the spillover phenomena. As an example, consider a simple reaction $A + B \rightarrow C$ which occurs with A, B, and C adsorbed and with the surface reaction ($A^* + B^*$, on a single type of active site *) controlling. The reaction rate can be written, following a Langmuir–Hinshelwood model, as

$$\frac{dP_C}{dt} = \text{rate} = \frac{k_r(C_s)^2 K_A K_B (P_A P_B - P_C/K)}{(1 + K_A P_A + K_B P_B + K_C P_C)^2} \tag{33}$$

where P_A, P_B, P_C are the partial pressures of A, B, and C; C_s is the surface concentration of active sites; K_A, K_B, K_C are the equilibrium constants for adsorption of A, B, and C; K is the reaction equilibrium constant; k_r is the surface reaction rate constant.

Spillover may have a dramatic effect on this analysis. If spillover of A exists and new active sites are created, C_s, the total concentration of active sites, may be divisible into C_{s0}, the initial concentration of sites on the surface, and C_{si}, the concentration of sites created by spillover. Even if the mechanism and specific activity are the same, a separate term in the equation may be included since $C_s \rightarrow C_{s0} + C_{si}$. As assumed, spillover can modify the effective surface concentration of A. This may involve including another term in addition to the product $K_A P_A$, or maybe just a modification of K_A is needed. The actual concentration of A on the surface would have to involve more than a single term: on the source of spillover, on the inert surface, and sorbed at the active site. If spillover induces activity on the surface, then a separate kinetic equation may be necessary. Again the questions concerning the sorption and kinetic relationships on these new sites need to be addressed.

One can hypothesize a complex set of possibilities: simple additive terms   (e.g., $C_s \to C_{s0} + C_{si}$)   and/or   cross   terms   (e.g., $k_r(C_s)\{[P_A(K_A + K'_A)]K_B P_B\} + \cdots$. To simplify our analysis, we will consider the contributions independently: the number of catalytic sites, the relative activity on induced sites, the availability of reacting species, and the contribution of transport effects.

The total number of active sites will change if spillover induces activity onto the support. Several questions are raised to ascertain the impact of these phenomena: is the reaction the same on the spillover induced sites, and what is the activity of these sites (relative specific activity and number of sites)?

For reactions that are the same on metal and other catalytic sites (e.g., hydrogenation or total oxidation), the reaction may seem to proceed in a similar fashion on the metallic source of spillover and on the "diluent" support. Some careful studies may be able to discriminate between activity on the metal and the spillover-induced sites. As an example, hydrogenation of ethylene occurs on Pt (or Ni) and on silica or alumina activated by spillover. The product (i.e., only ethane) is the same, as the kinetics often are (rate = $k[C_2H_4]^0[H_2]^1$), but the specific mechanism is different. Deuteration is able to discriminate between the relative rate of "alkyl reversal." Deuteration of ethylene on an "activated" silica produces d_2-ethane as the initial product (*137*), contrary to the results for metal-catalyzed ethylene hydrogenation (*1*).

The difference in selectivity for different sites was suggested by Sachtler and Bostelaar in their studies of the hydrogenation of molecules with a prochiral center (*190*). Catalysts were prepared that contained two types of sites: Ni metal and nickel-R-R'-tartrate complexes. Separately, the nickel is not enantioselective and the Ni complexes are inactive for the hydrogenation of methyl acetoacetate. In combination on a silica support the product contained enantiomeric excesses over 30% (= ([R] − [S]/[R] + [S]) × 100). The authors proposed a dual-site mechanism whereby hydrogen is dissociated on the nickel and then migrates across the silica to the sites involving the Ni–tartrate complexes. The spiltover hydrogen then reacts selectively with the adsorbed acetate to produce R-methyl hydroxybutyrate. They further suggested that spillover may similarly be involved in a dual-site mechanism for Fischer–Tropsch catalysis. It may also be possible that hydrogen spiltover from the metal is able to induce activity involving the adsorbed complexes and is then not needed to provide hydrogen as a reactant. This is the first intriguing suggestion that spiltover hydrogen is involved in asymmetric synthesis.

For reactions that occur with different selectivities or kinetics on the induced sites, the implications are far more complex. Teichner has found that the selectivities for the isomerization of methylcyclopropane and for cracking of benzene and cyclohexadienes may be substantially different on "activated" silica or alumina than on the "normal" oxides (see above). In these cases, a separate rate expression may need to be included. Depending on the rate constants and relative kinetics, this can substantially change the reaction-rate expression. Further, differences in the activation energies may affect the contribution at different temperatures. If the reaction temperature is sufficient, the activation of the support may be able to occur during the course of the experiment.

The crucial question is the relative activity on the metal surface and the activity induced on the "support." The number of sites potentially induced on the support is low per unit area for most refractory supports (SiO_2 and Al_2O_3). The dispersion of the support and ratio of solid diluent to supported catalyst will dictate the potential effect of activity on the support. A 1% Pt catalyst on a 200 m^2/g support will have a maximum (100% dispersion) of 1.5×10^{13} Pt/cm^2. As discussed above, the estimated maximum number of spiltover hydrogen atoms on silica or alumina is $10^{12}/cm^2$. If we assume that one site is produced for each spiltover hydrogen atom and that the specific activity on the support and on the Pt is the same, the ratio of activity on the metal to that on the support would be 15:1. Obviously, the relative activity for a supported metal catalyst will depend on the percentage and dispersion of metal, the number of active sites of all types, and their relative activity. For the example cited, under these constraints the relative activity of the support would probably be minimal. For systems where the concentration of induced active sites is larger or systems where a solid diluent is present, the contribution of the support can be significant. Studies that focus on the relative activity (concentration of active sites and their specific activity) are needed to *quantify* the potential contribution of the induced activity compared to the inherent activity.

As mentioned above, reactions with differing selectivities on the sources of spillover (usually a metal) and on the support can have a more profound effect. Again the relative activity and, therefore, the potential effect needs to be studied and quantified.

The availability of potential reactant species is changed by spillover from one site, where reaction does not occur, to another site, where the reaction may occur. We believe the most significant expression of these phenomena is the retention or induction of activity. Secondary reaction involving the spillover species can retain or induce activity on other sites. As discussed above, this may reverse or prevent the deactivation of catalytic sites, for example, spillover may be involved in the removal of coke. Further, spillover can promote the indirect activation of one metal by another in bimetallic catalyst systems.

Vannice has calculated that the direct participation of spillover species in hydrogenation reactions over supported metal catalysts is not significant (191), provided that the spiltover species reacts easily with the organic acceptor. This was not the case, as shown above, for ethylene hydrogenation on SiO_2 or Al_2O_3, where one form of the spiltover hydrogen gives rise to an induction period for the reaction. Thus the indirect contribution of spillover to catalysis seems to be more important. This involves the induction of activity on seemingly inert supports or other otherwise inactive metals and the retention of activity on sites that would deactivate (by coke formation, by oxidation and/or reduction, etc.) without spillover. Conversely, the spillover may be detrimental by contributing to undesirable side reactions like hydrogen transfer in catalytic hydrotreating. Also, the behavior of an activated support toward poisons is of interest since the catalytic activity of the support may be modified by the poison in a manner that differs from the effect of the poison on the metal.

The analysis of the effects of transport on catalysis has focused on a comparison of the availability of reacting species by diffusion to the rate of reaction on the catalytic sites. High-surface-area catalysts are usually porous. Comparison of transport to reaction rates has usually been based on Knudsen diffusion (by constricted collision with the pore walls) as the dominant mode of transport. DeBoer has noted that for small pores surface diffusion may dominate transport (192). Thiele modulus calculations may therefore not be valid if they are applied to systems where surface diffusion can be significant. This may mean that the direct participation of spillover species in catalysis becomes more important if the catalysts are more "microporous." Generalized interpretations of catalyst effectiveness may need to be modified for systems where one of the reactants can spill over and diffuse across the catalyst surface.

D. CONCLUDING REMARKS

This article has attempted to put spillover in perspective. At this time, there seem to be more questions than definitive conclusions. As with any newly discovered phenomenon, "spillover" has been used to explain a variety of phenomena on heterogeneous solids. Even if limited to interphase phenomena the picture is clouded. The influence is well documented, but the extent of the *relative* influence has been studied in less quantitative terms. Further, it is at this time necessary to focus the discussion of spillover to gain understanding and not to limit the approaches. The authors hope this article provides stimulation to further studies of spillover: how spillover occurs and what may be its potential significance.

ACKNOWLEDGMENTS

The authors gratefully thank NATO for support in this collaborative work. One of us, WCC, thanks NSF (CPE 81-21800) and PRF (13703-AC7) for their generous support of the research that contributed to these studies. The other authors, SJT and GMP, thank the CNRS for their support in this research.

REFERENCES

1. Ozaki, A., "Isotopic Studies of Heterogeneous Catalysis," Academic Press, New York, 1977.
2. Kuriacose, J., *Ind. J. Chem.* **5**, 646 (1957).
3. Taylor, H. S., *Actes Congr. Int. Catal., 2nd, 1960*, Vol. 1, p. 159 (1961).
4. Khoobiar, S., *J. Phys. Chem.* **68**, 411 (1964); Khoobiar, S., Peck, R. E., and Reitzer, B. J., *Proc. Int. Cong. Catal., 3rd, 1964*, p. 338 (1965).
5. Sinfelt, J. M., and Lucchesi, P. J., *J. Am. Chem. Soc.* **85**, 3365 (1963).
6. Pajonk, G. M., Teichner, S. J., and Germain, J. E., eds., "Spillover of Adsorbed Species." Elsevier, Amsterdam, 1983; also a volume of "Discussions." University of Claude Bernard-Lyon 1, Villeurbanne, 1984.
7. Sermon, P. A., and Bond, G. C., *Catal. Rev.* **8**, 211 (1973).
8. Dowden, D. A., "Catalysis," Vol. VIII, Chapter 6. Chem. Soc., London, 1980.
9. Bond, G. C., *in* "Spillover of Adsorbed Species" (G. M. Pajonk, S. J. Teichner, and J. E. Germain, eds.), p. 1. Elsevier, Amsterdam, 1983.
10. Conner, W. C., "Discussions," p. 71. University of Claude Bernard-Lyon 1, Villeurbanne, 1984.
11. Sermon, P. A., and Bond, G. C., *J. Chem. Soc., Faraday Trans.* **76**, 889 (1980).
12. Pajonk, G. M., and Teichner, S. J., *in* "Adsorption at the Gas–Solid and Liquid–Solid Interface" (J. Rouquerol and K. S. W. Sing, eds.), p. 281. Elsevier, Amsterdam, 1982.
13. Altham, J. A., and Webb, G., *J. Catal.* **18**, 133 (1970).
14. Fujimoto, K., and Toyoshi, S., *Proc. Int. Congr. Catal., 7th, 1980* p. 235 (1981).
15. Keren, E., and Soffer, A., *J. Catal.* **50**, 43 (1977).
16. Fujimoto, K., Ohno, A., and Kunugi, T., *in* "Spillover of Adsorbed Species" (G. M. Pajonk, S. J. Teichner, and J. E. Germain, eds.) p. 241. Elsevier, Amsterdam, 1983.
17. Robell, A. J., Ballou, E. V., and Boudart, M., *J. Phys. Chem.* **68**, 2748 (1964).
18. Kramer, R., and Andre, M., *J. Catal.* **58**, 287 (1979).
19. Berzins, A. R., Lau Vong, M. S. W., Sermon, P. A., and Wurie, A. T., *Adsorpt. Sci. Technol.*, **1**, 51 (1984).
20. Anderson, J. R., Foger, K., and Breakspere, R. J., *J. Catal.* **57**, 458 (1979).
21. Ostermaier, J. J., Katzer, J. R., and Manogue, W. H., *J. Catal.* **41**, 277 (1976).
22. Gajardo, F., Gleason, E. F., Katzer, J. R., and Sleight, A. W., *Proc. Int. Congr. Catal., 7th, 1980*, p. 1462 (1981).
23. Apple, T. M., Gajardo, P., and Dybowski, C., *J. Catal.* **68**, 103 (1981).
24. Apple, T. M., and Dybowski, C., *J. Catal.* **71**, 316 (1981).

25. Jiang, X. Z., Hayden, T. F., and Dumesic, J. A., *J. Catal.* **83**, 168 (1983).
26. Van Meerbeck, A., Jelli, A., and Fripiat, J. J., *J. Catal.* **46**, 320 (1977).
27. Morterra, C., and Low, M. J. D., *Ann. N.Y. Acad. Sci.* **220**, 133 (1973).
28. Engels, S., Malsch, R., and Wilde, M., *Z. Chem.* **16**, 416 (1976).
29. Sinfelt, J. H., and Via, G. H., *J. Catal.* **56**, 1 (1978).
30. Dowden, D. A., Haining, I. H. B., Irving, J. D. N., and Whan, D. A., *J. Chem. Soc., Chem. Commun.* p. 631 (1977).
31. Barbier, J. Charcosset, H., Periera, G., and Riviere, J., *Appl. Catal.* **1**, 71 (1981).
32. Paryjczak, T., and Zielinski, P., *React. Kinet. Catal. Lett.* **23**, 171 (1983).
33. Paryjczak, T., and Zielinski, P., *React. Kinet. Catal. Lett.* **23**, 165 (1983).
34. Takeuchi, T., Matsuyama, M., and Yashiki, M., *J. Res. Inst. Catal., Hokkaido Univ.* **28**, 335 (1980).
35. Engels, S., Khue, N. P., and Wilde, M., *Z. Anorg. Allg. Chem.* **461**, 155 (1980).
36. Inui, T., Ueno, K., Funabiki, M., Suehiro, M., Sezume, T., and Takegami, Y., *J. Chem. Soc., Faraday Trans.* **75**, 1495 (1979).
37. Inui, T., Funabiki, M., and Takegami, Y., *J. Chem. Soc., Faraday Trans.* **76**, 2237 (1980).
38. Crucq, A., Degols, L., Lienard, G., and Frennet, A., *in* "Spillover of Adsorbed Species" (G. M. Pajonk, S. J. Teichner, and J. E. Germain, eds.), p. 137. Elsevier, Amsterdam, 1983.
39. Levy, R. B., and Boudart, M., *J. Catal.* **32**, 304 (1974).
40. Boudart, M. Aldag, A. W., and Vannice, M. A., *J. Catal.* **18**, 46 (1970).
41. Sermon, P. A., and Bond, G. C., *J. Chem. Soc., Faraday Trans.* **76**, 8 (1976).
42. Parera, J. M., Figoli, N. S., Jablonski, E. L., Sad, M. R., and Beltraimini, J. N., "Catalyst Deactivation," Stud. Surf. Sci. Catal., Vol. 6. Elsevier, Amsterdam, 1980.
43. Figoli, N. S., Sad, M. R., Beltramini, J. N., Jablonski, E. L., and Parera, J. M., *Ind. Eng. Chem. Prod. Res. Dev.* **19**, 545 (1980).
44. Beck, D. D., Bawagan, A. O., and White, J. M., *J. Phys. Chem.* **88**, 2771 (1984).
45. Bond, G. C., and Tripathi, J. B. P., *J. Less-Common. Met.* **36**, 31 (1974).
46. Bond, G. C., and Mallat, T., *J. Chem. Soc., Faraday Trans.*, **77**, 1743 (1981).
47. Carter, J. L., Lucchesi, P. J., Corneil, P., Yates, D. J. C., and Sinfelt, J. H., *J. Phys. Chem.* **69**, 3070 (1965).
48. Cavanagh, R. R., and Yates, J. T., Jr., *J. Catal.* **68**, 22 (1981).
49. Ambs, W. J., and Mitchell, M. M., Jr., *J. Catal.* **82**, 226 (1983).
50. Scott, K. F., and Phillips, C. S. G., *J. Catal.* **51**, 131 (1978).
51. Mazabrard, A. R., unpublished results.
52. Tiller, H. J., Kuhn, W., and Meyer, K., *Z. Chem.* **14**, 450 (1974).
53. Bianchi, D., Maret, D., Pajonk, G. M., and Teichner, S. J., *in* "Spillover of Adsorbed Species" (G. M. Pajonk, S. J. Teichner, and J. E. Germain, eds.), p. 45. Elsevier, Amsterdam, 1983.
54. Conner, W. C., Jr., Cevallos-Candau, J. F., Shah, N., and Haensel, V., *in* "Spillover of Adsorbed Species" (G. M. Pajonk, S. J. Teichner, and J. E. Germain, eds.), p. 31. Elsevier, Amsterdam, 1983.
55. Sheng, T. C., and Gay, I. D., *J. Catal.* **71**, 119 (1981).
56. Finlayson-Pitts, B. J., *J. Phys. Chem.* **86**, 3499 (1982).
57. Steinberg, K. H., Hofmann, F., Bremer, H., Dmitriev, R. V., Detjuk, A. N., and Minachev, C. M., *Z. Chem.* **19**, 34 (1979).
58. Dmitriev, R. V., Steinberg, K. H., Detjuk, A. N., Hofmann, F., Bremer, H., and Minachev, C. M., *J. Catal.* **65**, 105 (1980).
59. Dmitriev, R. V., Detjuk, A. N., Minachev, C. M., and Steinberg, K. H., *in* "Spillover of Adsorbed Species" (G. M. Pajonk, S. J. Teichner, and J. E. Germain, eds.), p. 17. Elsevier, Amsterdam, 1983.

60. Dalla-Betta, R. A., and Boudart, M., *J. Chem. Soc., Faraday Trans.* **72**, 1723 (1976).
61. Kazanski, V. B., Borokov, V. Yu., and Kustov, L. M., *Proc. Int. Congr. Catal., 8th, 1984.* Vol. 3, p. 3 (1984).
62. Il'Chenko, N. I., *Russ. Chem. Rev. (Engl. Transl.)* **41**, 47 (1972).
63. Ekstrom, A., Batley, G. E., and Johnson, D. A., *J. Catal.* **34**, 106 (1974).
64. L'Homme, G. A., Boudart, M., and D'Or, L., *Bull. Acad. R. Belg. Cl. Sci.* **52**, 1206 and 1249 (1966).
65. McKee, D. W., *Carbon* **8**, 623 (1970).
66. Sermon, P. A., and Bond, G. C., *J. Chem. Soc., Faraday Trans.* **72**, 730 (1976).
67. Gerand, B., and Figlarz, M., *in* "Spillover of Adsorbed Species" (G. M. Pajonk, S. J. Teichner, and J. E. Germain, eds.), p. 275. Elsevier, Amsterdam, 1983.
68. Tinet, D., Estrade-Szwarckopf, H., and Fripiat, J. J., *in* "Metal-Hydrogen Systems." Pergamon, Oxford, 1982.
69. Bond, G. C., and Tripathi, J. B. P., *J. Chem. Soc., Faraday Trans.* **72**, 933 (1976).
70. Batley, G. E., Ekström, A., and Johnson, D. A., *J. Catal.* **34**, 368 (1974).
71. Che, M., Canosa, B., and Gonzales-Elipe, A. R., *J. Chem. Soc., Faraday Trans.* **78**, 1043 (1982).
72. Hurst, N. W., Gentry, S. J., Jones, A., and McNicol, B. D., *Catal. Rev.* **24**, 233 (1982).
73. Gentry, S. J., Hurst, N. W., and Jones, A., *J. Chem. Soc., Faraday Trans.* **77**, 603 (1981).
74. Szabo, Z. G., and Konkoly-Thege, I., "Thermal Analysis," p. 229 (1980).
75. Mieville, R. L., *J. Catal.* **87**, 437 (1984).
76. Isaacs, B. H., and Petersen, E. E., *J. Catal.* **77**, 43 (1982).
77. Praliaud, H., and Martin, G. A., *in* "Spillover of Adsorbed Species" (G. M. Pajonk, S. J. Teichner, and J. E. Germain, eds.), p. 191. Elsevier, Amsterdam, 1983.
78. Dalmon, J. A., Mirodatos, C., Turlier, P., and Martin, G. A., *in* "Spillover of Adsorbed Species" (G. M. Pajonk, S. J. Teichner, and J. E. Germain, eds.), p. 169. Elsevier, Amsterdam, 1983.
79. Grange, P., Van, T. T., and Delmon, B., *C. R. Hebd. Seances Acad. Sci.* **229C**, 1007 (1981).
80. Mutombo, H., Grange, P., and Delmon, B., *J. Chim. Phys.* **75**, 518 (1978).
81. Fraser, D., Moyes, R. B., and Wells, P. B., *Proc. Int. Congr. Catal., 7th, 1980,* p. 1424 (1981).
82. Imamura, H., and Tsuchiya, S., *J. Chem. Soc. Faraday Trans.* **79**, 1461 (1983).
83. Imamura, H., Takahashi, T., and Tsuchiya, S., *J. Catal.* **77**, 289 (1982).
84. Vannice, M. A., and Neikam, W. C., *J. Catal.* **20**, 260 (1971).
85. Neikam, W. C., and Vannice, M. A., *J. Catal.* **27**, 207 (1972).
86. Erre, R., and Fripiat, J. J., *in* "Spillover of Adsorbed Species" (G. M. Pajonk, S. J. Teichner, and J. E. Germain, eds.), p. 285. Elsevier, Amsterdam, 1983.
87. Erre, R., Van Damme, H., and Fripiat, J. J., *Surf. Sci.* **127**, 48 (1983).
88. Erre, R., Legay, M. H., and Fripiat, J. J., *Surf. Sci.* **127**, 69 (1983).
89. Lacroix, M., Pajonk, G. M., and Teichner, S. J., *Bull. Soc. Chim. Fr.* p. 101 (1980).
90. Teichner, S. J., Pajonk, G. M., and Lacroix, M., *in* "Surface Properties and Catalysis by Non Metals" (J. P. Bonnelle *et al.*, eds.), p. 457. Reidel Publ., Dordrecht, Netherlands, 1983.
91. Tinet, D., Partyka, S., Rouquerol, J., and Fripiat, J. J., *Mater. res. Bull.* **17**, 561 (1982).
92. Taylor, R. E. Silva-Crawford, M. M., and Gerstein, B. C., *J. Catal.* **62**, 401 (1980).
93. Bond, G. C., *in* "Metal–Support and Metal–Additive Effects in Catalysis," Stud. Surf. Sci. Catal., Vol. 11, p. 1. Elsevier, Amsterdam, 1982.
94. Tauster, S. J., Fung, S. C., and Garten, R. L., *J. Am. Chem. Soc.* **100**, 170 (1978).
95. Sexton, B. A., Hughes, A. E., and Foger, K., *J. Catal.* **77**, 85 (1982).
96. Tauster, S. J., and Fung, S. C., *J. Catal.* **55**, 29 (1978).
97. Chen, B.-H., and White, J. M., *J. Phys. Chem.* **86**, 3534 (1982).
98. Baker, R. T. K., Prestridge, E. B., and Murrell, L. L., *J. Catal.* **80**, 348 (1983); Baker, R. T. K., Prestridge, E. B., and Garten, R. L., *ibid.* **56**, 390 (1979).

99. Baker, R. T. K., Prestridge, E. B., and Garten, R. L., *J. Catal.* **59**, 293 (1979).
100. Huizinga, T., and Prins, R., *J. Phys. Chem.* **85**, 2156 (1981).
101. Conesa, J. C., and Soria, J., *J. Phys. Chem.* **86**, 1392 (1982); Conesa, J. C., Munuera, G., Munoz, A., Rives, V., Sanz, J., and Soria, J. in "Spillover of Adsorbed Species" (G. M. Pajonk, S. J. Teichner, and J. E. Germain, eds.), p. 149. Elsevier, Amsterdam, 1983.
102. Herrmann, J. M., Disdier, J., and Pichat, P., in "Metal-Support and Metal-Additive Effects in Catalysis," Stud. Surf. Sci. Catal., Vol. 11, p. 27. Elsevier, Amsterdam, 1982; Hermann, J. M., and Pichat, P., *J. Catal.* **78**, 425 (1982); Horsley, J. A., *J. Am. Chem. Soc.* **101**, 2870 (1979).
103. Herrmann, J. M., and Pichat, P., *J. Catal.* **78**, 425 (1982).
104. Herrmann, J. M., and Pichat, P., in "Spillover of Adsorbed Species" (G. M. Pajonk, S. J. Teichner, and J. E. Germain, eds.), p. 77. Elsevier, Amsterdam, 1983.
105. Disdier, J., Herrmann, J. M., and Pichat, P., *J. Chem. Soc., Faraday Trans.* **79**, 651 (1983).
106. Kunimori, K., and Uchijima, T., in "Spillover of Adsorbed Species" (G. M. Pajonk, S. J. Teichner, and J. E. Germain, eds.), p 197. Elsevier, Amsterdam, 1983.
107. Kunimori, K., Metsui, S., and Uchijima, T., *J. Catal.* **85**, 253 (1984).
108. Den Otter, G. J., and Dautzenberg, F. M., *J. Catal.* **53**, 116 (1978).
109. Beck, D. D., and White, J. M., *J. Phys. Chem.* **88**, 174 (1984).
110. Foger, K., *J. Catal.* **78**, 406 (1982).
111. Sadeghi, H. R., and Henrich, V. E., *J. Catal.* **87**, 279 (1984).
112. Resasco, D. E., and Haller, G. L., *J. Catal.* **82**, 279 (1983).
113. Burch, R., and Flambard, A. R., *J. Catal.* **78**, 389 (1982).
114. Boudart, M., Benson, J., and Kohn, H., *J. Catal.* **5**, 307 (1966).
115. Levy, R., Ph.D. Thesis, Stanford University, Stanford, California (1974).
116. Jiehan, H., Zupei, H., Yonzi, S., and Hangli, W., in "Spillover of Adsorbed Species" (G. M. Pajonk, S. J. Teichner, and J. E. Germain, eds.), p. 53. Elsevier, Amsterdam, 1983.
117. Gadgil, K., and Gonzalez, R. D., *J. Catal.*, 40, 190 (1975).
118. Che, M., *Proc. Int. Congr. Catal., 7th, 1980.* Discuss. Pap. p. 287 (1981).
119. Nogier, J. P., Bonardet, J., and Fraissard, J., in "Spillover of Adsorbed Species" (G. M. Pajonk, S. J. Teichner, and J. E. Germain, eds.), p. 233. Elsevier, Amsterdam, 1983.
120. Minachev, C., Dimitriev, R., Steinberg, K., Detjuk, A., and Bremer, H., *Izv. Akad. Nauk, Ser. Khim.* p. 2670 (1975).
121. Minachev, C., Dimitriev,, R., Steinberg, K., and Bremer, H., *Izv. Akad. Nauk, Ser. Khim.* p. 2682 (1978).
122. Duprez, D., and Miloudi, A., in "Spillover of Adsorbed Species" (G. M. Pajonk, S. J. Teichner, and J. E. Germain, ads.), p. 163. Elsevier, Amsterdam, 1983.
123. Bianchi, D., Lacroix, M., Pajonk, G. M., and Teichner, S. J., *J. Catal.*, **68**, 411 (1981).
124. Bianchi, D., Lacroix, M., Pajonk, G. M., and Teichner, S. J., *J. Catal.* **59**, 467 (1979).
125. Herzberg, G., *J. Chem. Phys.* **70**, 4806 (1979).
126. Daage, M., and Bonnelle, J., in "Spillover of Adsorbed Species" (G. M. Pajonk, S. J. Teichner, and J. E. Germain, eds.), p. 261. Elsevier, Amsterdam, 1983.
127. Bond, G., Sermon, P., and Tripathi, J., *Ind. Chim. Belge* **38**, 506 (1973).
128. Tomita, D., and Tamai, Y., *J. Catal.* **27**, 293 (1973).
129. Gates, B., Katzer, J., and Schuit, G., "Chemistry of Catalytic Processes," p. 289. McGraw-Hill, New York, 1979.
130. Parera, J., Traffano, E., Masso, J., and Pieck, C., in "Spillover of Adsorbed Species"(G. M. Pajonk, S. J. Teichner, and J. E. Germain, eds.), p. 101. Elsevier, Amsterdam, 1983.
131. Schwabe, U., and Bechtold, E., *J. Catal.* **26**, 427 (1972).
132. Fleisch, T., and Abermann, R., *J. Catal.* **50**, 268 (1977).
133. Selwood, P., *Proc. Int. Congr. Catal., 4th, 1968*, p. 248 (1971)
134. Lacroix, M., Pajonk, G. M., and Teichner, S. J., *Bull. Soc. Chim. Fr.* p. 84 (1981).

135. Maret, D., Pajonk, G. M., and Teichner, S. J., *in* "Spillover of Adsorbed Species" (G. M. Pajonk, S. J. Teichner, and J. E. Germain, eds.), p. 215. Elsevier, Amsterdam, 1983.

136. Maret, D., Pajonk, G. M., and Teichner, S. J., "Catalysis on the Energy Scene" Stud. Sur. Sci. Catal., Vol. 19, p. 347. Elsevier, Amsterdam, 1984.

137. Conner, W. C., Cevallos-Candau, J., and Lenz, D., unpublished results.

138. Beck, D. D., and White, J. M., *J. Phys. Chem.* **88**, 2764 (1984).

139. Bianchi, D., Gardes, G. E. E., Pajonk, G. M., and Teichner, S. J., *J. Catal.* **38**, 135 (1975).

140. Lacroix, M., Pajonk, G. M., and Teichner, S. J., *Bull. Soc. Chim. Fr.* p. 258 (1981).

141. Gardes, G. E. E., Pajonk, G. M., and Teichner, S. J., *J. Catal.* **33**, 145 (1974).

142. Mestdagh, M., Stone, W., and Fripiat, J., *J. Phys. Chem.* **76**, 1220 (1972).

143. Cirillo, A., Ph.D. Dissertation, University of Wisconsin, Milwaukee (1979).

144. Lobashina, N. E., Savvin, N. N., and Myasnikov, I. A., *Dokl. Akad. Nauk SSSR* **268**, 1434 (1983); *Kinet. Katal.* **24**, 747 (1983).

145. Yang, R. T., and Wong, C., *J. Chem. Phys.* **75**, 4471 (1981).

146. Yang, R. T., and Wong, C., *J. Catal.* **85**, 154 (1984).

147. Radovic, L. R., Walker, P. L., and Jenkins, R. G., *J. Catal.* **82**, 382 (1983).

148. Inui, T., Otowa, T., Tsutchiyashi, K., and Takegami, Y., *Carbon* **20**, 382 (1983).

149. Ravick, R. T., Wentrcek, P. R., and Wise, H., *Fuel* **53**, 274 (1974).

150. Bond, G. C., Fuller, M. J., and Molloy, L. R., *Proc. Int. Congr. Catal., 6th, 1976* p, 356 (1977),

151. Tascon, J. M. D., Mestdagh, M. M., Grange, P., and Delmon, B., *Iberoam. Symp. Catal., 8th, 1982*, p. 710 (1983).

152. Masai, M., Nakahara, K., and Yabashi, M., *Chem. Lett. Chem. Soc. Jpn.* p. 503 (1979).

153. Masai, M., Yabashi, M., and Kobasyashi, H., *Chem. Lett., Chem. Soc. Jpn.* p. 833 (1979).

154. Masai, M., Nakahara, K., Yabashi, M., Murata, K., Nishiyama, S., and Tsuruya, S., *in* "Spillover of Adsorbed Species" (G. M. Pajonk, S. J. Teichner, and J. E. Germain, eds.), p. 89. Elsevier, Amsterdam, 1983.

155. Batley, G. E., and Ekström, A. *J. Catal.* **34**, 360 (1974).

156. Chadwick, D., and Christie, A. B., *Proc. Ecoss Cannes, 3rd*, 1, p. 423 (1980).

157. Solymosi, F., Völgyesi, L., and Sarkany, J., *J. Catal.* **54**, 336 (1978).

158. Bansagi, T., Rasko, J., and Solymosi, F., *in* "Spillover of Adsorbed Species" (G. M. Pajonk, S. J. Teichner, and J. E. Germain, eds.), p. 109. Elsevier, Amsterdam, 1983.

159. Solymosi, F., and Rasko, J., *Appl. Catal.* **10**, 19 (1984).

160. Hecker, W. C., and Bell, A. T., *J. Catal.* **85**, 389 (1984).

161. Webb, G., and MacNab, J. I., *J. Catal.* **26**, 226 (1972).

162. Solymosi, F., Erdöhelyi, A., and Bansagi, T., *J. Chem. Soc., Faraday Trans.* **77**, 2645 (1981); *J. Catal.* **68**, 371 (1981); Solymosi, F., Tombacz, I., and Kocsis, M., *ibid.* **75**, 78 (1982).

163. Palazov, A., Kadinov, G., Bonev, H., and Shopov, D. *J. Catal.* **74**, 44 (1982).

164. Bracey, J. D., and Burch, R., *J. Catal.* **86**, 384 (1984).

165. Vannice, M. A., and Sudhaka, C., *J. Phys. Chem.* **88**, 2429 (1984).

166. Kosaki, Y., Miyamoto, A., and Murakami, Y., *Chem. Lett. Chem. Soc. Jpn.* p. 935 (1975).

167. Antonucci, P., van Truong, N. Giordano, N., and Maggiore, R., *J. Catal.* **75**, 140 (1982).

168. Lau, M. S. W., and Sermon, P. A., *J. Chem. Soc., Chem. Commun.* p. 891 (1978).

169. Fujimoto, K., Masamizu, K., Asaoka, S., and Kunigi, T., *J. Chem. Soc. Jpn.* p. 1062 (1976).

170. Marcq, J. P., Wispenninckx, X., Poncelet, G., Keravis, D., and Fripiat, J. J., *J. Catal.* **73**, 309 (1982).

171. Marcq, J. P., Poncelet, G., and Fripiat, J. J., *J. Catal.* **87**, 339 (1984).

172. Benali, R., Hoang-Van, C., and Vergnon, P., *Bull. Soc. Chim. Fr.* p. 417 (1985).

173. Cirillo, A. C., Bryan, L., Gerstein, B., and Fripiat, J. J., *J. Chem. Phys.* **73**, 3060 (1980).

174. Delmon, P., *React. Kinet. Catal. Lett.* **13**, 203 (1980).

175. Pirote, D., Grange, P., and Delmon, P., *Proc. Int. Congr. Catal., 7th, 1980* p. 1422 (1981).

176. Carter, J. L., Lucchesi, P. J., Sinfelt, J. H., and Yates, D. J. C., *Proc. Int. Congr. Catal., 3rd, 1964* Vol. 1, p. 644 (1965).

177. Schlatter, J. C., and Boudart, M., *J. Catal.* **24**, 482 (1972).

178. Gardes, G. E. E., Pajonk, G. M., and Teichner, S. J., *C.R. Hebd. Seances Acad. Sci.* **227C**, 191 (1973).

179. Gardes, G. E. E., Pajonk, G. M., and Teichner, S. J., *C.R. Hebd. Seances Acad. Sci.* **278C**, 659 (1974).

180. Teichner, S. J., Mazabrard, A. R., Pajonk, G. M., Gardes, G. E. E., and Hoang-Van, C., *J. Colloid Interface Sci.* **58**, 88 (1977).

181. Lacroix, M., Pajonk, G. M., and Teichner, S. J., *Bull. Soc. Chim. Fr.* p. 94 (1981).

182. Lacroix, M., Pajonk, G. M., and Teichner, S. J., *C.R. Hebd. Seances Acad. Sci.* **287C**, 499 (1978); *Bull. Soc. Chim. Fr.* p. 265 (1981).

183. Lacroix, M., Pajonk, G. M., and Teichner, S. J., *Proc. Int. Congr. Catal., 7th, 1980* p. 279 (1981).

184. Lacroix, M., Pajonk, G. M., and Teichner, S. J., *J. Catal.*, in press (1986).

185. Hoang-Van, C., Mazabrard, A. M., Michel, C., Pajonk, G. M., and Teichner, S. J., *C.R. Hebd. Seances Acad. Sci.* **281C**, 211 (1975).

186. Lacroix, M., Pajonk, G. M., and Teichner, S. J., *React. Kinet. Catal. Lett.* **12**, 369 (1979).

187. Hindin, S. G., and Weller, S. W., *J. Phys. Chem.* **60**, 1501 (1956); *Adv. Catal.* **9**, 70 (1957).

188. Hilaire, P., French At. Energy Comm., Paris, Reprint No. 2260 (1963); Thesis, Lyons (1963).

189. Kokes, R. J., and Dent, A. L., *Adv. Catal.* **22**, 1 (1972).

190. Sachtler, W. M. H., and Bostelaar, L.-J., *in* "Spillover of Adsorbed Species" (G. M. Pajonk, S. J. Teichner, and J. E. Germain, eds.), p. 207. Elsevier, Amsterdam, 1983.

191. Vannice, M. A., personal correspondence.

192. deBoer, J. H., *Proc. Int. Congr. Catal., 4th, 1968* p. 9 (1971).

Mechanistic Aspects of Transition-Metal-Catalyzed Alcohol Carbonylations

THOMAS W. DEKLEVA

Electronics Group
ICI Americas Inc.
Wilmington, Delaware 19897

AND

DENIS FORSTER

Central Research Laboratories
Monsanto Company
800 N. Lindbergh Boulevard
St. Louis, Missouri 63167

I. Introduction

The metal-catalyzed carbonylation reactions of olefins have been well documented and extensively reviewed. The complementary area of alcohol carbonylation reactions has been a slower developing field, but over the past 10 years it has been the subject of many investigations and we feel that it is an appropriate time to review our understanding of the chemistry involved. In doing so, we also hope to expose areas in which further study is necessary.

In principle, the reaction of an alcohol with carbon monoxide or synthesis gas can lead to several different products:

$$ROH \longrightarrow \begin{cases} \xrightarrow{\ CO\ } RCOOH & (1a) \\[2mm] \xrightarrow{\ CO/H_2\ } RCHO \longrightarrow RCH_2OH & (1b) \\[2mm] \xrightarrow{\ CO/H_2\ } HCO_2R & (1c) \end{cases}$$

81

In the presence of carbon monoxide and an oxidizing agent, a coupling reaction can be catalyzed:

$$2ROH + 2CO + [O] \longrightarrow R-O-\overset{\overset{O}{\|}}{C}-\overset{\overset{O}{\|}}{C}-O-R + H_2O \qquad (2)$$

Practically, all of the above reactions have been realized, with different metals and conditions. In determining the scope of this review, we have attempted to focus our attention on the nature of the transformations at the metal center, especially with regard to oxidation state and formation of the initial alkyl-, alkoxy-, or carboalkoxy-metal bond from saturated precursors. Therefore, while it appears that hydrocarboxylation reactions make some contribution to the total reactivity in a variety of alcohol carbonylation systems, we feel that the mechanistic aspects of this topic would be better covered separately. So, except for noting where this chemistry makes probable contributions, it will not be discussed here. Similarly, homologation reactions, which are believed to usually proceed by way of aldehyde intermediates, will be discussed only as they pertain to the incorporation of the CO into the metal–carbon bonds, that is, the factors governing the subsequent hydrogenation reactions will not be covered.

Several key mechanistic steps appear to be necessary to allow the various catalytic cycles to occur. A short review of some of these fundamental steps involving transition-metal centers appears to be in order.

A. General Reactivity of Organometallic Complexes

Prerequisite to any catalytic activity is the ability of the metal center to interact effectively with alcohols or alcohol-derived precursors. There are several ways in which this can occur, and most of these have been observed or postulated in at least one catalytic scheme. In order to understand the specific reactivities, though, the reader should be familiar with some fundamental aspects of organo-transition-metal chemistry. These will be discussed only very briefly. The reader should also recognize that in order for catalysis to occur, a balance of reactivities is required.

The first of these reactions is oxidative addition. This is the initial step in many of these catalytic reactions, accompanying the formation of the initial metal–substrate bond. The relative reactivity toward oxidative addition is favored by the presence of electron density on the metal center, so, for a given oxidation state and set of ligands, the rates generally increase in going down a given triad and from left to right in the periodic table (1–3). Similarly, ligands which act as π acids, drawing electron densities from the metal center, generally diminish reactivity. This is easily rationalized by examining the mechanisms of this reaction type. The reactivities of transition-metal complexes with a variety of substrates have been the subjet of intense investigation over the past 20 years. In this article, we are most interested in the reactivities with alkyl halides, because of their importance in carbonylation chemistry. While the specifics of a given system can be controversial, it is now generally accepted that there are two mechanistic classes which have been demonstrated, these being nucleophilic displacement and free-radical reactions. We shall not go into the criteria by which these distinctions are made, since this has been adequately covered elsewhere and the interested reader is referred there (1–3). However, some brief comments are in order. It is often difficult to determine the mechanistic reaction type even under ideal conditions, making it almost impossible to do so under the conditions employed in the systems discussed in this review. Perhaps for this reason alone, the nucleophilic displacement reaction is the one invoked almost exclusively for the systems to be discussed here. While there is no doubt that such S_N2-type reactivity is important, the possible contributions for free-radical mechanisms, either by initial electron transfer or halogen abstraction, should not be ignored. Because of this, we consider it important to note the conditions which favor the intermediacy of radical species. The singularly most important factor governing the reaction type is the nature of the substrate. Free-radical mechanisms become more likely as R varies, primary < secondary < tertiary < benzyl, allyl, and as X varies, $RSO_3^- < Cl^- < Br^- < I^-$ (2, 3). The importance of the metal is less defined, since the oxidative addition reaction generally involves an overall two-electron change in the metal regardless of particular pathway, and predictions based on the accessibility of a particular oxidation state can be very misleading. In fact, many transition-metal complexes exhibit different reactivities with different substrates. Nevertheless, in the study of alcohol carbonylations, it is not uncommon to generalize the findings of one system (usually methanol) to all others in an attempt to present a single scheme for a particular system. This should be done with extreme care; in the absence of rigorous supporting data, this generalization can probably not be justified.

The rearrangement of a metal–alkyl to metal–acyl species, required for incorporation of the CO into the organic substrate, is termed migratory insertion. The reaction is obviously thermodynamically favorable, since it has been observed for a wide variety of metal systems (4), with an estimated reaction enthalpy of -33 kJ/mol (5). The reactivity of transition-metal alkyls toward carbonyl insertion usually decreases in going down a given group (6, 7), probably because of the corresponding increase in the metal–carbon bond energies [$5d > 4d > 3d$ (8)].

B. Specific Reactivities Involved in Alcohol Carbonylations

Metal hydride species exhibit a wide range of reactivity, depending on the nature of the M–H bond. In those cases where the bond is very hydridic in nature, alcohols normally react to form alkoxides, with liberation of H_2:

$$L_xM—H + HOR \longrightarrow L_xM—OR + H_2 \tag{3}$$

At the other end of the spectrum, reactions of alcohols with acidic M–H bonds [$HCo(CO)_4$ being the most notable example] can result in protonation of the alcohol. This protonation then can be followed potentially by several pathways:

$$L_xM—H + ROH \longrightarrow \{ROH_2^+\}\{L_xM^-\} \tag{4}$$

$$\{ROH_2^+\}\{L_xM^-\} \longrightarrow L_xMR + H_2O \tag{5}$$

$$ROH_2^+ \xrightarrow{-H_2O} R^+ \xrightarrow{+CO} RCO^+, \text{etc.}) \tag{6}$$

$$ROH_2^+ \longrightarrow R^+ \longrightarrow \text{olefin} + H^+ \tag{7}$$

$$\text{olefin} + HML_x \longrightarrow RML_x, \text{ etc.} \tag{8}$$

The reactions of alcohols with carbonyl-containing species are also well documented to occur by alcohol attack on the coordinated CO to yield carboalkoxy moieties, with concomitant loss of a proton:

$$M—CO + ROH \longrightarrow M—\overset{\overset{\displaystyle O}{\displaystyle \|}}{C}—OR^- + H^+ \tag{9}$$

Coupling of two such units results in the formation of oxalates.

C. Effects of Halide Promoters

A frequent theme in alcohol carbonylations by transition metals is the use of a halide or pseudo-halide promoters or cocatalysts. Despite major problems of corrosion associated with its use, iodide is almost always found to be most effective in this capacity. This is because the halide serves several purposes, for each of which iodide is ideally suited. One of the most important roles of these anion promoters can be that of facilitating the formation of metal–carbon bonds via the formation of intermediate alkyl halides. Under typical catalytic conditions for a variety of systems, at least some portion of the added halide is converted to the corresponding strong halo-acid. In fact, conditions are generally set so that this event is maximized.

These halo-acids and the alcohols are in equilibrium with, and preferentially form, the corresponding alkyl halide:

$$HX + ROH \;\rightleftharpoons\; RX + H_2O \qquad (10)$$

The alkyl halides are much more susceptible to a nucleophilic attack by an electron-rich metal center than are the neutral or protonated alcohols. The relative importance of the halides in this role follows the classical halide sequence, that is, $RI > RBr > RCl$.

The tendencies of these halides to form strong protic acids allow an alternative pathway to occur. In cases where the formation of the alkyl halide is less efficient, alcohol dehydration can become more significant, resulting in high levels of olefins. At the same time, the reaction of HX with the metal centers is an effective way of generating (possibly catalytically active) metal hydrides. When this occurs, the stage is set for hydrocarboxylation. As discussed earlier, hydrocarboxylation is considered to be outside the scope of this present review. Nevertheless, it is an important pathway for generating metal–alkyl bonds, and frequently is competitive in this way with longer chain alcohol carbonylations.

A third effect of halide promoters is to enhance the relative nucleophilicities of the metal center. This generally occurs by halide displacement of a coordinated neutral ligand or by bridge-splitting reactions to generate anionic metal species. In cases where this type of halide promotion has been independently demonstrated, the effect of the increase in electron density on the reactivity is very large.

Finally, nucleophilic halide ions can enhance the reactivity by promoting the removal of generated acyl moieties from the metal centers:

$$X^- + R\overset{\displaystyle O}{\overset{\displaystyle \|}{C}}-M \xrightarrow{-M^-} R\overset{\displaystyle O}{\overset{\displaystyle \|}{C}}-X \xrightarrow{R'OH,\,-HX} RCOOR' \qquad (11)$$

Iodide, being a much better nucleophile than H_2O or alcohol, is especially efficient in this regard.

II. Rhodium-Catalyzed Carbonylations

In the late 1960s, workers at Monsanto began studies into the carbonylation of methanol to acetic acid. The process they developed (9–11), now known worldwide as the Monsanto acetic acid process, is based on an iodide-promoted rhodium catalyst system. Because of the high efficiency and selectivity of the reaction (typical commercial operating conditions are 150–200°C and 30–100 atm, giving selectivities >99% based on CH_3OH),

TABLE I

Relative Rate and Product Distribution Data for the Iodide-Promoted
Rhodium-Catalyzed Carbonylations of Higher Alcohols at 175°C
and 48 atm CO Pressure[a]

Alcohol	Major products (selectivity)	Relative rates
Methanol	Acetic acid (>99%)	1
Ethanol	Propanoic acid (>99%)	0.1
1-Propanol	Butanoic acid (80%)	0.2
	2-Methyl propanoic acid (20%)	
2-Propanol	Butanoic acid (50%)	0.5
	2-Methyl propanoic acid (50%)	
1-Butanol	Pentanoic acid (80%)	0.1
	2-Methyl butanoic acid (20%)	
1-Hexanol	Heptanoic acid (87%)	0.1
	2-Methyl hexanoic acid (10%)	
	2-Ethyl pentanoic acid (3%)	
3-Hexanol	Heptanoic acid (24%)	0.5
	2-Methyl hexanoic acid (48%)	
	2-Ethyl pentanoic acid (28%)	
Benzyl alcohol	Phenylacetic acid (>85%)	5
Allyl alcohol	Butanoic acid (50%)	0.2
	2-Methyl propanoic acid (40%)	

[a] From Ref. 31.

more than 90% of all new facilities built in the past 10 years for acetic acid
production have been based on this chemistry (12). The technology is
currently licensed to 9 major organizations worldwide and at present
accounts for the production of more than a million metric tons of acetic acid
annually, or 40% of the market share (12, 13).

The reaction conditions for the methanol carbonylation reaction are even
milder than the commercial conditions indicate. Indeed, the complete catalytic cycle can be realized at room temperature and 1 atm of carbon monoxide!
This allowed Forster to specifically interpret the macroscopic kinetic observations in mechanistic terms (14) by demonstrating the component reactions
under ambient conditions. Since then, a great deal of effort has been directed
to determining the generality of this mechanism with other alcohols. In fact, a
wide array of alcohols can be carbonylated by this catalyst system (e.g.,
Table I), though not necessarily by the same mechanistic pathways.

A. NATURE OF CATALYST SPECIES

One of the beauties of the catalytic system is that a variety of rhodium
compounds can be used as precursors for the methanol carbonylation

reaction since most are converted to the most active species under the reaction conditions. In particular, rhodium(III) halides and rhodium(I) phosphine complexes give almost identical reaction rates, after initial induction periods, indicative of a common catalyst. Related kinetic and spectroscopic studies have shown that the carbonylation of rhodium(III) halides in alcoholic (15) and aqueous (16) media under much milder conditions result in the formation of the dicarbonyldihalorhodate(I) anions:

$$RhX_3 + 3CO + H_2O \longrightarrow RhX_2(CO)_2^- + H^+ + CO_2 + HX \qquad (12)$$

This has also been verified in the catalytic system. When rhodium(I) or rhodium(III) phosphine complexes are charged into the reaction media as the catalyst precursors, the phosphines are quaternized to give isolable phosphonium salts, and in situ spectroscopic studies show that the same diiododicarbonylrhodium(I) anion is the main species generated in the reaction mixture (17). Pannetier et al. (18) observed that rhodium complexes containing chelating ligands gave much poorer methanol conversion rates over a limited reaction period than did the reactions starting with rhodium complexes containing monodentate ligands. This probably reflects the slower removal and quarternization of the chelating phosphines.

The $RhI_2(CO)_2^-$ species considered to be the active catalyst for methanol carbonylation has also been observed as the only carbonyl-containing rhodium complex during the carbonylation of other alcohols (Fig. 1; See Sections II,C and D).

A very recent report by Hickey and Maitlis (19) suggests that, in aprotic solvents, a five-coordinate species such as $\{Rh(CO)_2I_3\}_2^-$ may be acting as a "supernucleophile," at least toward CH_3I. This suggestion is based on the observation that salts such as $\{AsPh_4\}\{X\}$ (where $X = I^-$, Cl^-) promote the oxidative addition of methyl iodide to the anionic $RhI_2(CO)_2^-$ species and extends the concept that additional electron density enhances the nucleophilicity of the metal center (20). It is interesting to note, however, that salts such as $\{NBu_4\}\{I\}$, which do not have countercations with aromatic addenda, do not exhibit this extra-ordinate promotional effect. More work is necessary to fully understand this phenomenon, but one possible explanation for it will be discussed in the next section.

B. Carbonylation of Methanol

The mechanism of the rhodium-catalyzed methanol carbonylation has been extensively reviewed elsewhere (17, 21, 22). The reaction is characterized by kinetics which are independent of CO pressure and are first order both in rhodium and methyl iodide. Taken with the spectroscopic data, these are

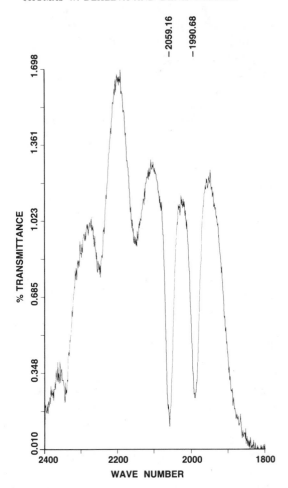

FIG. 1.   Representative *in situ* IR spectrum obtained during the carbonylation of *n*-alcohols; taken during the carbonylation of ethanol at 170°C and 31 atm. From Refs. *24* and *31*.

generally discussed in terms of the mechanism summarized in Scheme 1. This system is the first to be discussed in which organic iodide (as HI or methyl iodide) is necessary for efficient catalysis. The importance of this promoter should be obvious from Scheme 1, for the reasons given in Section I,C. Critical to catalysis is the *in situ* formation of methyl iodide, since the oxidative addition of methyl iodide to the rhodium(I) center is the rate-determining step. Because of this, iodide sources incapable of generating this organic intermediate in significant quantities (e.g., alkali metal iodides) are ineffectual promoters for this reaction.

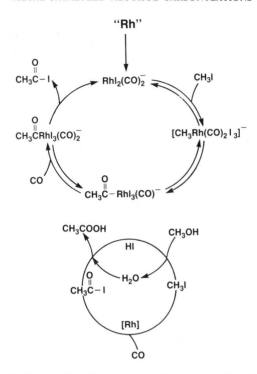

SCHEME 1. Proposed mechanism for the Monsanto acetic acid synthesis.

Another key feature of the metal complex cycle in which the iodide acts most effectively is the nature of the active catalyst itself. The oxidative addition step is considered to be nucleophilic in nature, based on activation parameters and relative rate data (23, 24a) (Section II,C), and the presence of a negative charge on the metal center appears to significantly enhance the nucleophilicity (and hence reactivity toward methyl iodide) of the metal relative to neutral rhodium(I) species (20). Extrapolations of available data (24–26) indicate that, at 25°C, the diiododicarbonylrhodium(I) species has a Pearson nucleophilicity parameter (25) toward methyl iodide of ~5.5. In relation to other common nucleophiles, this value corresponds to nucleophilic reactivity toward methyl iodide comparable to that of pyridine ($n = 5.2$), an order of magnitude greater than chloride ($n = 4.4$), and two orders of magnitude slower than iodide ($n = 7.4$).

In terms of generating the diiododicarbonylrhodium(I) anion, under the catalytic conditions, HI (or some strong acid) is specifically required to stabilize the +1 oxidation state. In the absence of the acid component (e.g., with NaI alone), the catalytic conditions are sufficiently reducing to remove the rhodium as metal.

It is interesting to note here, based on the findings of Hickey and Maitlis (*19*) (Section II,A), that the hydroxylic acidity of the catalytic system may be important in another regard. The ability of Lewis acids to catalyze migratory insertion reactions is well known (*27*). The first step in the accepted mechanism is generally written as in Scheme 1. The alkyldicarbonyl-rhodium(III) has never been observed and is considered to have only a very transitory existence. Nevertheless, under the catalytic conditions, the data indicate that in the following equations k_1 is rate limiting, k_{-1} is kinetically insignificant, and k_2 is very fast:

$$\{RhI_2(CO)_2\}^- + CH_3I \underset{k_{-1}}{\overset{k_1}{\rightleftharpoons}} RhI_3(CH_3)(CO)_2^- \tag{13}$$

$$\{RhI_3(CH_3)(CO)_2\}^- \overset{k_2}{\longrightarrow} \{RhI_3(\overset{\overset{\text{O}}{\|}}{C}CH_3)(CO)\}^- \tag{14}$$

While the effect of the hydroxylic solvent on k_2 cannot be quantified, it would be expected to accelerate it. The work of Hickey and Maitlis (*19*) involved the use of nonhydroxylic solvents, in which this accelerating effect is not available, and the assumption that k_2 is fast under these conditions may not be justified. It is even possible that it may be slowed to the extent that it contributes to the rate-limiting step. If this is the case, the rate enhancements observed in their work might also be attributable to the acceleration of the k_2 step by the Lewis acid countercations. Such effects of cations on migratory insertions are expected to be more important under their experimental conditions, where ion pairing was acknowledged to be significant. However, while this alternative explanation may appear to be consistent with the available data, we must emphasize that it is speculative. Certainly, more detailed studies are required to determine definitively the reason for this acceleratory effect.

C. CARBONYLATION OF PRIMARY LINEAR ALCOHOLS

The general reaction scheme used to describe methanol carbonylation has been extended to include the carbonylations of benzyl alcohol (*28*), ethanol (*24a, 29, 30*) and 1-propanol (*24a, 31*). In fact, the study with the last system has made it necessary to expand the original scheme to account for the production of mixtures of isomeric products (*vide infra*).

The relative rate data and activation parameters (Tables II and III) obtained for these systems suggest very convincingly that the oxidative addition of the corresponding alkyl iodide to the rhodium(I) center is nucleophilic in nature. There appear to be slight variances in the absolute,

TABLE II
Relative Rate Data for the Iodide-Promoted Rhodium-Catalyzed Carbonylation of a Variety of Alcohols[a]

Alcohol	Relative rate (*31*)	S_N2 displacement rate for organic halides (*57*)
Methanol	21	30
Ethanol	1.0	1.0
1-Propanol	0.47	0.4
2-Propanol	1.2–3.8	0.02

[a] From Refs. *31* and *57*.

TABLE III
Activation Parameters for the Carbonylation of Linear Primary Alcohols with the Rhodium–Iodide System[a]

Alcohol	E_a (kJ mol^{-1})	$H^{\ddagger}$ (kJ mol^{-1})	$S^{\ddagger}$ (J mol^{-1} K^{-1})
Methanol	67.1	63.6	−115.9
Ethanol	80.1	76.4	−110.7
1-Propanol	83.1	79.5	−109.2

[a] Data from Ref. *31*.

but not relative, rates from the data presented by the three groups, which may simply be a reflection of the differences in experimental procedures. The kinetic profiles for these systems again indicate that the reactions are first order each in rhodium and alkyl iodide and independent of CO pressure. Dake and co-workers (*30*) reported a more complicated rate law for their study on ethanol carbonylation. This appears not to be the result of any difference in mechanism, but rather the result of perturbations of the equilibrium affecting the concentration of ethyl iodide:

$$CH_3CH_2OH + HI \; \rightleftharpoons \; CH_3CH_2I + H_2O \qquad (15)$$

Labeling studies (*24, 31*) indicated that there was no significant kinetic isotope effect when either the substrate (EtOH system) or protic solvent (*n*-PrOH system) was replaced by deuterium-substituted species, again consistent with the S_N2-type reactivity.

Studies involving the carbonylation of 1-propanol (*24a*) and specifically labeled [1-^{13}C]ethanol (*24c*) give mutually supportive information as to the stability and reactivity of one of the key organometallic intermediates. Initial

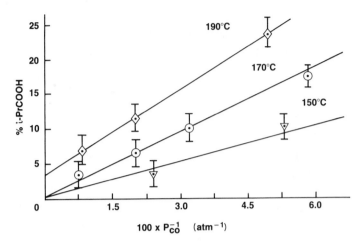

FIG. 2. Variation of product distribution with pressure during the carbonylation of 1-propanol. From Refs. *24* and *31*.

screening runs done previously at Monsanto (*31*) (Table I) indicated that the carbonylations of longer linear alcohols always resulted in such isomeric mixtures of the homologated acids, although the source of this behavior had not been determined. During the 1-propanol study, it was found that while the absolute rate was independent of CO pressure (and thus, concentration), the product distribution was not. Increasing the CO pressure (ca. 20–130 atm) resulted in mixtures which contained decreasing amounts of isobutyric acid. Extrapolation to infinite CO pressure [as $1/P(CO)$ approached zero, Fig. 2] indicated that in this limit only *n*-butyric acid was formed. Again, since the reaction exhibited rate parameters consistent with the anticipated S_N2-type reactivity, it was concluded that the product selectivity was determined after this single rate-determining step. This was also borne out by labeling studies, in which the products, *n*- and isobutyric acids, both had the same isotopic composition as the recovered *n*-PrI, indicating a similar history. The data are most consistent with the reactions shown in Scheme 2. In this model, nucleophilic oxidative addition is rate determining and gives rise to a short-lived alkyl dicarbonyl rhodium(III) species. By all accounts, *cis*-dicarbonyl rhodium(III) species are very unstable. This instability is relieved, at least in this case, by two possible "decomposition" routes. The first involves the familiar migratory insertion reaction to generate an acyl monocarbonyl rhodium(III) center, which then proceeds to regenerate the $RhI_2(CO)_2^-$ and *n*-butyric acid, by a route identical to that proposed for the methanol system (vide supra). Alternatively, the alkyl dicarbonyl rhodium(III) intermediate can dissociate a CO

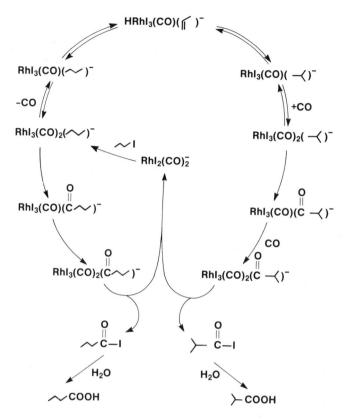

SCHEME 2. Mechanism for the iodide-promoted rhodium-catalyzed carbonylation of 1-propanol to account for the formation of isomeric butyric acids. From Refs. *24* and *31.*

ligand to form a more stable, but formally coordinatively unsaturated, monocarbonyl derivative. Facile β-hydride elimination generates an intermediate hydrido-olefin complex, which can then reinsert to form either the same *n*-propyl or new isopropyl moiety. Further reaction of the isopropyl rhodium(III) species gives rise to the isobutyric acid. The retention of isotopic integrity discussed earlier also indicated that the intermediate hydride was not rapidly exchanged with the solvent (this is important in analyzing the results of a later study). Since the action responsible for isomerization is the β-hydride elimination to form the hydrido-olefin complex, it seemed likely that the same sort of process occurs during the carbonylation of all linear alcohols higher than methanol. This adequately explains the production of isomeric acid mixtures in the carbonylations of longer chain linear alcohols.

In the case of ethanol, however, the presumed intermediate is as hydrido–ethylene complex, which on reinsertion can give rise only to a single ethyl moiety.

By specifically labeling the precursor substrate with ^{13}C in the C-1 position, the degeneracy of the process was lifted, revealing the same product pressure dependence as above. That is, the mechanism outlined in Scheme 2 does appear to be a general one for this rhodium system.

D.   CARBONYLATION OF SECONDARY ALCOHOLS, MODELED BY 2-PROPANOL

To date, mechanistic studies into the carbonylations of secondary alcohols with the same type of rhodium/RI catalyst system have used 2-propanol as a model substrate. At least part of the reason for this has been to minimize the expected complexities of the product analyses. The carbonylation of 2-propanol gives mixtures of n- and isobutyric acids. Two studies have been (24b, 32) reported with this system. The first of these (32) concluded that the reactivity could be described in terms of the same nucleophilic mechanism as has been described above, despite the fact that the reaction rates at 200°C were approximately 140 times faster than predicted by this type of chemistry (24b). Other data also indicated that this S_N2-type reactivity was probably not the sole contributor to the reaction scheme. For example, the authors were not able to adequately explain either the effect of reaction conditions on product distribution or the activation parameters. They also did not consider the possible contribution of a hydrocarboxylation pathway, which is known to be extremely efficient in analogous systems (33). For these reasons, a second study into the carbonylation of 2-propanol was initiated (24b, 31).

In contrast to the findings of Hjortkjaer and Jorgensen (32), Dekleva and Forster (24b) found a rather dramatic inhibitory effect of CO pressure on the rates of reaction (Fig. 3). Even under their much milder reaction conditions, they found that the organic medium was capable of effectively generating large amounts of propylene. In the presence of rhodium catalysts, the levels of propylene were significantly less than in the blank experiments, suggesting that an efficient hydrocarboxylation pathway was indeed operative. Because of the apparent superposition of a second contributing pathway (vide infra) and the apparent buffering capacity of the system with respect to propylene production, detailed kinetic analyses did not yield easily interpretable information. However, further information was obtained by examining the product (iso- vs. n-butyric acids and their isopropyl esters) both in the absence and presence of added propylene as a function of pressure. It was

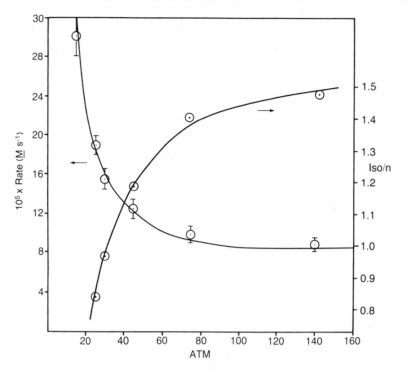

FIG. 3. Effect of CO pressure on the rate and product distribution for the carbonylation of 2-propanol at 170°C. From Refs. *24* and *31*.

concluded that this reaction pathway was responsible for the predominant production of *n*-butyric acid.

Based mainly on the inhibition of CO and known reactivities (*33*), the active catalyst of the hydrocarboxylation was considered to be the hydride, $\{HRhI_3(CO)\}^-$, which arose from the addition of free HI to $\{RhI_2(CO)_2\}^{-1}$:

$$\{RhI_2(CO)_2\}^- + HI \rightleftharpoons \{HRhI_3(CO)_2\}^- \qquad (16)$$

$$\{HRhI_3(CO)_2\}^- \rightleftharpoons \{HRhI_3(CO)\}^- + CO \qquad (17)$$

Studies with specifically deuterium labeled 2-propanols also indicated that this was a contributing pathway. *In situ* generation of propylene necessarily must result in loss of one β deuterium from the isopropyl precursor. Under the conditions employed for these labeling studies, the hydride would be derived from HI (as opposed to DI), so that such an overall reaction would be expected to result in loss of one β deuterium in going from isopropyl substrate to butyric acid. Some loss was observed. However, there was also very significant retention of isotopic integrity in these reactions, suggesting a second pathway which does not involve breaking the β C–D (or C–H) bonds. Even assuming a negligible kinetic isotope effect for this second contributing pathway, the production of acids containing the retained deuterium was ~ 20 times too fast to be attributable to S_N2-type reactivity. Product analyses similar to those described earlier suggested that this second pathway was responsible for mainly isobutyric acids. The relatively fast rates for the reactions yielding retention of isotopic integrities, and predominance of initial iso-alkyl suggested that some type of radical pathway is involved. This suggestion was based mainly on literature precedents for the oxidative additions of secondary alkyl iodides to transition-metal complexes because no experimental evidence other than the accelerated rate was accessible to support or refute this suggestion.

E. SUMMARY

From a mechanistic point of view, the iodide-promoted rhodium-catalyzed carbonylation of methanol, involving rate-limiting oxidative addition of methyl iodide to the rhodium(I) center, is the best understood of those to be presented in this review. Similarly, it has been possible to extend this interpretation to other systems, specifically those of linear primary alcohols. In the course of doing so, the nucleophilic nature of this oxidative addition reaction has been confirmed and characterized under operating conditions. From this knowledge, it is now possible to extract further information as to the reactivities of the intermediates involved. Also, because of the apparent predictability of reactivities for these reactions, it has been possible to suggest with some confidence that at least one system has contributions of a mechanism involving oxidative additions of alkyl halides by a radical pathway. This is not without precedent in the literature for systems at or near ambient conditions, but represents the first and only case, at least to our knowledge, where this has been invoked in catalytic carbonylation reactions. The validity of this postulate has yet to be fully confirmed.

Finally, it should be apparent that the nature of the reaction media has a profound effect on the reactivity of this system, and that, particularly for secondary alcohols, generation of olefins (and metal hydrides) occurs quite easily. Since this involves only the organic equilibria, this situation is not unique to rhodium chemistry. Unless great care has been taken to eliminate possible contributions of the hydrido/olefin pathway to the total reaction scheme, then, the hydrocarboxylation route should probably be considered to be a contributing reaction with other catalytic systems.

III. Iridium-Catalyzed Carbonylations

Iridium complexes in conjunction with iodide promoters are also excellent alcohol carbonylation catalysts (9, 34, 35). However, insofar as achieving a mechanistic understanding of the chemistry involved, methanol is the only substrate to have been studied in any depth (36). The iridium system apparently resembles the rhodium system, as far as fundamental steps are concerned, though in the former case inorganic iodide plays a more significant role and there appear to be contributions of neutral metal centers. Also, the relative rates of many of the steps involved differ, the differences reflecting the greater strengths of the iridium–carbon bond. As a result, interpretation of the overall reaction in the iridium system is considerably more complex. Nevertheless, a thorough combination of *in situ* spectroscopy and kinetic measurements, as well as studies of the chemistry of the observed intermediates, has allowed an understanding of the system.

A.   OVERALL SCHEME—IDENTIFICATION OF
THREE DISCRETE DOMAINS

The chemistry involved in this system has been reviewed recently (17, 37), and so treatment here will be abbreviated (the reader is referred to Ref. 36 for more specific information). Suffice it to say that the complexities of the system made it necessary to break down the discussion into recognizably different regimes. The resulting interpretations are given in Scheme 3.

Under conditions of low methyl/iridium ratios, in media with low levels of water and ionic iodide, the major species in solution was found to be $Ir(CO)_3I$, and the carbonylation reaction was found to be inhibited by increasing CO pressure. In separate experiments involving reaction of this complex with methyl iodide, the product was found to be the dicarbonyl iridium(III) species, $Ir(CH_3)(CO)_2I_2$. Apparently, the presence of three carbonyl ligands on the iridium(I) center is sufficient to completely inhibit any nucleophilic behavior by the tricarbonyl complex. Prior dissociation of one of the carbonyl ligands produces a metal complex capable of this type of reactivity.

When the concentration of ionic iodide in the catalytic system is increased, there is a change in the predominant form of the iridium to $\{Ir(CH_3)(CO)_2I_3\}^-$. The reaction rate increases with increasing CO pressure and is inhibited by increasing iodide concentration. Under these conditions, although the overall reaction rate is only slightly faster, it appears that there has been a shift in the rate-determining step. This is probably attributable to the greater nucleophilicity possible for the anionic $\{IrI_2(CO)_2\}^-$ complex

98   THOMAS W. DEKLEVA AND DENIS FORSTER

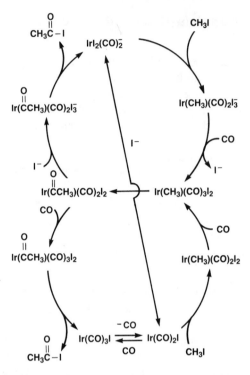

SCHEME 3.   Proposed mechanism for the iridium-catalyzed carbonylations of methanol.
From Ref. *36*.

relative to the neutral species, so that the initial formation of the $Ir-CH_3$ bond is now much faster than the formation of the neutral methyl tricarbonyl iridate(III), $Ir(CH_3)(CO)_3I_2$:

$$\{CH_3Ir(CO)_2I_3\}^- \longrightarrow CH_3Ir(CO)_2I_2 + I^- \qquad (18)$$

$$\downarrow CO$$

$$CH_3Ir(CO)_3I_2$$

Only when this tricarbonyl species is formed does the rearrangement to the acetyl-iridium complex occur. The observation that the neutral tricarbonyl complex rearranges to form an acetyl species much more readily than the anionic dicarbonyl species undoubtedly can be related to the relative strengths of the individual $Ir-CO$ bonds in the two complexes. Similarly, the marked contrast to this type of reactivity with that of the rhodium system, where the analogous $\{CH_3Rh(CO)_2I_3\}^-$ has never been detected because of the extreme facility of the corresponding isomerization, can also be rationalized in terms of the relative metal–alkyl bond strengths.

Under a wider range of conditions, particularly at higher methanol and water concentrations, the main species observed in solution is the hydride $\{HIr(CO)_2I_3\}^-$. The carbonylation reaction, under these conditions, is independent of CO pressure and approximately first order in methanol. It is perhaps not surprising, given the presence of high hydride concentrations, that the catalyst solutions concurrently catalyze the water–gas shift (WGS) reaction (Scheme 4), though at rates less than those of the carbonylations. To explain the observation that the hydridoiridium species was the predominant metal complex when methanol carbonylation was the predominant catalytic reaction, it was suggested that this species is in rapid equilibrium with the true active catalyst, the $\{IrI_2(CO)_2\}^-$ species discussed previously. Such an equilibrium appears to rationalize all of the available data and is consistent with the propensity of iridium to form oxidative addition adducts.

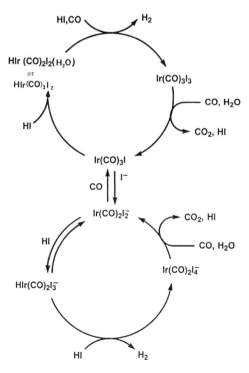

SCHEME 4. Proposed mechanisms for the iridium-catalyzed water–gas shift reaction. From Ref. *36*.

B. Summary

Despite the greater complexity involved with the solution chemistry of iridium with respect to that of rhodium, it is interesting to note that the overall reaction rates for the two are comparable (34). This appears to be due to the fact that different steps in the analogous catalytic cycles are rate limiting, reflecting the expected differences in reactivities of the two metals. As a result, the chemistries involved in the two systems complement one another particularly well. For example, extension of the simple behavior of the rhodium system can be used to interpret the more complicated iridium chemistry. The analogous $IrI_2(CO)_2^-$, expected to be more nucleophilic than the rhodium cogener, is apparently so much more so that it cannot be observed. Yet, despite its very low concentrations, under certain conditions it contributes significantly to the total reactivity. On the other hand, studies with the latter $5d$ metal have allowed the observation of several key high-valent species. Although the corresponding rhodium compounds have not been observed, they are also mechanistically very important. Taken together, the two systems present a unified picture of the chemistry involved in each.

IV. Cobalt-Catalyzed Carbonylations

The cobalt-catalyzed carbonylation of methanol to acetic acid has been developed as a commercial process and practiced on a large scale for more than 20 years (38, 39). Relative to the Monsanto acetic acid synthesis, it has two key disadvantages: (i) it requires significantly more severe operating conditions, and (ii) it is less selective on both major reactants (CO and CH_3OH). As with rhodium and iridium, the chemistry of the cobalt-catalyzed reactions has been well reviewed (40–43). Yet, despite the massive amount of attention which experimentalists have given this system, many aspects are still mechanistically unclear. In particular, the effects of added promoters have yet to be adequately explained. This is due, primarily, to the complicated behavior exhibited by cobalt–carbonyl systems, and thus the sensitivity of the system to reaction conditions. Some very recent studies may have introduced new information in these reactivities, which might be useful in interpreting some of the otherwise confusing data.

A. Nature of the Species Involved

The active species in the cobalt-catalyzed carbonylation of methanol to acetic acid are generally considered to be either the tetracarbonyl cobalt anion or hydride (*40–43*). These species are related according to the following:

$$2CoX_2 + 2H_2O + 10CO \;\rightleftharpoons\; Co_2(CO)_8 + 4HX + 2CO_2 \qquad (19)$$

$$Co_2(CO)_8 + H_2O + CO \;\rightleftharpoons\; 2HCo(CO)_4 + CO_2 \qquad (20)$$

$$HCo(CO)_4 + base \;\rightleftharpoons\; \{Co(CO)_4\}^- + baseH^+ \qquad (21)$$

In methanol solutions, the cobalt dimer also undergoes disproportionation according to

$$3Co_2(CO)_8 + nCH_3OH \;\rightleftharpoons\; 2\{Co(CH_3OH)_n\}^{2+} + 4\{Co(CO)_4\}^- \qquad (22)$$

Variations in reaction conditions, particularly with respect to acid and iodide concentrations, have profound effects on the positions of these equilibria, and these have, for the most part, added significant complexity to the behavior of the systems. It has also been determined that the catalytic activity of the cobalt catalyst system is increased with the introduction of H_2 (*44*) or halides (*vide infra*), though these increases in rate are quite often at the expense of selectivity. The presence of added H_2 serves as an alternate route to the formation of the hydride species:

$$Co_2(CO)_8 + H_2 \;\rightleftharpoons\; 2HCo(CO)_4 \qquad (23)$$

Unfortunately (at least from the point of view of carbonylation), higher levels of H_2 give rise to increased rates of formation of homologated products. The reasons for this will be discussed in Section IV,C,2. At very high levels of hydrogen, metallic cobalt deposits from the reaction mixtures.

Mechanistically, alcohol carbonylation reactions catalyzed by the $HCo(CO)_4/\{Co(CO)_4\}^-$ system appear to be governed by several features which are unique to this system. In particular, the high inherent acidity of the $HCo(CO)_4$ species (*45*), coupled with the nucleophilicity of the conjugate base (*35*), is responsible for the activation of the substrate and formation of the alkyl–cobalt bond. In addition, the facility of homolytic cleavage of cobalt–carbon bonds (*46, 47*) may be responsible for the complications in selectivity not normally observed with other systems.

B. NATURE OF THE CATALYTIC CYCLES

In the case of the unpromoted cobalt carbonylation catalyst, a relatively clear picture as to the nature of the transformations is available. The original proposal of Wender *et al.* that the first step in the mechanism is the protonation of methanol by the strongly acidic $HCo(CO)_4$ (*48*) has stood the test of time and is now generally accepted for this mechanism (Scheme 5). Subsequent migratory insertion yields the corresponding acyl derivative, which, when followed by hydrolysis by solvent water or alcohol, leads to the

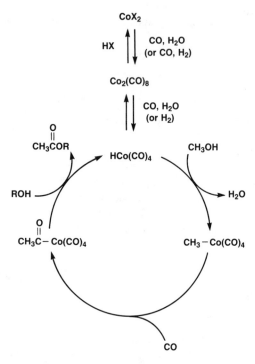

SCHEME 5.   Proposed mechanism for the unpromoted cobalt-catalyzed carbonylation of methanol.

formation of carboxylic acids or esters, respectively. This scheme includes all the salient features required for other closely related catalytic systems (*i.e.*, homologations, hydrocarbonylations) and can be used as a core scheme for future discussions. Because of the commonality of certain key steps in these reactions, information derived from one system can be extrapolated to the others. For example, although there are losses in selectivity due to secondary reactions of the cobalt–alkyl or cobalt–acyl species, each of these mechanisms has as a primary step, the formation of a common cobalt–alkyl bond. This is important for the present discussion, since much of the more recent work in this area has focused on the homologation reactions, and the findings can provide valuable information into the nature of the probable transformations. The nature of these secondary reactions gives some further insight into the nature of the intermediates.

As mentioned above, the cobalt system for methanol carbonylation to acetic acid usually involves the use of iodide promoters. However, the role of the iodide is still unclear from the published work in the area. There are several points in the catalytic cycle where the presence of iodide (organic or inorganic) can cause kinetic enhancements (*vide infra*). In fact, the intrinsic enhancements may be even greater than actually observed, since iodide can also reduce the concentrations of the presumed catalytic species by favoring the formation of inactive $\{CoI_4\}^{2-}$ (*49*). This balancing of effects adds to the uncertainties in interpretation of data with these systems.

Hohenschutz *et al.* (*38, 39*) originally proposed that the function of the iodide is to produce methyl iodide, and that the catalytic cycle involved the formation of the methyl–cobalt bond through the nonspecific interaction of this methyl iodide with $HCo(CO)_4$:

$$HI + CH_3OH \;\rightleftharpoons\; CH_3I + H_2O \tag{24}$$

$$HCo(CO)_4 + CH_3I \;\rightleftharpoons\; Co(CH_3)(CO)_4 + HI \tag{25}$$

However, it is more likely that the nucleophilic anion interacts preferentially with the methyl iodide to initiate the cycle:

$$\{Co(CO)_4\}^- + CH_3I \;\longrightarrow\; Co(CH_3)(CO)_4 + I^- \tag{26}$$

The rate enhancements observed when using inorganic iodide salts can be similarly rationalized, the methyl iodide being effectively generated, in these cases, by a combination of the acidic hydride and the salt [these equilibria are essentially equivalent with those in Eqs. (24) and (26)]:

$$HCo(CO)_4 + CH_3OH \;\longrightarrow\; CH_3OH_2^+ + Co(CO)_4^- \tag{27}$$

$$CH_3OH_2^+ + I^- \;\longrightarrow\; CH_3I + H_2O \tag{28}$$

$$CH_3I + Co(CO)_4^- \;\longrightarrow\; Co(CH_3)(CO)_4 + I^- \tag{29}$$

The role of methyl iodide as a promoter has also been defined during the early stages of the reaction by the work of Roper and Loevenich (*50*). They examined the products generated from homologation reaction mixtures containing either $CD_3I + CH_3OH$ or $CH_3I + CD_3OD$, and showed that the primary species involved was methyl iodide. Similarly, the very recent work of Gautheir-LeFaye *et al.* (*51*) implicated the *in situ* generation of methyl iodide from the tetracarbonyl cobalt hydride and inorganic iodide salts during the catalytic hydrocarbonylation of methanol to acetaldehyde.

In the study of Gauthier-LeFaye *et al.* (*51*), however, inorganic iodide was found to further enhance the reactivity synergistically with its role in forming methyl iodide. To explain this enhancement, these authors suggested that the iodide was catalyzing the disproportionation of the dimer $Co_2(CO)_8$ to form the active hydride according to

$$3Co_2(CO)_8 + 4A^+I^- \;\rightleftharpoons\; 2CoI_2 + 4A^+\{Co(CO)_4\}^+ + 8CO \tag{30}$$

$$2CoI_2 + 2H_2 + 8CO \;\rightleftharpoons\; Co_2(CO)_8 + 4HI \tag{31}$$

$$2Co_2(CO)_8 + 4A^+I^- + 2H_2 \;\rightleftharpoons\; 4A^+\{Co(CO)_4\}^- + 4HI \tag{32}$$

where, for example, $A^+ = Na^+, K^+, P(CH_3)(C_4H_9)_3^+$]. Also, it was suggested that the association of the $\{Co(CO)_4\}^-$ anion with the A^+ countercation enhances the reactivity of the anion, relative to the corresponding hydride or $Co\{Co(CO)_4\}_2$ species. Braterman et al. (52) also found that halide ions, in the presence of organic halides, markedly accelerated the disproportionation of cobalt carbonyls to the corresponding alkylated or acylated product. While this empiricism may be correct, the mechanistic reasons for this enhancement are as yet unclear. Mizoroki and Nakayama (53) previously suggested that iodide reacts with the cobalt carbonyls to generate an iodo-containing cobalt–carbonyl species which attacked methanol directly:

$$CH_3OH + HCo(CO)_3I \longrightarrow Co(CH_3)(CO)_3I + H_2O \qquad (33)$$

While other workers have similarly rationalized rate enhancements in terms of long-lived reactive iodo-containing complexes (54, 55), their presences have yet to be substantiated. Nevertheless, the suggestion of Mizoroki and Nakayama (53) that I^- acts to labilize CO ligands may be used to explain the promotional effect on the dissociation reactions.

In considering the effects of promoters in this multistep mechanism, attention has been given, thus far, only to the alkylation step. There is little doubt that this is important, given the parallel enhancements observed with carbonylations and homologations, which contain this as a common mechanistic step. However, in carbonylation chemistry, another promotional effect of iodide may be considered on the basis of some work by Imyanitov et al. (56). It is well known that the cobalt-catalyzed hydroformylation reaction can be conducted in the presence of alcohols or water without the observation of significant by-product ester or acid formation (41). Thus, the intermediate cobalt–acyl species is trapped much more efficiently by the cobalt hydride (either inter- or intramolecularly) than by these reagents. It was also known that the hydrocarboxylation of olefins by cobalt is strongly promoted by pyridine derivatives. Imyanitov et al. used these data to design experiments which demonstrated that pyridine accelerated the reaction between acetyl–cobalt tetracarbonyl and methanol by a factor of greater than 400, by nucleophilically displacing $\{Co(CO)_4\}^-$ anion to generate an intermediate acetyl pyridinium salt. This, in turn, is more rapidly hydrolyzed under the reaction conditions. Since iodide is much more nucleophilic than pyridine (25, 57), it seems likely that iodide can act in the same capacity. It is difficult to assess the quantitative ramifications of this enhancement, since no study has yet isolated this step under catalytic conditions. However, it may be significant that the reported turnover rates for methanol carbonylation to acetic acid with iodide promoters are comparable to those reported for homologation in the absence of iodide promoters.

Scheme 6 summarizes the possible effects of the iodide promoters.

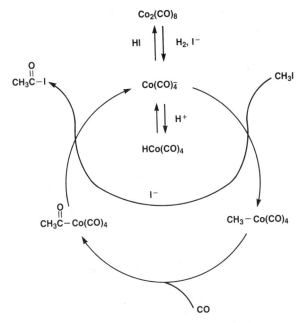

SCHEME 6. Proposed mechanism of the iodide-promoted cobalt-catalyzed carbonylation of alcohols encompassing the effects discussed in Section IV,B.

C. COMPETING REACTIONS

1. Water–Gas Shift Reaction

Under carbonylation conditions, the water–gas shift (WGS) reaction as well as alcohol hydrogenolysis and homologation reactions can be easily observed (38, 39). There have been no thorough mechanistic studies into the WGS reaction, although this reaction probably proceeds by a metal cycle involving shuttling between CoX_2 and $Co_2(CO)_8/HCo(CO)_4$, analogous to that found for rhodium.

2. Alcohol Hydrogenolysis Reactions

It was initially suggested (48) that the hydrogenolysis reaction involved a radical process without the formation of a metal–carbon bond, as in the following equations:

$$Co_2(CO)_8 \rightleftharpoons 2 \cdot Co(CO)_4 \tag{34}$$

$$\cdot Co(CO)_4 + H_2 \rightleftharpoons H \cdot + HCo(CO)_4 \tag{35}$$

$$H \cdot + ROH \rightleftharpoons R \cdot + H_2O \tag{36}$$

$$R \cdot + HCo(CO)_4 \longrightarrow RH + \cdot Co(CO)_4 \tag{37}$$

This is unlikely, however, since in methanol homologation studies, methane is generated preferentially, even when there are comparable amounts of methanol and ethanol present in the reaction media (58). Acetic acid decarboxylation has also been suggested as a pathway for methane formation:

$$CH_3COOH \longrightarrow CH_4 + CO_2 \qquad (38)$$

This also seems unlikely to be a major contributor, since the later study by Koermer and Slinkard (58) showed substantial methane formation in the absence of significant CO_2 generation.

A more probable pathway is hydrogenolysis of the methyl–cobalt bond, either by direct reaction with $HCo(CO)_4$ or via initial homolytic cleavage. This latter route, that is, initial homolytic dissociation, seems most likely, given the available data and the known reactivities of cobalt. In a study involving various aromatically substituted benzyl alcohols, under homologation reaction conditions, the predominant pathway was found to be greatly affected by the substitution pattern, in some case yielding hydrogenolysis products as the major reaction products (59). Electron-donating substituents on the aromatic ring increased the rate as well as the selectivity to homologated versus hydrogenolyzed product. In fact, this study gave much insight into the reaction mechanisms. Obviously, the enhanced rate points to initial protonation of the benzyl alcohols by the acidic $HCo(CO)_4$. The significance of the hydrogenolysis reaction in these systems, relative to other systems involving alkanols, can be rationalized in terms of the inherent stability of the benzyl radical (5, 60), and so the ease of bond homolysis. The presence of electron-donating substituents on the carbon center is known to weaken the cobalt–carbon bond (46, 47), making it more susceptible to both homolysis and migratory insertion. The relative increase of homologation under these conditions suggests that the latter reaction is appreciably more enhanced.

3. *Alcohol Homologation Reactions*

The catalysis of the homologation of methanol to ethanol by $Co_2(CO)_8$ was first described in the scientific literature 30 years ago (61). The reaction was largely ignored for many years, but has become the subject of intense investigation over the past 10 years as a route to ethanol from synthesis gas. The reaction appears to suffer an inherent selectivity problem, with the upper range of selectivity to ethanol being 60–70%. Numerous ligands and promoters have been explored without significant improvement in the area. It is difficult to compare the results of various investigating groups, since under ostensibly similar reaction conditions there are rather wide variations in rates and selectivities. Since almost all of the studies available have involved batchwise experiments, it may be suggested that the source of these variabilities are the differences in the heat-up and cool-down procedures. Probably the most reliable study of the unpromoted reaction is that of Koermer and Slinkard (58), who studied the reaction in the continuous mode. Examples of the product distribution observed after several of their reactions, which reflect the complexity of the system, are given in Table IV. In these homologation reactions, the methyl–cobalt moiety is believed to be generated as described in Section IV,B. Again, the selectivity is determined with the reaction of the cobalt–acyl species. Hydrogenolysis generates the intermediate acetaldehyde, which on hydrogenation gives the desired ethanol.

Although several studies have examined the effects of various promoters and ligands on the methanol homologation reaction, none has identified a system with substantially improved selectivity. However, there are many claims that iodide accelerates the rate of the reaction (62–64). While the possible sources of this enhancement have been discussed in Section IV,B, it should be noted that the systems from which these interpretations were extracted are by no means simple. Qualitative comparisons among the various studies of promoted and unpromoted systems are difficult for the reasons given above, but, in addition, because the variety of forms by which iodine is introduced (e.g., I_2, CH_3I, or iodide salts) apparently produce different effects (51, 63, 64). Also, many of the systems involve two promoter components (e.g., triphenylphosphine + methyl iodide or tri-p-tolylphosphine + I_2), which further complicates the interpretations as to the role(s) of the halide.

There have been only limited studies of the homologations of higher alcohols with unpromoted cobalt catalysts. The relative rates are shown in Table V (55, 62). Close examination of the products from the various alcohols indicates a greater complexity for the reactions than in the methanol case. For example, the reaction involving tert-butyl alcohol is rapid but gives 3-methylbutanol as the predominant product and not the 2,2-dimethylpropanol which would be expected from direct homologation. This is suggestive of dehydration of the alcohol and hydroformylation of the resulting isobutylene:

$$(CH_3)_2CHCH_2CHO \longrightarrow (CH_3)_2CHCH_2CH_2OH$$

$$(CH_3)_3COH \xrightarrow{-H_2O} (CH_3)_2C{=}CH_2 \tag{39a}$$

$$(CH_3)_3CCHO \longrightarrow (CH_3)_3CCH_2OH$$

$$\tag{39b}$$

TABLE IV

Representative Product Distributions for the Homologation of Methanol Using
Cobalt in Both Batchwise and Continuous Mode Experiments[a]

	Experiment number		
	1	2	3
Mode	Batch	Continuous	Continuous
Pressure (atm)	260–310	334	334
Temperature (°C)	190	190	190
Liquid reaction volume (ml)	34	400	400
Gas flow (l/min STP)	—	2.3	2.7
CO/H_2	1.0	0.91	1.1
Catalyst[b]	$Co_2(CO)_8$	$Co(acac)_2$	$Co(acac)_2$
[Co] (M, in methanol)	0.24	0.05	0.05
Liquid feed rate (ml/min)	—	3.1	3.3
Solvent[c]	None	None	25%
Molar selectivities (%)[d]			
Acetaldehyde	2.0	2.0	1.5
1,1-Dimethoxyethane	5.2	10.2	12.1
Ethanol	37.5	51.6	56.6
1-Methoxy-1-ethoxyethane	—	0.6	0.4
Methyl acetate	17.4	8.3	4.7
Methyl formate	2.8	1.4	4.2
Ethyl formate	3.1	1.7	2.8
1-Propanol	2.3	3.8	3.3
2-Propanol	—	0.4	0.5
1-Butanol	—	0.5	0.2
2-Butanol	—	0.3	0.2
2-Methyl-2-butanol	—	1.5	0.2
1-Methoxy-2-propanol	1.1	2.5	e, f
2-Methoxy-1-propanol	2.7	2.3	e, f
Dimethyl ether	11.3	4.7	2.7
Methyl ethyl ether	—	2.1	2.6
Diethyl ether	1.0	0.2	0.2
Methyl propyl ether	—	Trace	0.2
Methane	10.9	5.9	8.2
Carbon dioxide	2.7	—	—
Carbon monoxide conversion (%)	17	41	26
Methanol conversion (%)	30	31	25
Methanol accountability (%)	108	96	96

[a] Data from Ref. 58. Reprinted with permission from *Industrial and Engineering Chemistry Product Research and Development*. Copyright (1978) American Chemical Society.

[b] $Co(acac)_2$ is cobalt(II) acetylacetonate. Using $Co(acac)_2$ and the $Co_2(CO)_8$ gave essentially the same results.

[c] Solvent was 1,4-dioxane; percentage solvent by volume before mixing.

[d] Molar selectivity to product, i = (moles of product i recovered)/(total moles of all products recovered) × 100%. Water was excluded as a product.

[e] Analysis was obscured by solvent peak.

[f] Based on batch studies with 1,4-dioxane as a solvent where analysis of methoxypropanols was accomplished, the maximum contribution of methoxypropanol is estimated to be less than 5 relative % of the total ethanol potential. For experiment 3, if this estimate of methoxypropanols were included, the individual product selectivities would be reduced by a maximum of 3.5% and total ethanol potential would drop from 71.0% to 68.5%.

TABLE V

Relative Rates for the Cobalt-Catalyzed
Homologations of a Variety of Alcohols as
Measured by Gas Uptake[a]

Alcohol	Relative rate[b]
Methanol	41
Ethanol	1.0
2-Propanol	0.86
tert-Butyl alcohol	100
Glycol	3.6
Benzyl alcohol	14
p-Methoxy-benzyl alcohol	Very fast

[a] Data from Ref. *62*.
[b] Relative rates not compatible with data in Tables I or II.

A similar dehydration pathway was suggested from labeling studies involving ethanol homologation (*65*). Ethanol, initially containing a ^{14}C label only in the 2 position, was homologated under unspecified conditions. The product *n*-propanol contained essentially equal amounts of the ^{14}C label in the 2 and 3 positions. This was interpreted in terms of a mechanism involving the intermediacy of free ethylene. While this seems a plausible explanation, it is not the only one available. For example, if the ethyl–cobalt moiety formed initially with retention of isotopic integrity, an intramolecular scrambling (e.g., via β-hydride elimination) could also account for the equilibration of the label. It would be necessary to do the same labeling study under several reaction conditions to determine the cause of this equilibration.

Finally, many studies have also been reported in which a transition-metal cocatalyst has been used in the methanol homologation reaction (*66*). The intention, in some cases, was to improve the overall rate and selectivity to ethanol by using an aldehyde hydrogenation catalyst in conjunction with the cobalt. Ordinarily, many of the unwanted by-products which form in the unpromoted cobalt system arise from secondary reactions of the long-lived acetaldehyde. By efficiently removing this intermediate from the reaction medium, it is hoped that these problems would be reduced. Many metals have been used in this context and rate enhancements have been observed in several cases. It has been suggested that the underlying reason for these enhancements, at least with $RhCl_3$, $IrCl_3$, and $RuCl_3$, is that these species serve as sources of stoichiometric amounts of acid [reductions of these salts in hydroxylic media give stoichiometric amounts of HX (*67*)]. A recent study of the cobalt–ruthenium system by Moser *et al.*, involving *in situ* spectroscopy,

also challenges the role of ruthenium in reactions with the organic components (68). Rather, the enhancement by ruthenium is attributed to its ability
to enhance the WGS reaction, and, thus, lower the concentration of water in
the reaction system. This is entirely consistent with previous observations
(69) that the presence of water in the reaction media is detrimental to the
reactivity.

V. Ruthenium-Catalyzed Carbonylations

In sharp contrast to the massive effort which has been directed toward
rhodium- and cobalt-catalyzed alcohol carbonylations, the chemistry of
ruthenium has received less direct attention, the most notable work being
that of Braca and co-workers (70–73). This catalytic system is particularly
complicated by the fact that it serves also as an extremely efficient catalyst for
the water–gas shift and homologation reactions. In fact, most of the work
using analogous ruthenium-based homogeneous catalyst systems has been
directed toward CO-hydrogenation chemistry (74–77), particularly with
respect to alcohol and carboxylic acid homologations (78), with strikingly
similar operating conditions and promoters. Unfortunately, despite this,
there is still no clear picture of the actual species involved. Since Braca et al.
have attempted to separate the alcohol carbonylations from these other
processes, their work serves as a convenient point for beginning discussion,
from which we will attempt to include the relevant features of these other
systems.

A. NATURE OF THE SPECIES INVOLVED

The typical operating conditions used by Braca et al. in their study were
150 atm and 200°C (see Table VI). Because of the efficiency of the WGS
reaction, reactions starting with charged H_2O and CH_3OH were often
characterized by production of high levels of dimethyl ether, CO_2, and H_2
according to the following equations (71):

$$2CH_3OH \rightleftharpoons CH_3OCH_3 + H_2O \qquad (40)$$

$$H_2O + CO \longrightarrow CO_2 + H_2 \qquad (41)$$

For mechanistic studies, then, it was necessary to minimize the effects of the
WGS reaction (70, 72). As a result, the reaction conditions were essentially
anhydrous, using dimethyl ether as substrate. Even so, complications of
homologation reactions to ethyl-containing species were still significant.

TABLE VI

Carbonylation and Homologation of Dimethyl Ether and Homologation of Methanol and Methyl Acetate[a,b]

	Experiment						
	1	2	3	4	5	6	7
Catalyst	$RuI_2(CO)_4$	$Ru(acac)_3$					
Iodide	MeI, NaI	MeI	MeI	NaI	MeI	MeI	MeI
Ru/MeI/NaI	1/1.2/10	1/10	1/8.5	1/10	1/12	1/10	1/5
Reagents	MeOH	Me_2O	Me_2O	Me_2O	AcOMe	Me_2O	Me_2O
(mmol)	(550)	(123)	(260)	(100)	(223)	(100)	(95)
	—	Toluene	AcOH	AcOH	AcOH	AcOMe	AcOMe
		(235)	(530)	(430)	(218)	(194)	(94)
	—	—	—	—	—	AcOH	AcOH
						(250)	(142)
H_2/CO, initial	1.0	0.5	1.0	0.5	0.22	0.22	1.0
H_2/CO, final	1.75	0.77	0.5	0.55	0.19	0.2	1.55
Time (h)	2	28	12	28	8	14	20
Conversion	95	Me_2O	Me_2O	Me_2O	AcOMe	Me_2O	Me_2O
(%)		(78)	(96)	(70)	(23)	(58)	(16)
	—	—	AcOH	AcOH	AcOH	AcOMe	AcOMe
			(25)	(30)	(−)	(−)	(61)
	—	—	—	—	—	AcOH	AcOH
						(8)	(5)
Selectivity[b] (%)							
C_1	14	1.8	1.5	0.2	2.1	5.4	0.4
C_2	43	2.0	4.5	3	1.8	3.6	2.0
AcOMe	7.5	34.7	37	12	—	18	—
AcOH	—	14	—	—	38	—	—
AcOEt	20.2	20	35	79	49.2	64	73
Hydrocarbon	3.3	22.8	18	4.5	5.7	7.5	23
Heavy ends	12	4.7	4	1.3	3.2	1.5	1.6
Acetyl groups							
(+/−) (mmol)	+66	+80.2	+109	−7	+2	+35	−15
Ethyl groups							
produced (mmol)	258	24	160	150	21	45	52

[a] From Ref. 70.

[b] Reaction conditions were 200°C, 150 atm at room temperature; catalyst concentration, 1.5×10^{-2} M; CO/H_2 (1/1) was fed to the reaction to maintain the pressure at an approximately constant value.

[c] Calculated as $\{$(product, mol) $\times$ (no. of methyl groups)$\}/\{\sum$ (reagents converted, mol) $\times$ (no. methyl groups)$\}$; C_1, MeOH, Me_2O, 1/2 MeOEt; C_2, EtOH, Et_2O, 1/2 MeOEt; hydrocarbon, $CH_4 + C_2H_6$; heavy ends, n-PrOH, EtCOOH, and their C_4 or C_5 esters.

However, several observations of note were made. First and foremost was that H_2 was necessary for catalysis. This fact is not incorporated into the catalytic cycle usually cited (Scheme 7), but rather, H_2 is considered to be involved only in activating the precursors. Reduction of Ru(III) (and higher valent species) to Ru(II) is known to be preceded by heterolytic cleavage of H_2 (*79, 80*), and under the anhydrous conditions of the system, it is possible that H_2 is constantly necessary to maintain the ruthenium in a lower oxidation state (the catalytic cycle may generate an otherwise inactive high-valent species, either directly or by some side reaction). The presence of H^+ and an iodide promoter is also requisite for reactivity. It has also been concluded that for catalysis, the I/Ru ratio must equal or exceed 3 (*73, 78*).

Braca and co-workers have recovered or observed several iodocarbonyl-ruthenium(II) complexes, including *cis*-$RuI_2(CO)_4$[ν_{CO} 2161m, 2105s, 2096s, and 2066m cm^{-1} (*81*)], $\{RuI_3(CO)_3\}^-$ [ν_{CO} 2116s and 2042s cm^{-1} (*82*)], and $\{RuI_2(CO)_4\}_3$ [ν_{CO} 2124s, 2052s, and 2008m cm^{-1} (*83*)], from catalyzed reaction mixtures, starting either from $Ru_3(CO)_{12}$ or Ru(acac)$_3$. Knifton has recovered similar product mixtures from acid homologation reactions (*78*).

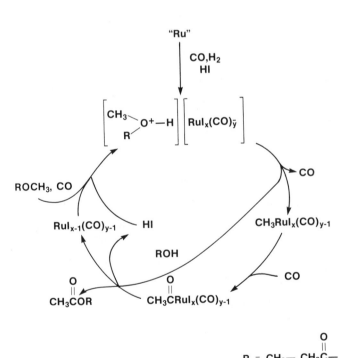

SCHEME 7. Suggested mechanism for the ruthenium-catalyzed carbonylations of alcohols.

Nevertheless, mechanistic schemes which involve this system are nonspecific about even the oxidation state, much less the coordination sphere, of the active catalyst. Generally, though, anions of the type $\{RuI_x(CO)_y\}^-$ are listed in the catalytic cycle (Scheme 7). Unfortunately, neither of these workers have taken advantage of *in situ* spectroscopy to determine more directly the nature of the species involved, specifically with regard to the ruthenium composition at various operating conditions and reactivities. Although the large number of ruthenium–carbonyl complexes and associated stretching frequencies could conceivably complicate spectral analyses, such experiments would appear to be useful in further understanding the system, given the discrepancies in the presently available indirect information. For examples, Braca *et al.* (70, 72) observed unquantified mixtures of $\{RuI_3(CO)_3\}^-$ and cis-$RuI_2(CO)_4$ when using a nonpolar heptane/CH_3I solvent system (150°C, 100 atm, 1:1 H_2/CO). This is quite different from the results of a study of Dombeck (74), in which he observed significant amounts of $\{HRu_3(CO)_{11}\}^-$ in the more polar sulfolane solvent under more severe conditions (230°C, 550 atm, 1:1 H_2/CO, from $Ru_3(CO)_{12}$/KI). In fact, in the latter case, maximum activity for CO hydrogenation was, in fact, observed with a $\{HRu_3(CO)_{11}\}^-$/$\{RuI_3(CO)_3\}^-$ ratio of 2/1). Similarly, the high efficiency with which the systems catalyze the WGS reaction also suggests the presence of highly reactive anionic cluster hydrides (77, 84). Since these systems would appear to be intimately related, it is entirely probable that the same clusters are present in Braca's work. Their effect on the carbonylation has yet to be determined, but the view that the active catalysts are single monomeric ruthenium species may be overly simplistic. However, on the other hand, it is interesting to note that workers at Texaco have found that the selectivity toward carbonylation (compared with homologation) can approach 90% by dispersing the ruthenium on a heterogeneous carrier (85). This observation might be used to support the idea that monomeric species are involved.

B. Nature of the Catalytic Cycle

The catalytic activity of the methanol carbonylation is very dependent on the nature of the iodide promoter, and different chemistry appears to follow using HI or NaI in this regard (72). However, under otherwise identical conditions, the catalytic activity increased in the order NaI < CH_3I < HI. Contrary to what is observed for the rhodium/iodide catalyst, Braca *et al.* did not consider CH_3I to be directly involved in the catalytic carbonylation cycle (70–73). This conclusion is based on the observation that CH_3I was not carbonylated under their reaction conditions. Instead, because of the necessity of a proton supplier and the promoting effect of NaI, these authors suggested the intermediacy of a long-lived protonated dimethyl ether or methyl acetate as the methyl transfer reagent $(70-73)$, which acts also as a counterion for the anionic $\{RuI_x(CO)_y\}^-$ species (Scheme 7). Such a non-specific methyl transfer has been used to explain the unpromoted cobalt carbonylation system [cf. Eq. (25)]. However, the presence of significant levels of such a protonated species discounts the facile formation of CH_3I under the reaction conditions. Given the leaving-group capability of I^- vs. ROH, it would seem reasonable that if this methyl transfer is at all nucleophilic in nature, CH_3I would serve as a better substrate for such a reaction than would the protonated ether or ester (Scheme 7). An alternate explanation as to the necessity of H^+ may be that this reagent triggers CO dissociation, perhaps by oxidizing the ruthenium to a higher valent state, to generate the active catalyst, for example,

$$\{RuI_x(CO)_y\}^- + H^+ \ \rightleftharpoons \ HRuI_x(CO)_{y-z} + zCO \tag{42}$$

$$HRuI_x(CO)_{y-z} \ \rightleftharpoons \ \{RuI_x(CO)_{y-z}\}^- + H^+ \tag{43}$$

The same type of reactivity has been observed in the rhodium system (Section II,D). The dependence on CO pressure for this ruthenium system (essentially independent, but showing a slight maximum at 20–40 atm) is consistent with the existence of several competing equilibria (72).

In mechanistic terms, the reactions shown in Scheme 7 are otherwise necessarily vague. Braca et al. (73) have also noted that the iodide counter-cations $(AI; A = Na^+, K^+, N(CH_3)_4^+)$ promote the carbonylation reaction rather than the homologation, improving the yields of carboxylic acid. This has been interpreted in terms of two effects. The first involves the Lewis acid $(Li^+ \approx Na^+ \gg Cs^+)$ assisted nucleophilic attack of the coordinated acyl species by alcohol or water:

$$A^+ \cdots O{=}\overset{\overset{\displaystyle R}{\displaystyle |}}{C}{-}RuI_x(CO)_{y-1} + HOR' \quad (R' = H, alkyl)$$

$$\longrightarrow \ \{RuI_x(CO)_{y-1}\}^- + A^+ + H^+ + RCOOR' \quad (A^+ = Li^+, Na^+, Cs^+) \tag{44}$$

The promotional effect of the $N(CH_3)_4^+$ is rationalized differently. In this case, the cation is capable of generating $N(CH_3)_3$, which in turn can intercept the ruthenium–acetyl species to form N-acetyl ammonium salts, for example,

$$N(CH_3)_4^+ I^- \ \rightleftharpoons \ N(CH_3)_3 + CH_3I \tag{45}$$

$$N(CH_3)_3 + (R\overset{\overset{\displaystyle O}{\displaystyle \|}}{C})RuI_x(CO)_{y-1} \ \rightleftharpoons \ \{(CH_3)_3N{-}\overset{\overset{\displaystyle O}{\displaystyle \|}}{C}{-}R\}^+ + \{RuI_x(CO)_{y-1}\}^- \tag{46}$$

$$\{(CH_3)_3N{-}\overset{\overset{\displaystyle O}{\displaystyle \|}}{C}{-}R\}^+ + R'OH \ \longrightarrow \ RCOOR' + (CH_3)_3NH^+ \tag{47}$$

Such a reactivity has been observed previously during the pyridine-promoted solvolysis of acetyltetracarbonyl cobalt(I) (56). Both effects in the case of ruthenium suggest that the carbonylated product does not leave the metal center by reductive elimination of acyl iodide, but rather by nucleophilic displacement.

C. Homologation Reactions

The same difficulties encountered in determining the nature of the species or cycles involved in the carbonylation reaction exist for the homologation reaction, and so little is actually known about the specifics of the latter reaction. However, most workers in the area of methanol homologation seem to believe that the resulting ethanol is not derived from hydrogenating free acetaldehyde. Rather, the carbonylated species is reduced before leaving the metal center, for example,

$$
\begin{array}{cc}
\overset{O}{\overset{\|}{CH_3C}}-RuI_x(CO)_{y-1} + H_2 \longrightarrow & \overset{OH}{\overset{|}{CH_3CH}}-RuI_x(CO)_{y-1} \qquad (48)
\end{array}
$$

$$
\overset{OH}{\overset{|}{CH_3CH}}-RuI_x(CO)_{y-1} + H_2 \longrightarrow CH_3CH_2OH + HI + RuI_{x-1}(CO)_{y-1} \qquad (49)
$$

Since the secondary reactions involved in alcohol homologation reactions, strictly speaking, are outside the scope of this review, the effects of various promoters on this reactivity will not be discussed. The presented material is given only to serve as a comparison with the analogous cobalt systems, in which the weakness of the metal–carbon bonds governs this aspect of reactivity.

D. Summary

The ruthenium–iodide-catalyzed carbonylation of alcohols is greatly complicated by the facility with which the same system catalyzes the competitive water–gas shift and homologation reactions. The resulting inability to totally isolate the reaction(s) of interest necessitates that conclusions are based on observations which are less direct than in other systems discussed in this article. Further work, aimed at determining the nature of the proposed transformations, perhaps through the use of model compounds, would appear to be required to unravel the finer mechanistic details of the system.

VI. Carbonylations Catalyzed by Nickel, Palladium, and Platinum

The carbonylation of alcohols can proceed with formation of carboxylic acid by catalytic insertion of CO into the carbon–oxygen bond. An alternative reaction gives rise to oxalate or formate esters, when the CO is inserted into the oxygen–hydrogen bond. The members of the nickel triad carbonylate alcohols to give each of these products, and they will be discussed separately.

A. NICKEL-CATALYZED CARBOXYLIC ACID SYNTHESES

The carbonylation of alcohols to give the next higher carboxylic acid is catalyzed by nickel. Catalytic systems are based on $Ni(CO)_4$, NiX_2, or metallic nickel in the presence of a halogen (86, 87). In all of these cases, though, the active species can be derived from nickel(0) species, for example (42),

$$NiI_2 + H_2O + 5CO \rightleftharpoons Ni(CO)_4 + 2HI + CO_2 \qquad (50)$$

Reaction conditions are rather severe, typically 200 atm and 250–300°C (86, 87). Systems having iodine or iodide have the greatest activity. Recently, several groups have found that the severity of the reaction conditions necessary to prepare acetic acid can be reduced significantly by addition of tin, Group VIB, or phosphine promoters, leading to a number of patents in this area (88–94). Unfortunately, no systematic kinetic or spectroscopic work has been done, and only very limited amounts of data are available. Mechanistic studies are limited to a few reported product analyses, possibly because the toxicity of $Ni(CO)_4$ deterred earlier experimentalists from studying the system in greater depth. However, discrepancies have been reported even with these data, attributable to the earlier difficulties in separating, quantifying, and characterizing the components of the complex product mixtures (95, 96). Adkins and Rosenthal (95) reported that carbonylation of linear-chained alcohols gave exclusively branched-chain acids as products. However, later studies of Reppe and co-workers (96) showed that such reactions led to mixtures of isomeric products. The carboxyl group generally adds to the carbon atom initially carrying the hydroxyl group. The same nickel systems also catalyze the formation of diacids from linear diols. However, in both cases, the severe reaction conditions caused significant alcohol dehydrations following the sequence tertiary > secondary > primary. Secondary alcohols reacted faster than did primary alcohols, which in turn reacted faster than methanol (95–97).

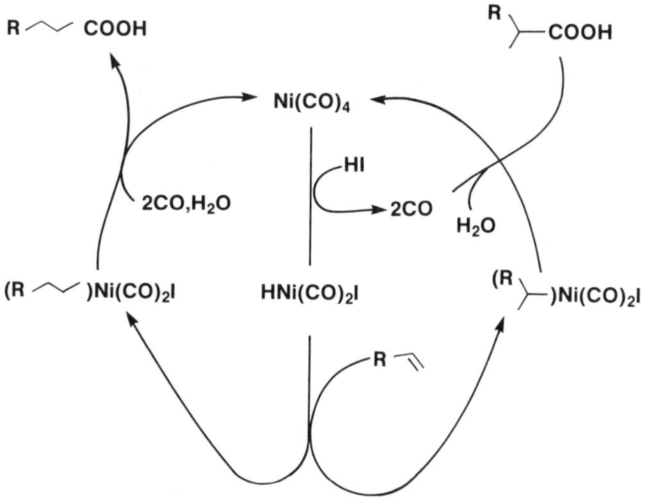

SCHEME 8. Suggested mechanism for the contribution of hydrocarboxylation during the nickel-catalyzed carbonylation of higher alcohols.

These observations are easily incorporated into mechanistic schemes essentially identical to those described previously for rhodium (Section II), in which the major pathways for carbonylation chemistry are governed by the organic pre-equilibria involved. In cases where secondary alcohols are present, dehydrations are facile, and, presumably, hydrocarboxylation gives at least part of the product in these cases (Scheme 8). Despite the fact that nickel is a moderately good catalyst for this type of reactivity (86, 87), there are several observations which mitigate against the singularity of this type of reaction (42). First, the high-yield carbonylation of methanol to acetic acid cannot proceed through an olefin intermediate. Also, both 1- and 2-propanol give propylene on dehydration but yield different amounts of isomeric butyric acids (96, 97). A second pathway, then, involves the oxidative addition of the corresponding alkyl iodide to the $Ni(CO)_4$ catalyst to generate the requisite Ni–alkyl species (Scheme 9). This was originally suggested by Reppe (96). There is insufficient evidence to differentiate the nature of this oxidative addition step in the catalytic system. The enhanced reactivity of the nickel in the presence of phosphines and SnI_2 suggests that the reaction is accelerated by the presence of more electron density on the metal center. In some cases, under milder conditions, Ni(0) complexes are known to react with aromatic halides by initial single-electron transfer reactions (98). The reactions of alkyl halides with similar nickel species are known to require more forcing conditions, although the catalytic conditions are certainly forcing enough. Foa and Cassar have also suggested that

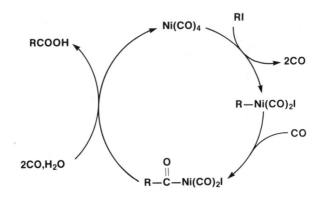

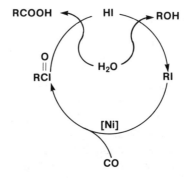

SCHEME 9.   Suggested mechanism for the carbonylation of alcohols by nickel.

$\{Ni(CO)_3X\}^-$ species may be more active than the tetracarbonyl species for nucleophilic displacement reactions (99, 100). In a later study on the carboxylation of aryl halides, these same workers suggested that more complex anions (e.g., $\{Ni_2(CO)_6\}_2^-$ and $\{Ni_3(CO)_8\}_2^-$) may be contributing to the total reactivity under certain conditions (101). Obviously, then, in view of the complicated behavior which has been suggested, more detailed studies are required to differentiate the contributions of the possible reactions for a given substrate.

B.   PALLADIUM-CATALYZED OXALATE SYNTHESES

Palladium-based catalysts can be used for the oxidative coupling of CO to dialkyl oxalates and/or carbonates (102–104), and sufficient information is available that mechanistic schemes can be considered. The organic transformation occurs with reduction of Pd(II) to Pd(0):

$$Pd(2+) + 2CO + 2ROH \longrightarrow Pd(0) + RO\overset{\overset{O}{\|}}{C}-\overset{\overset{O}{\|}}{C}OR + 2H^+ \qquad (51)$$

Reoxidation of the palladium back to the divalent state is most commonly accomplished with $CuCl_2/O_2$. The overall catalytic cycle then becomes

$$2ROH + 2CO + 0.5O_2 \longrightarrow RO\overset{\overset{O}{\|}}{C}-\overset{\overset{O}{\|}}{C}OR + H_2O \qquad (52)$$

and is accompanied by the formation of H_2O. In the absence of suitable drying agents, the CO is oxidized to CO_2 and the formation of oxalates is greatly reduced. Generally, the problem is overcome by carbonylating *ortho*-esters. In this case, the resulting coproduct is the corresponding ester:

$$2CO + R'C(OR)_3 \overset{(cat)}{\longrightarrow} R'COR + RO\overset{\overset{O}{\|}}{C}-\overset{\overset{O}{\|}}{C}OR \qquad (53)$$

Additionally, $Fe(NO_3)_3$ (*105*), $LiCl/NH_3$ (*106*), and KNO_3 (*107*) have been used to overcome the problem. These same palladium systems also catalyze the competitive formation of dialkyl carbonates, and this observation must be incorporated into the mechanistic scheme. Typical reaction conditions are 60–140°C and 80–100 atm (*105–108*).

   Mechanistic aspects of the stoichiometric palladium-catalyzed portion of the reaction have been studied by Rivetti and Romano (*109–112*) using phosphine model complexes of the form $PdX_2(P)_2$ (where $X = Cl^-$ or OAc^- and P = tertiary phosphine, generally PPh_3). Reaction conditions in these systems are much milder, typically less than 80°C. These systems effectively mimic the features observed in the catalytic cycle and appear to be good models for them. The overall reaction is

$$2ROH + 3CO + PdX_2P_2 + P + 2 \, base$$

$$\longrightarrow Pd(CO)P_3 + 2 \, base\text{-}HX + RO\overset{\overset{O}{\|}}{C}-\overset{\overset{O}{\|}}{C}OR \qquad (54)$$

When $X = OAc^-$, the initially coordinated anion is sufficiently basic to neutralize the liberated H^+ [Eq. (54)]; however, $X = Cl^-$ requires the presence of external (normally amine) base to effect the same transformation. In both cases the palladium is recovered as the carbonyl derivative, $Pd(CO)(PPh_3)_3$ [and/or the related $Pd_3(CO)_3(PPh_3)_4$]. Alkoxycarbonyl derivatives such as $Pd(OAc)(COOMe)(PPh_3)_2$ and $Pd(COOMe)_2(PPh_3)_2$ have been isolated from these systems and are invoked as catalytic intermediates, the total mechanism being described in the Scheme 10. The salient

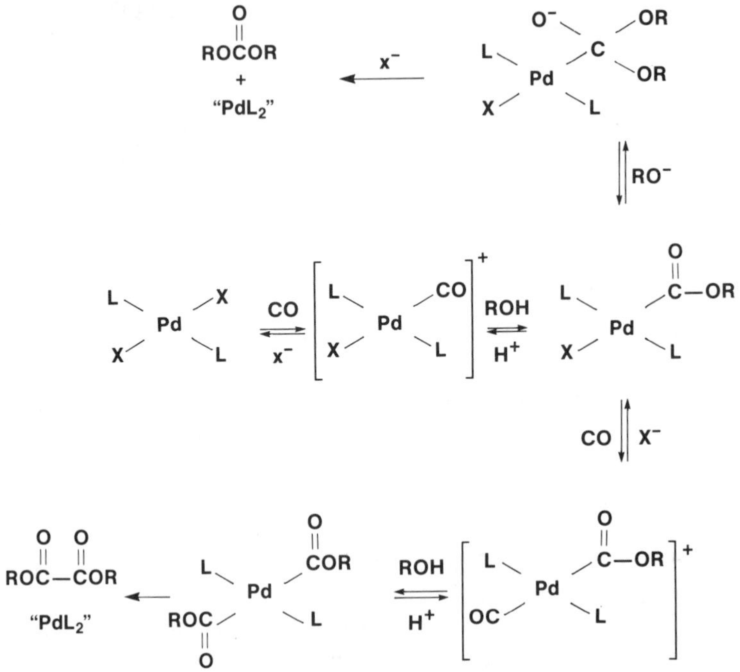

SCHEME 10.   Proposed mechanism for the palladium-catalyzed syntheses of oxalates. From Ref. *110*.

feature in this scheme is the attack of the nucleophilic alcohol on coordinated CO. Ionization of the starting PdX_2L_2 complex in the polar MeOH solvent generates the electrophilic cation. Nucleophilic attack by the available alcohol, rather than the corresponding alkoxide, on the coordinated CO ligand has as its precedent the analogous Pt system

$$\{PtCl(CO)(PEt_3)_2\}^+ + ROH \;\rightleftharpoons\; PtCl(\overset{\overset{\textstyle O}{\textstyle \|}}{C}OR)(PEt_3)_2 + H^+ \qquad (55)$$

first observed by Clark and co-workers (*113–115*) and studied kinetically by Byrd and Halpern (*116*). The mechanism also appears to account quite well for the other available data. Increasing the CO pressure increases the yield of oxalates, both absolutely and relative to the carbonates, up to approximately 8 atm, after which point the effects are negligible (Fig. 4). The initial rate increase with CO may be attributable to the increased formation of the cationic intermediate with CO pressure. Similarly, added bases (e.g., triethylamine, diisopropylamine, pyridine) favor the formation of the carbonates at

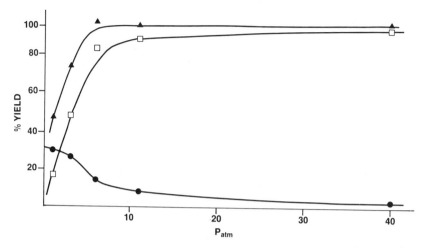

FIG. 4. Effect of CO pressure on dimethyl oxalate (□), dimethyl carbonate (●), and dimethyl oxalate + dimethyl carbonate (▲) yields for the palladium-catalyzed oxalate synthesis. From Ref. *110*.

low CO pressures but have little effect at the higher CO pressures. Both of these observations are consistent with the mechanisms shown in that they imply a competition for the cation intermediate.

Another possible route into the carboalkoxy species which may be considered is the formal insertion of CO into a preformed Pd–alkoxy species. Such a mechanism was considered a viable possibility in the catalytic nonphosphine system of Fenton and Steinwand (*102*). Given that in the absence of added CO but under otherwise identical reaction conditions alcohols are oxidized to the corresponding aldehyde or ketone (*117, 118*), this may be reasonable. Presumably this latter reactivity can be attributed to initial alkoxy formation, followed by β-hydride elimination:

$$\underset{H}{\overset{H}{\underset{|}{\overset{|}{RC}}}}-O-Pd-X \quad \rightleftharpoons \quad \underset{H}{\overset{H}{\underset{|}{\overset{|}{RC}}}}=O-Pd-X \quad \longrightarrow \quad RCHO + Pd(0) + HX \qquad (56)$$

One might envisage that CO is able to intercept the alkoxy group before this elimination occurs. A similar set of reactions may also operate in the phosphine systems of Rivetti and Romano, despite the facility with which β-hydride eliminations occur in this type of system. The available information would not appear to be capable of determining the possible contribution of this second pathway to the total reactivity.

A recent patent has described the carbonylation of methanol to give acetic acid using a palladium-based system (119). The system requires alkyl halide promoters and electron-rich nitrogen ligands (e.g., 2,2'-bipyridine) and operates in the ranges 125-250°C and 20-210 atm. There is insufficient information available to allow discussion of pathways involved.

C. PLATINUM-CATALYZED FORMATE SYNTHESIS

Recently, Head and Tabb reported that methanol could be carbonylated to give mainly methyl formate using a platinum phosphine catalyst system (120). Typical reaction conditions were 110-180°C and 35-150 atm CO, with the reaction yielding up to 500 turnovers after an 18 h period. The most significant difference from the formally analogous palladium system (vide supra, Section VI,B), aside from the product, was that the platinum system was catalytic, that is, the latter system was capable of re-oxidizing the zero-valent complex without external reagents. To account for this and their limited rate and spectroscopic data, these authors proposed the mechanistic cycle shown in Scheme 11. Reoxidation of the Pt(0) was achieved by the oxidative addition of CH_3OH. The reaction invoked CO insertion into a $Pt-OCH_3$ bond [Eq. (57)] instead of the attack of an external nucleophile (CH_3O^- or CH_3OH) on precoordinated CO [Eq. (58)]:

$$(P)_2PtH(OCH_3) + CO \longrightarrow (P)_2PtH(CO_2CH_3) \qquad (57)$$

$$(P)_2Pt(CO)_2 + CH_3O^- \longrightarrow \{(P)_2Pt(CO)(CO_2CH_3)\}^- \qquad (58)$$

The former mechanism was modeled after the analogous reactions proposed by Otsuka et al. (121) for the Pt-catalyzed water-gas shift (WGS) reaction ($H_2O + CO \rightarrow CO_2 + H_2$). However, aside from the fact that the reactivity pattern for the alcohol system [$P = P(p\text{-tolyl})_3 > PEt_3 > P(i\text{-Pr})_3$] was exactly opposite that seen in the WGS system [$L = P(i\text{-Pr})_3 > PEt_3 \gg PPh_3$], there would appear to be an anomaly in using this mechanism for the present case. The predominant species in solution, as observed by IR, was $Pt(CO)_2P_2$ (for $P = PEt_3$). To generate a species from this with sufficient electron density to oxidatively add ROH to any significant extent, the platinum would be required to lose at least one CO ligand (Scheme 11). It is surprising, then, that the reaction was approximately first order in CO pressure (pressures greater than 35 atm), since any pre-equilibrium involving CO dissociation would rather be expected to be inhibited by the higher pressures (alternatively, if CO loss were rate determining to give an immediately intercepted intermediate, the reaction might be expected to be in independent of CO). The explanation that the rate acceleration is due to the increase in rate of reduction elimination seems reasonable, except that it

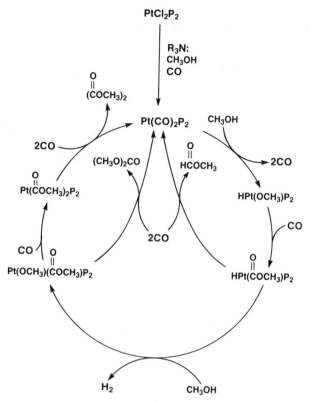

SCHEME 11. Proposed mechanism for the platinum-catalyzed syntheses of formates. From Ref. *120*.

implies that this occurs before or is the rate-limiting step, and this conclusion would be inconsistent with the observation that the $Pt(CO)_2(PEt_3)_2$ species was the most predominant species under the reaction conditions. One might imagine, based on formate production, that species such as $\{HPtI_{2,3}\}^+$, $HPt(OCH_3)I_{2,3}$ or $HPt(COOCH_3)I_{2,3}$ are all likely intermediates in the catalytic cycle. Since all of these species are expected to be infrared inactive over the energy range available, they may be present in larger concentrations than recognized. However, this would appear to be inconsistent with the available ^{31}P-NMR data, obtained from depressurized reaction solutions, and may not be the required explanation. The fact that charged *cis*-$Pt(CO)_2\{P(p\text{-tolyl})_3\}_2$ catalyzes the reaction would suggest, though, that Pt(0) is at least capable of being transformed into the active species.

Again, as in many of the systems discussed here, more work is necessary to unravel the mechanisms involved.

VII.   Other Metal-Catalyzed Carbonylations

A.   IRON-CATALYZED CARBONYLATIONS

Workers at Argonne Laboratory have recently discovered a catalytic method for the homologation of methanol to ethanol (122–124). The workers consider the mechanism to involve initial formation of acetaldehyde, followed by its hydrogenation.

The reaction, which has been operated at $200°C$ and 300 atm ($3:1$) CO/H_2, is given by

$$CH_3OH + 2CO + H_2 \longrightarrow CH_3CH_2OH + CO \qquad (59)$$

and has as an initial pre-equilibrium, the base (B; e.g., trimethylamine or 1-methylpiperidine) catalyzed formation of methyl formate (125):

$$CH_3OH + CO \overset{B}{\rightleftharpoons} HCO_2CH_3 \qquad (60)$$

$$HCO_2CH_3 + B \rightleftharpoons CH_3B^+ + HCO_2^- \qquad (61)$$

In contrast to other systems, particularly the acidic $HCo(CO)_4$ catalyst reaction, this methodology takes advantage of general reactivities of metal carbonyls and uses the quaternary ammonium cation as the methyl transfer reagent. The method's characteristic preference for methanol vs. ethanol homologation is attributed to the differences in reactivity for methyl vs. ethyl formate as a substrate in S_N2-type reactivity [Eq. (61)]. In such displacement reactions involving poor leaving groups (HCO_2^-), rates for methyl transfers can be two orders of magnitude larger than those for ethyl transfers (126).

As in the cobalt system, the reaction generates significant quantities of CH_4 as a by-product (in some cases, the methane is the major product). The selectivity of homologated alcohol to hydrocarbon appears to be independent of the partial pressures of either CO or H_2 (127), and the authors suggest that this could be attributable, at least in the Mn system, to the relative rates of methyl migration to homolytic bond dissociation. In the iron system, methane is generated by simple reductive elimination from the $HFe(CH_3)(CO)_4$ intermediate.

A variety of metals (e.g., Fe, Ru, Rh, Mn) have been used to effect the reaction; when the proper cocatalysts are added, a manganese catalyst (apparently $\{Mn(CO)_5\}^-$) is most effective. However, the initial discovery was made with, and the work is generally discussed in terms of, the $Fe(CO)_5$/amine catalysts. The corresponding mechanistic interpretation is given in Scheme 12. Under the conditions of the reaction, the iron exists almost entirely as the anionic $HFe(CO)_4^-$ (in general, the metals exist as the

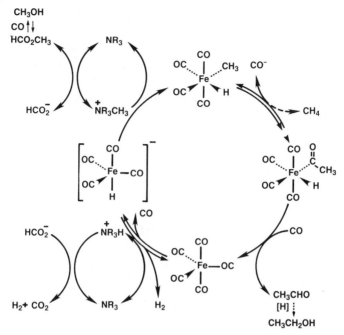

SCHEME 12.   Proposed mechanism for the amine-promoted iron-catalyzed homologations of alcohols. From Ref. *127*.

corresponding metallate anions), formed during the decomposition of the formate. The key requirements. of this system, as far as the metals are concerned, are the ability to facilitate this formate decomposition as well as to act as a good nucleophile toward the substrate $NR_3(CH_3)^+$. This iron hydride species, $HFe(CO)_4^-$, functions well in both regards, although it shows poor selectivity, yielding approximately equal amounts of methane and ethanol. The anionic $Mn(CO)_5^-$ acts as a better nucleophile [estimated, in separate experiments, to be 2.7 times faster than iron (*127*)] and is more selective to form ethanol (> 80%; the greater selectivity with this metal is attributed to the more favorable formation of Mn–acyl complexes) but is much poorer in effecting the formate decomposition reaction. This last property causes the overall reaction rate to be relatively low. However, by adding $Fe(CO)_5$ (to accelerate the decomposition reaction) to the $Mn_2(CO)_{10}$ system, the overall rate of the latter reaction increases by a factor of three, with full retention of the higher selectivity. The rates are reportedly comparable with current homologation technologies based on cobalt carbonyls (*123*).

B. COPPER-CATALYZED CARBONYLATIONS

Copper(I) carbonyls in the presence of H_2SO_4 ($>85\%$) catalyze the carbonylation of alcohols under ambient conditions (128). In this case, yields of up to 80% have been reported. The necessity of such high acid concentrations suggest that the chemistry involved may be described as a modified Koch reaction:

$$ROH + H^+ \rightleftharpoons ROH_2^+ \rightleftharpoons R^+ + H_2O \xrightleftharpoons{+CO} R\overset{\overset{\displaystyle O}{\|}}{C}{}^+ \xrightarrow{+H_2O} RCOOH$$

(62)

Consistent with this explanation are the observations that tertiary alcohols react much faster than primary and secondary alcohols and that even when linear alcohols are carbonylated the predominant products are the branched tertiary acids. Both of these are consistent with the formation of carbonium ions, which then rearrange to form the favored tertiary structure, for example,

$$CH_3CH_2CH_2CH_2OH + H^+ \longrightarrow CH_3CH_2CH_2CH_2^+ + H_2O$$

$$\downarrow$$

(63)

$$\underset{\underset{\displaystyle CH_3}{|}}{\overset{\overset{\displaystyle CH_3}{|}}{H_3C-C^+}}$$

The copper salts (typically Cu_2O) are transformed into the cationic carbonyl derivatives under the strongly acidic conditions:

$$Cu_2O + 2H^+ \longrightarrow 2Cu^+ + H_2O \xrightarrow[n=3,4]{n\ CO} \{Cu(CO)_n\}^+ \qquad (64)$$

By increasing the effective concentration of CO in the liquid phase, these act as CO suppliers to the generated organic cations:

$$R^+ + \{Cu(CO)_n\} \longrightarrow R\overset{\overset{\displaystyle O}{\|}}{C}{}^+ + \{Cu(CO)_{n-1}\}^+ \qquad (65)$$

$$R\overset{\overset{\displaystyle O}{\|}}{C}{}^+ + H_2O \longrightarrow RCOOH + H^+ \qquad (66)$$

VIII. Concluding Remarks

In this overview, we have attempted to present the various aspects of metal-catalyzed carbonylation reactions largely within a mechanistic framework. In some cases, a detailed understanding of the reaction pathways is now available, while in others the situations are such that only a rudimentary understanding has been accessed. This situation certainly arises in large measure because the severe conditions required for carbonylation reactions give rise to interfering, competitive secondary and side reactions, making it difficult to examine the system of interest directly. As a result, in many cases this lack of understanding results simply from the absence of any useful kinetic and/or spectroscopic data. Toward this end, it is expected that advances in *in situ* spectroscopic techniques should be most helpful in enhancing our understanding of these reactions.

REFERENCES

1. Collman, J. P., *Acc. Chem. Res.* **1**, 136 (1968).
2. Halpern, J., *Acc. Chem. Res.* **3**, 386 (1970).
3. Stille, J. K., and Lau, K. S. Y., *Acc. Chem. Res.* **10**, 434 (1977).
4. Calderazzo, F., *Angew. Chem., Int. Ed. Engl.* **16**, 299 (1977).
5. Halpern, J., *Acc. Chem. Res.* **15**, 238 (1982).
6. Wojcicki, A., *Adv. Organomet. Chem.* **11**, 88 (1972).
7. Wojcicki, A., *Adv. Organomet. Chem.* **12**, 33 (1974).
8. Berke, H., and Hoffmann, R., *J. Am. Chem. Soc.* **100**, 7224 (1978).
9. Paulik, F. E., and Roth, J. F., *J. Chem. Soc., Chem. Commun.* p. 1578 (1968).
10. Roth, J. F., Craddock, J. H. Hershman, A., and Paulik, F. E., *Chem. Technol.* p. 600 (1971).
11. Paulik, F. E., Hershman, A., Knox, W. R., and Roth, J. F., to Monsanto Company, U.S. Patent 3,769,329 (1973).
12. Eby, R. T., and Singleton, T. C., *in* "Applied Industrial Catalysis" (B. E. Leach, ed.), Vol. 1, pp. 275–299. Academic Press, New York, 1983.
13. Dickson, S. E., Bakker, J., and Kitai, A., "Chemical Economics Handbook," 602.5020A-.5021L, SRI International, Menlo Park, California, 1982.
14. Forster, D., *J. Am. Chem. Soc.* **98**, 846 (1976).
15. Forster, D., *Inorg. Chem.* **8**, 2556 (1969).
16. James, B. R., and Rempel, G. L., *J. Chem. Soc., Chem. Commun.* p. 158 (1967).
17. Forster, D., *Adv. Organomet. Chem.* **17**, 255 (1979).
18. Brodzki, D., Leclere, C., Denise, B., and Pannetier, G., *Bull. Soc. Chim. Fr.* p. 61 (1976).
19. Hickey, C. E., and Maitlis, P. M., *J. Chem. Soc., Chem. Commun.* p. 1609 (1984).
20. Forster, D., *J. Am. Chem. Soc.* **97**, 951 (1975).
21. Forster, D., *Ann. N.Y. Acad. Sci.* **295**, 79 (1977).
22. Mullen, A., *in* "New Syntheses with Carbon Monoxide" (J. Falbe, ed.), Chapter 3.2. Springer-Verlag, Berlin and New York, 1980.
23. Hjortkjaer, J., and Jensen, V. W., *Ind. Eng. Chem. Prod. Res. Dev.* **15**, 46 (1976).
24a. Dekleva, T. W., and Forster, D., *J. Am. Chem. Soc.*, **107**, 3565 (1985).
24b. Dekleva, T. W., and Forster, D., *J. Am. Chem. Soc.*, **107**, 3568 (1985).

128   THOMAS W. DEKLEVA AND DENIS FORSTER

24c. Dekleva, T. W., and Forster, D., *J. Mol. Catalysis* **33**, 269 (1985).
25. Pearson, R. G., Sobel, H., and Songstad, J., *J. Am. Chem. Soc.* **90**, 319 (1968).
26. Moore, J. W., and Pearson, R. G., "Kinetics and Mechanism," p. 267. Wiley, New York, 1981.
27. Collman, J. P., and Hegedus, L. S., "Principles and Applications of Organotransition Metal Chemistry," p. 270. University Science Books, Mill Valley, California, 1980.
28. Masuda, A., Mitani, H., Oku, K., and Yamazaki, Y., *Nippon Kagaku Kaishi* **2**, 249 (1982).
29. Hjortkjaer, J., and Jorgensen, J. C., *J. Mol. Catal.* **4**, 199 (1978).
30. Dake, S. B., Kohle, D. S., and Chaudhari, R. V., *J. Mol. Catal.* **24**, 99 (1984).
31. Dekleva, T. W., and Forster, D., *187th Nat. Meet., Am. Chem. Soc., 1984* INDE 0108 (1984).
32. Hjortkjaer, J., and Jorgensen, J. C. E., *J. Chem. Soc., Dalton Trans.* 2 p. 763 (1978).
33. Forster, D., Hershman, A., and Morris, D. E., *Catal. Rev.—Sci. Eng.* **23**, (1 & 2), 89 (1981).
34. Brodzki, D., Denise, B., and Pannetier, G., *J. Mol. Catal.* **2**, 149 (1977).
35. Matsumoto, T., Mizoroki, T., and Ozaki, A., *J. Catal.* **51**, 96 (1978).
36. Forster, D., *J. Chem. Soc., Dalton Trans.* p. 1639 (1979).
37. Forster, D., and Singleton, T. C., *J. Mol. Catal.* **17**, 299 (1982).
38. Hohenschutz, H., von Kutepow, N., and Himmele, W., *Hydrocarbon Process.* **45** (11), 141 (1966).
39. von Kutepow, N., Himmele, W., and Hohenschutz, H., *Chem.-Ing.-Tech.* **37**, 383 (1965).
40. Falbe, J., ed., "New Syntheses with Carbon Monoxide." Springer-Verlag, Berlin and New York, 1980.
41. Falbe, J., "Carbon Monoxide in Organic Synthesis." Springer-Verlag, Berlin and New York, 1970.
42. Piacenti, F., and Bianchi, M., *in* "Organic Syntheses via Metal Carbonyls" (I. Wender and P. Pino, eds.), Vol. 2, Chapter 1. Wiley, New York, 1977.
43. Forster, D., and Singleton, T. C., *J. Organomet. Chem.* **17**, 299 (1982).
44. Reppe, W., and Friederich, H. to BASF, DE Offen. 902,495 (1951); Friederich, H., BASF, De Offen 933,148 (1953).
45. Vidal, J. L., and Walker, W.E., *Inorg. Chem.* **20**, 249 (1981).
46. Halpern, J., *Ann. N. Y. Acad. Sci.* **239**, 2 (1974).
47. Halpern, J., *Adv. Chem. Ser.* **191**, 165 (1980).
48. Wender, I., Metlin, S., and Orchin, M., *J. Am. Chem. Soc.* **73**, 5704 (1951).
49. Mizoroki, T., and Nakayama, M., *Bull. Chem. Soc. Jpn.* **38**, 1876 (1965).
50. Roper, M., and Loevenich, H., *J. Organomet. Chem.* **255**, 95 (1983).
51. Gauthier-LeFaye, J., Perron, R., and Colleuille, Y., *J. Mol. Catal.* **17**, 339 (1982).
52. Braterman, P. S., Walker, B. S., and Robertson, T. H., *J. Chem. Soc., Chem. Commun.* p. 651 (1977).
53. Mizoroki, T., and Nakayama, M., *Bull Chem. Soc. Jpn.* **41**, 1628 (1968).
54. Mizoroki, T., Matsumo, T., and Ozaki, A., *Bull. Chem. Soc. Jpn.* **52**, 479 (1979).
55. Bahrmann, H. and Cornils, B., *in* "New Syntheses with Carbon Monoxide" (J. Falbe, ed.), p. 226. Springer-Verlag, Berlin and New York, 1980.
56. Imyanitov, N. S., Bogoradoskaya, N. M., and Semenova, T. A., *Kinet. Katal.* **19**, 573 (1978).
57. Hendrickson, J. B., Cram, D. J., and Hammond, G. S., "Organic Chemistry," p. 393. McGraw-Hill, New York, 1970.
58. Koermer, G. S., and Slinkard, W. E., *Ind. Eng. Chem. Prod. Res. Dev.* **17**, 231 (1978).
59. Wender, I., Greenfield, H., Metlin, S., and Orchin, M., *J. Am. Chem. Soc.* **74**, 4079 (1952).
60. O'Neal, H. E., and Benson, S. W., *in* "Free Radicals" (J. K. Kochi, ed.), Vol. 2, Chapter 17. Wiley, New York, 1973.
61. Wender, I., Friedel, R. A., and Orchin, M., *Science* **113**, 206 (1951).
62. Berty, J., Marko, L., and Kollo, D., *Chem. Tech. (Leipzig)* **8**, 260 (1956).

63. Pretzer, W. R., and Kobylinski, T. P., *Ann. N. Y. Acad. Sci.* **333**, 58 (1956).
64. Mizoroki, T., and Nakayama, M., *Bull. Chem. Soc. Jpn.* **37**, 236 (1964).
65. Burns, G. R., *J. Am. Chem. Soc.* **77**, 6615 (1955).
66. Fakley, M. E., and Head, R. A., *Appl. Catal.* **5**, 3 (1983).
67. Steinmetz, G. R., and Larkins, T. H., *Organometallics* **2**, 1879 (1983).
68. Crossen, J. E., Krouse, S. A., and Moser, W. R., to be published; *187th Nat. Meet., Am. Chem. Soc., 1984* INDE 0109.
69. Forster, D., and Schaefer, G., unpublished observations.
70. Braca, G., Sbrana, G., Valentini, G., Andrich, G., and Gregorio, G., *J. Am. Chem. Soc.* **100**, 6238 (1978).
71. Braca, G., Sbrana, G., Valentini, G., Andrich, G., and Gregorio, G., *Fundam. Res. Homogeneous Catal.* **3**, 221 (1979).
72. Braca, G., Paladini, L., Sbrana, G., Valentini, G., Andrich, G., and Gregorio, G., *Ind. Eng. Chem. Prod. Res. Dev.* **20**, 115 (1981).
73. Braca, G., Sbrana, G., Valentini, G., and Cini, M., *J. Mol. Catal.* **17**, 323 (1982).
74. Dombeck, B. D., *J. Organomet. Chem.* **250**, 467 (1983).
75. Dombeck, B. D., *J. Am. Chem. Soc.* **103**, 6508 (1981).
76. Dombeck, B. D., *J. Am. Chem. Soc.* **102**, 6855 (1980).
77. Ford, P. C., ed., "Catalytic Activation of Carbon Monoxide," ACS Sympo. Ser. 152. Am. Chem. Soc., Washington, D.C., 1981.
78. Knifton, J. F., *J. Mol. Catal.* **11**, 91 (1981).
79. Halpern, J., and James, B. R., *Can. J. Chem.* **44**, 671 (1966).
80. Brothers, P. J., *Prog. Inorg. Chem.* **28**, 1 (1981).
81. Calderazzo, F., and L'Eplattenier, F., *Inorg. Chem.* **6**, 1220 (1967).
82. Colton, R., and Farthing, R. H., *Aust. J. Chem.* **20**, 1283 (1967); **24**, 903 (1971); Cleare, M. J., and Griffith, W. P., *J. Chem. Soc. A* p. 792 (1969).
83. Johnson, B. F. G., Johnston, R. D., and Lewis, J., *J. Chem. Soc. A* p. 372 (1969).
84. Laine, L. M., Rinker, R. G., and Ford, P. C., *J. Am. Chem. Soc.* **99**, 252 (1977).
85. Knifton, J. F., Vanderpool, S. H., Estes, J. H., and Buinicky, E. P., to Texaco Development Corporation, WO 81 00856 (1979).
86. Jolly, P. W., and Wilke, G., "The Organic Chemistry of Nickel," Vol. 2, Chapter 6. Academic Press, New York, 1975.
87. Mullen, A., *in* "New Syntheses with Carbon Monoxide" (J. Falbe, ed.), Chapter 3.6.3.3. Springer-Verlag, Berlin and New York, 1980.
88. S. P. Current, to Chevron Research Co., European Patent 99,665 Al, (1984).
89. Mitsubishi Gas Chemical Co., Ltd., Japanese Patent 57/175140 A2 (82/175140) (1982).
90. Isshiki, T., Kijimi, Y., Miyauchi, Y., and Kondo, T., to Mitsubishi Gas Chemical Co., Inc., European Patent 65,817 A1, (1982).
91. Naglieri, A., and Rizkalla, N., to Halcon SD Group, Inc., U. S. Patent 4,356,320A (1982).
92. Rizkalla, N., to Halcon SD Group, Inc., French Patent 2,496,643 A1 (1982).
93. Naglieri, A. N., and Rizkalla, N., to Halcon International, Inc., U. S. Patent 4,134,912 (1979).
94. Rizkalla, N., to Halcon International, Inc., British Patent 2,128,609 (1984).
95. Adkins, H., and Rosenthal, R. W., *J. Am. Chem. Soc.* **72**, 4550 (1950).
96. Reppe, W., Kroper, H., von Kutepow, N., and Pistor, H. J., *Justus Liebigs Ann. Chem.* **582**, 72 (1953).
97. Simon, R. H., Ph.D. Dissertation, Yale University, New Haven, Connecticut (1957).
98. Tsou, T. T., and Kochi, J. K., *J. Am. Chem. Soc.* **101**, 6319 (1979).
99. Cassar, L., and Foa, M., *Inorg. Nucl. Chem. Lett.* **6**, 291 (1970).
100. Foa, M., and Cassar, L., *Gazz. Chim. Ital.* **102**, 85 (1972).

101. Cassar, L., and Fao, M., *J. Organomet. Chem* **51**, 381 (1973).
102. Fenton, D. M., and Steinwand, P. J., *J. Org. Chem.* **39**, 701 (1974).
103. Fenton, D. M., and Steinwand, P. J., U. S. Patent 3,393,136 (1972).
104. Fenton, D. M., and Olivier, K. L., *Chem. Technol.* p. 220 (1972).
105. Yamazaki, T., Eguchi, M., Uchiumi, S., Takahashi, M., and Kurahashi, M., (to UBE Industries, Ltd.), U. S. Patent 3,994,960 (1976).
106. Cassar, L., and Gardano, A., to Montedison Spa.), U. S. Patent 4,118,589 (1978).
107. Yamazaki, T., to UBE Industries Ltd., Japan Kokai 73-29,428 (1974).
108. Montedison Spa., DE Offen. 2,601,139 (1975).
109. Revetti, F., and Romano, U., *Chim. Ind. (Milan)* **67**, 7 (1980).
110. Rivetti, F., and Romano, U., *J. Organomet. Chem.* **174**, 221 (1979).
111. Rivetti, F., and Romano, U., *J. Organomet. Chem.* **154**, 323 (1978).
112. Hidal, M., Kokura, M., and Uchida, Y., *J. Organomet. Chem.* **52**, 431 (1973).
113. Clark, H. C., and von Werner, K., *Synth. React. Inorg. Met.-Org. Chem.* **4**, 355 (1974).
114. Clark, H. C., Dixon, K. R., and Jacobs, W. J., *J. Am. Chem. Soc.* **91**, 1346 (1969).
115. Cherwinski, W. J., and Clark, H. C., *Can. J. Chem.* **47**, 2665 (1969).
116. Byrd, J. E., and Halpern, J., *J. Am. Chem. Soc.* **93**, 1634 (1971).
117. Lloyd, W. G., *J. Org. Chem.* **32**, 2816 (1967).
118. Tsuji, J., "Organic Synthesis with Palladium Compounds," p. 77. Springer-Verlag, Berlin and New York, 1980.
119. Van Leeuwen, P., to Shell International Research, European Patent Appl. 0090443 (1983).
120. Head, R. A., and Tabb, M. I., *J. Mol. Catal.* **26**, 149 (1984).
121. Yoshida, T., Ueda, Y., and Otsuka, S., *J. Am. Chem. Soc.* **100**, 3941 (1978).
122. Feder, H. M., and Chen, M. J., to U. S. Department of Energy, U. S. Patent 4,301,312 (1981).
123. Chen, M. J., Feder, H. M., and Rathke, J. W., *J. Am. Chem. Soc.* **104**, 7346 (1982).
124. Chen, M. J., Feder, H. M., and Rathke, J. W., *J. Mol. Catal.* **17**, 331 (1982).
125. Colquhoun, H. M., Holton, J., Thompson, D. J., and Twigg, M. V. T., "New Pathways for Organic Synthesis," p. 212. Plenum, New York, 1984.
126. De la Mare, P. B. D., Fowden, L., Hughes, E. D., Ingold, C. K., and Mackie, J. D. H., *J. Chem. Soc.* p. 3200 (1955), and references therein.
127. Roth, S. A., Stucky, G. D., Feder, H. M., Chen, M. J., and Rathke, J. W., *Organometallics* **3**, 108 (1984).
128. Souma, Y., and Sano, H., *Bull. Chem. Soc. Jpn.* **46**, 3237 (11973).

Quantum-Chemical Cluster Models of Acid–Base Sites of Oxide Catalysts

G. M. ZHIDOMIROV

Institute of Catalysis
Siberian Branch of the Academy of Sciences of the USSR
630090 Novosibirsk, USSR

AND

V. B. KAZANSKY

Zelinsky Institute of Organic Chemistry
Academy of Sciences of the USSR
117334 Moscow, USSR

I. Introduction

The development of the theory of heterogeneous catalysis is characterized by the following two clearly defined trends. In the first approach a catalyst is considered as an infinite periodic crystal lattice where electrons can be transferred over considerably large distances. The chemisorption and chemical activation of adsorbed molecules in such a system are then treated by taking into account the symmetry of the regular crystal lattice and the ability of free electrons or holes to be localized within the adsorption region. This collective approach originates from the band theory of the electron structure of solids. For metal catalysts it was most consistently developed by Dowden in his theory of partially occupied *d*- electron bands (*1*). For semiconducting systems including oxides, it was discussed in detail at a qualitative level by Vol'kenshtein (*2*).

An opposite local approach consists of considering the interactions of adsorbed molecules only with the nearest atoms of the surface of a catalyst. Key importance is ascribed in this case to defects and various imperfections of the crystal lattice at the surface of a solid. This approach was first formulated for metals in Taylor's theory of active sites and, in a more general form, in Balandin's multiplet theory as early as the late 1920s.

In terms of atomistic description of the phenomenon of catalysis, the border between these two approaches is determined by the rate of attenuation of the electron perturbation in a solid with the distance from an adsorbed molecule as well as by the degree of similarity between electron structures of the whole surface of a solid and of a fragment considered as a model of the adsorption or of an active site. For a long time this problem lacked an unambiguous solution. Therefore both approaches were equally widely used to describe catalytic phenomena, often without proper regard to specific features of the systems considered. Thus, active sites on metal catalyst surfaces were frequently modeled by individual metal atoms. On the contrary, catalysis on insulator oxide surfaces was sometimes discussed in terms of their cooperative electron properties.

This problem was certainly clarified after introduction of so-called cluster quantum-chemical models of heterogeneous catalysts. Their surface is simulated in this case by a relatively small fragment known as a cluster. It is desirable, on the one hand, that the size of such a cluster should be as small as possible to permit a sufficiently rigorous quantum-chemical description of its electron properties and the interaction with adsorbed molecules. On the other hand, the cluster should be large enough to produce both the real geometry of the whole crystal lattice of a catalyst and its electron structure, which are important for chemisorption. The numerous quantum-chemical computations carried out for such clusters have demonstrated that chemisorption interactions are indeed sufficiently localized, even in the case of metals where maximal cooperative effects should be anticipated. Thus quantum chemistry quite unambiguously advocates the use of the local approach in describing the catalytic phenomena (3). We shall return to this question below in discussing the cluster approximation.

This conclusion is also in accordance with the rich chemical experience obtained over the past 10–20 years. Close analogy was revealed between homogeneous and heterogeneous catalysis for a great variety of reaction types. Methods have been developed for the goal syntheses aimed at preparing the local active sites on oxide catalyst surfaces by means of anchoring and different complexes. The data on chemisorption and catalysis on alloys (4) also support the local approach. A number of various spectral and physicochemical methods of studying the surfaces of solids were used for investigation of catalysis. The results of these studies have also pointed to a dominant role of local factors in heterogeneous catalysis and have revealed the nature of the following principal types of active sites of oxide catalysts.

(1) Brönsted acid sites (BASs) usually represented by rather strongly acidic hydroxyl groups formed due to the partial hydrolysis of oxide surfaces. They are believed to act as active sites in the acid heterogeneous catalysis involving intermediate protonation of substrates.

(2) Lewis acid sites (LASs) and/or acid–base pairs consisting of low-coordinated nontransition-metal cations produced by dehydrogenation of oxide surfaces and neighboring oxygen ions. Characteristic of these active centers is the heterolytic dissociation of adsorbed molecules, resulting in the activation of hydrocarbons or hydrogen involved in catalytic hydrogenation, dehydrogenation, hydrogen isotopic exchange reactions, etc.

(3) Low-coordinated transition-metal ions acting as active sites both in catalytic redox reactions of organic and inorganic compounds and in coordination-type catalysis.

This classification certainly does not cover all possible types of the centers capable of exhibiting the catalytic activity in oxides. In addition, the active sites can be represented by strained structures in which an increased coordination ability should be expected for cations because of weaker oxygen "binding." Of certain importance may also be charged or radical surface defects. Rather low surface concentrations of such active sites should, however, be anticipated in all these cases. Another interesting problem is that of possible catalytic effects of various functional groups, both of biographical nature and formed as a result of the reactant adsorption or chemical reaction. It is, however, quite obvious that they should be separately discussed with regard to the specific features of a system and reaction considered.

The cluster approach opens the way for constructing the cluster models and performing quantum-chemical calculations for active sites of any of the above types. However, in different cases the problem is of different complexity. One could mention, for example, purely methodological problems arising in quantum-chemical computations of transition-element compounds. Another difficulty is the necessity of a proper description of rather large clusters, for instance, in the case of oxides with closely packed crystalline structure. Finally, very important for the adequate choosing of a suitable cluster model is the level of experimental knowledge of an object of a study. In heterogeneous catalysis this means the identification of the type of an active site, its structural characteristics, and possible ways of its modification in the course of a reaction. A variety of quite different experimental methods were applied to investigate the structure and chemical behavior of different sites. BASs, for instance, were most studied in this respect due to their highly characteristic OH stretching vibrations, making IR spectroscopy exceptionally informative for these sites. In addition, these sites possess the most simple structure.

Experimental information on LASs is poorer and complicated by uncertainties and contradictions of experimental data. Thus, even in the case of zeolites
which were investigated in a great number of experimental works, there is still
no certainty about the nature of LASs. Our knowledge of active sites formed
by low-coordinated transition-metal ions and other surface structural defects
is even less definite.

It therefore seems quite natural to choose silica, silica aluminas, and
aluminium oxide as the objects of the first systematical quantum-chemical
calculations. These compounds do not contain transition elements. They are
built of the individual structural fragments: primary, secondary, etc. This
enables one to find the most suitable cluster models for quantum-chemical
computations. The covalent nature of these structures again makes quite
efficient a comparatively simple method of taking into account the boundary
conditions in the cluster calculations. Finally, these systems demonstrate
clearly defined Brönsted and Lewis acidity. This range of questions comprises
the subject of the present review. This does not by any means imply that there
are no quantum-chemical computations on the cluster models of the surface
active sites of transition element oxides. It would be more proper to say that
the few works of this type represent rather preliminary attempts, being far
from systematic studies. Also, many of them unfortunately include some
disputable points both in the statement of the problem and in the procedure
of calculations. In our opinion, the situation is such that it is still unreasonable to try to summarize the results obtained, and therefore this matter is not
reviewed in the present article.

II. Cluster Quantum-Chemical Models

The cluster approach opens the way for a direct application of quantum-
chemical methods to the problems of chemisorption and catalysis. Quantum
chemistry has a rich experience in describing the local chemical interactions
in molecular systems. Its methods are continuously refined and improved,
and the computational techniques have become more and more accurate and
predictive.

Certainly, the cluster calculations require some modification of quantum-
chemical computational schemes. This is primarily connected with the main
drawback of the cluster approximation, consisting of artificial scission of the
chemical bonds between the cluster and the rest of the lattice. In most cases,
however, such modifications do not involve the principal points, and
therefore widely used quantum-chemical computer programs can serve as the
basis for the cluster calculations. Some quantum-chemical methods used in

modern practice to describe chemical and physical properties of surface active sites are discussed below, together with the versions of the cluster approximation used in modeling a separate site on oxides.

A. QUANTUM-CHEMICAL METHODS

Practically the whole range of quantum-chemical methods, from the simplest semiempirical to rather sophisticated variants of *ab initio* approaches, is currently used in quantum-chemical studies of chemisorption and surface structures. Such a situation is quite justified. This is stimulated both by the characteristic features of the object of the studies and by the peculiarities of the problems to be solved.

Cluster structures modeling a surface usually comprise rather large molecular systems, especially when considering several adsorption centers or when taking into account surface structure transformations accompanying the chemisorption. Moreover, because of some unavoidable uncertainty in cluster simulations, it is desirable to perform a series of comparative computations for several enlarged clusters in order to test the stability of the results. For this reason the less time-consuming semiempirical methods are usually preferred, since all the problems to be solved (examinations of various chemisorption structures, calculations of potential energy surfaces of chemical rearrangements, comparative treatment of wide series, etc.) imply a great body of computations.

At the same time, a conclusive and sufficiently reliable answer is frequently required. We may be interested in, for example, the question of the possibility of dissociative adsorption, or the problem of the existence of some chemisorption structures as in the discussion (see below) on the coordinative binding of water molecules by silicon atoms, etc. *Ab initio* calculations are required in these cases. They are needed as well to check some principal conclusions based on semiempirical schemes. Also, they are useful in providing the basis for proper choice and improving the parametrization of semiempirical methods. Therefore the nonempirical approach is finding ever-increasing application to the surface problems.

The computational schemes of quantum chemistry are based either on molecular orbital (MO) or valence bond (VB) methods (5). MO φ_i is defined as an one-electron wave function describing an electron in some effective field of nuclei and other electrons. It is usually represented as a linear combination of atomic orbitals χ_k (LCAO):

$$\varphi_i = \sum_k C_{ik} \chi_k$$

The many-electron wave function Ψ of a molecular system is taken as the antisymmetrized product of φ_i, and for closed-shell systems it is convenient to represent it by a Slater determinant. Such an approach is known as the restricted Hartree–Fock (RHF) method and is the most widely used method in chemisorption calculations. Its principal drawback is the neglect of Coulomb electron correlation, which is of crucial importance for adequate treatment of chemical rearrangements with varying numbers of electron pairs.

The scheme of calculations can be improved by making use of the configuration interaction (CI) method, where the total wave function Φ is represented by a superposition of Slater determinants Ψ of different configurations. If the AO basis is large enough and a sufficiently great number of configurations are taken into account, good accuracy can be attained. Actually, modern computational results for small molecules already compete in accuracy with the experimental data.

A somewhat modified MO LCAO scheme, without restriction on the identity of spin orbitals φ_i^α and φ_i^β for electrons with spins α and β, is known as the unrestricted Hartree–Fock (UHF) method and is usually used to treat open-shell systems (free radicals, triplet states, etc.). Electron correlation is partially taken into account in this method, and therfore it can be expected to be more efficient than the RHF method when applied to calculate potential energy surfaces of chemical rearrangements whose intermediate or final stages may involve the formation of free- or bi-radical structures. The potentialities of the UHF method are now under active study in organic reaction calculations. Also, it is successfully coming into use in chemisorption computations (6).

Semiempirical quantum-chemical methods can be subdivided into two groups. The first one covers so-called simple MO LCAO methods, of which extended Hückel treatment (EHT) (7) and its modification by Anderson (ASED) (8), additionally taking into account core repulsion, are most widely used in chemisorption computations. Anderson's improvement of EHT was aimed at obtaining more reliable values of the total energy and at gaining an opportunity to optimize the geometry of chemisorption structures. The method was mainly used in calculations of chemisorption and catalysis on metals. Its validity for oxide systems, with their rather highly ionic bonds, is formally less justified.

The second group of semiempirical MO LCAO methods is constituted by zero-differential overlap methods (9). Two subgroups can be specified here which somewhat conventionally may be called physical and chemical. The former involves CNDO/2, INDO, and some of their modifications, for example, CNDO/S. These methods are directed to the calculations of the electron characteristics: charge distributions, dipole moments, polarizabilities, spin densities, electron spectra, etc.; that is, mainly physical properties. Specific features of the parametrization scheme of these methods permits its relatively simple extension to the heavy-element compounds, including $3d$, $4d$, and $5d$ transition elements (*10*). The latter subgroup covers CNDO/BW and MINDO/3 methods adapted for calculations of total energy, heat of atomization, geometry of molecular systems, and conformational barriers. These are just the characteristics that are of especial importance and interest in chemisorption and catalysis. However, the parametrization of the methods presents a considerably more difficult problem. Only second-row element parametrization (up to F) is now available. A simplified procedure of estimating the binding CNDO/BW parameters has been proposed by Mikheikin *et al.* (*11*) based on a certain correlation with CNDO/2 parameters. Pelmenshchikov (*12*), already within the scope of a general scheme, has suggested CNDO/BW parametrization for the computations of the chemisorption on aluminosilicates of compounds containing H, C, N, and O.

CNDO/BW has well proven itself in the calculations of the chemisorption on oxides since it provided a relatively good description of hydrogen bonds. However, the method suffers from a poor description of conjugated bonds in organic compounds that requires special corrections to be introduced. This drawback is successfully avoided in MINDO/3. However, to apply it to catalysis, one needs to extend it parametrization so as to cover third-row and subsequent elements. Pelmenshchikov *et al.* (*13*) have solved this problem for Si and Al with the aim to calculate the chemisorption on aluminosilicates. The next important problem is to extend parametrization to $3d$ transition elements. It is now being used to solve several groups simultaneously, and the first advances have already been made in this area. Blyholder *et al.* have proposed a parametrization for the computations of the chemisorption of hydrogen on iron (*14*) and nickel (*15*). MINDO/3, however, does not give a sufficient description of donor–acceptor interactions, particularly of hydrogen bonds. In view of these limitations of CNDO/BW and MINDO/3, it seems suitable to have them in complex (*13*) and sometimes to carry out parallel calculations in order to check the reliability of results. Subsidiary testing by *ab initio* computations would also be of great value. Also, MINDO-type procedures seem quite useful in this respect.

A separated group of quantum-chemical methods consists of the schemes using local approximation of the exchange potential, X_α-SW and X_α-DV. These methods take an intermediate place between semiempirical and *ab initio* techniques. X_α-SW appears quite efficient in computing ionization potentials and electron spectra and is now widely used in cluster calculations of photoelectron and X-ray photoelectron spectra of chemisorption structures. It faces difficulties, however, in computing total energy, geometry, etc. X_α-DV is also promising for electron spectra calculations, though it is now less economical than X_α-SW. The main attractive feature of X_α-DV is that it provides sufficiently good values of energetic and geometrical parameters. This method is therefore finding ever increasing use in calculations of surface structures and chemisorption.

Ab initio methods are the most rigorous and predictive. Their principal advantage consists of the possibility of a systematic refinement of the results of computation absent, in general, in the case of semiempirical methods. However, ad initio techniques are rather time consuming, the time being sharply increased with increasing the size of a system and/or inclusion of heavy atoms. Therefore the development of the pseudo-potential approximation for core electrons is of crucial importance in extending the applicability of nonempirical quantum-chemical approaches to catalysis. Most ad initio cluster computations on metals are now carried out under this approximation. The first such calculations for aluminosilicates were performed by Zelenkovskii (16). Another important problem of ab initio computations is the electron correlation. Alongside the traditional CI treatment, the schemes taking into account only the most important part of the interactions are worth mentioning here. Among them, the generalized valence bond (GVB) technique developed by Goddard (17) has found rather wide application, being aimed at surface and chemisorption calculations.

B. CLUSTER APPROXIMATION

The main shortcoming of the cluster approach consists of the scission of the chemical bonds between terminal atoms of a cluster and the rest of a lattice. As a result, so-called dangling bonds occur at the terminal atoms of a cluster, artificial electron "surface states" appear in the partially occupied band, and the charge distribution is disturbed. A cluster in this case possesses too many "surface" atoms. Unfortunately, to obtain a better surface/bulk ratio, one should consider such large clusters that the approach becomes useless.

Some characteristics of the electron structure (Fermi level, ionization potentials, electron affinity, atomization energy) which are sensitive to cooperative properties of a solid converge rather slowly upon extending the cluster. However, in the chemisorption calculations the local interactions are of primary importance, and the cluster approach appears quite satisfactory. Thus even in the most adverse case of metals where the cluster approach incorrectly predicts the existence of the band gaps it provides a good description of the chemisorption even for relatively small clusters. Sufficiently good predictions can also be obtained for the priorities of adsorption sites, geometry, energetics, vibrational parameters of chemisorption bonds, etc.

For instance, in their comparative computations of the chemisorption of atomic hydrogen on Be, Bagus et al. (18) have demonstrated that, when passing from the Be_{22} to the Be_{36} cluster, both equilibrium length and frequency of stretching vibrations of the Be–H bond changed by only 4 % and the chemisorption energy varied by $\sim 20 \%$. The correct hierarchy of priorities of the chemisorption states was attained even for as small cluster as Be_{13}.

An obvious way to improve the results of a cluster calculation is to enlarge the cluster. A similar result could be achieved in another way, however, by taking into account in some form the influence of the neglected resting part of a lattice. A great number of such approaches were recently suggested and examined. Roughly, two trends can be specified. The first one is to consider the periodic structure of a lattice. It can be incorporated into a cluster calculation merely by imposing a periodic boundary condition. Such a scheme was used by Bennett et al. (19) in chemisorption calculations and is known as the periodic cluster model. Its consistent generalization by Evarestov (20, 21) has led to the so-called quasi-molecular enlarged unit cell (QEUC) model.

Another way to treat the influence of the surrounding lattice on a cluster is to use the "embedded" or "immersed" cluster technique. A number of realizations of this scheme have been proposed. We shall merely mention the directions of the evolution of these schemes rather than considering the question in detail. One way is to match the cluster and band solutions. This method, generally based on the early works of Grimley, was realized in Pisani (22) within the scope of CNDO/2. The whole system being considered (support and adsorbate) was subdivided into the cluster (adsorbate and a part of adsorbent) and the environment. Then the self-consistent calculation was carried out at a fixed environment. This approach was used to compute the adsorption of H (22) and CO (23) on graphite.

Some other "embedded cluster" schemes have also been developed. Thus in Muscat (24) the chemisorption of hydrogen on transition metals was simulated by a cluster of atoms immersed into a homogeneous medium described by the pseudo-potential approximation (jellium model). A promising scheme, involving the computation of "surface states" in a sufficiently large cluster, was proposed in Whitten and Pakkanen (25). These states may then be used, generally speaking, in a more advanced chemisorption calculation. Some other interesting approaches to the "embedded cluster" problem (26–28) are also worth mentioning.

Most of the suggested refined methods of treating the environment require considerable complication of the scheme of calculations and much computer time. Therefore they were mainly used in the case of sufficiently homogeneous systems of rather simple structures: graphite, metals, and oxides with cubic crystal lattices. In contrast, the real surface of most oxides is characterized by high inhomogeneity: various functional groups, impurities, and structural defects. Even higher inhomogeneity should be anticipated for supported catalysts and amorphous supports. Simpler but more versatile models are needed in this case to describe the environment effects. It is just these methods that are mainly used in chemically oriented calculations, and therefore they are considered below in more detail.

The first of such procedures was the method of terminal atoms saturating the broken outer bonds of a cluster. It was initially used in cluster computations of point defects in homo-atomic crystals of diamond and silicon. The hydrogen atoms were used as saturating atoms. In Hayns (29) this approach was successfully extended to calculate the chemisorption of hydrogen on a graphite surface.

Mikheikin *et al.* (11) have formulated an alternative approach where terminal valencies are saturated by monovalent atoms whose quantum-chemical parameters (the shape of AO, electronegativity, etc.) are specially adjusted for the better reproduction of given characteristics of the electron structure of the solid (the stoichiometry of the charge distribution, the band gap, the valence band structure, some experimental properties of the surface groups, etc.). Such atoms were termed "pseudo-atoms" and the procedure itself was called the method of a cluster with terminal pseudo-atoms (CTP). The corresponding scheme of quantum-chemical calculations was realized within the frames of CNDO/BW (11), MINDO/3 (13), and CNDO/2 (30) as well as within the scope of the nonempirical approach (16).

A particular scheme of introducing the pseudo-atom A depends both on the type of a quantum-chemical method and on a specified set of experimental properties whose reproduction is assumed to be the most important in the forthcoming investigations. This set of properties can also include, instead of experimental values, the values obtained by a more rigorous theoretical method (extended cluster, embedded or periodic cluster, band approach). In the scope of ZDO-type methods, the following parameters of pseudo-atom A can in principle be adjusted (11).

(1) n, the principal quantum number of χ_A — AO of atom A;

(2) $l = s, p, \ldots$, the shape of χ_A;

(3) ζ, the exponent of STO χ_A;

(4) VOIP, the valence orbital ionization potential, and $U_{\mu\mu}$, the one-center core parameter of atom A;

(5) γ_{AA} and γ_{BB}, the integrals of electron–electron interaction (subscript B refers to an atom different from A);

(6) α_{AB}, the exponent entering the expression for the core repulsion;

(7) location of atoms A in a cluster, usually determined by the distance R_{AB} to the nearest atom B.

At the same time, it seems reasonable to reduce considerably the set of adjustable parameters. Actually, certain relations exist between them, originating both from their physical nature and the method of their determination. So the overlap intergrals S_{AB}, which alongside the VOIP of atom A affect essentially the main characteristics of the electron structure of a cluster, depend on n_A, l_A, ξ_A, and R_{AB}. The arbitrariness in choosing these parameters is therefore obvious. Taking into account, in addition, the model nature of the very cluster scheme which is directed to make some qualitative conclusions, it seems inappropriate to attach too much importance to this maximal optimization.

Let us illustrate this by the following result, obtained for bulk silica. In the initial realization of the cluster scheme (11) it was adoped for pseudo-atom A saturating dangling O–Si bonds that $n_A = 1$, $l_A = 0$, and $\xi_A = 1.2$. The only free parameters were $VOIP_A$ and R_{OA}. The value of R_{OA} was estimated from the condition of the identity of two integrals: the overlap integral between the sp^2 hybrid AO of atom O and the sp^3 hybrid AO of atom SI and the overlap integral between the sp^2 hybrid AO of atom O and the χ_A AO of atom A. This gave $R_{OA} = 1.02$ Å. Parameters γ_{AA} and γ_{BB} were calculated using the respective formulas of CNDO/BW. Then the $VOIP_A$ was adjusted in such a way as to reproduce properly the stoichiometry of charges in the bulk SiO_2. For this purpose, a set of computations with different $VOIP_A$ was performed for a tetrahedral $Si(OA)_4$ cluster (**1a**). The stoichiometric condition $q_{Si} = -2q_O$ was achieved at $VOIP_A = 10.68$ eV, giving $q_{Si} = 2.14$ eV. The observed dependence of q_{Si} and q_O on $VOIP_A$ was practically linear for $VOIP_A$ ranging from 9 to 12 eV.

Further ·improvement of the above scheme of calculations should be expected primarily from another choice of χ_A. Indeed, since A was taken as a model of a silicon atom within its nearest environment (SiO_3 fragment), it would be more consistent to take the $3s$ AO of Si as χ_A. The next steps in this direction imply the use of the sp^3 hybrid AO of Si or even certain effective orbital χ_A which better represents the interaction of the SiO_3 fragment with the neighboring O atom (or with the whole cluster). The latter already bridges to the pseudopotential description of the influence of environment on a cluster. At the same time, in view of the above reasoning, this should not lead to any considerable improvement of the whole scheme of calculations.

Although relatively small, the experience of the comparative computations available from the literature generally agrees with this conclusion. In this respect the MINDO/3 realization of the CTP scheme (13) is quite representative. Although, as in the case of CNDO/BW, the hydrogen $1s$ AO was chosen for χ_A, the exponent of the STO was adopted to be (as for the sp^3 AO of Si)

$$\zeta_A = \tfrac{1}{4}\zeta_{Si(3s)} + \tfrac{3}{4}\zeta_{Si(3p)} \tag{2}$$

and the Coulomb parameter for the valence AO for atom A was taken equal to that of the sp^3 hybrid AO of Si. Two-center parameters were also determined by

$$\beta_{AO} = \beta_{SiO}, \qquad \alpha_{AO} = \alpha_{SiO}, \qquad \beta_{ASi} = \beta_{SiSi}, \qquad \alpha_{ASi} = \alpha_{SiSi}$$

The distance R_{AO} was assumed to be equal to R_{SiO} (1.63 Å). The ionization potential I_A and the diagonal matrix element U_{ss} for atom A were adjusted both to satisfy the condition of the charge stoichiometry of a bulk cluster and to reproduce the energy gap $\Delta\mathscr{E}$ between the highest occupied (HOMO) and lowest unoccupied (LUMO) MOs of silica gel [8 eV (31)]. The final values were $I_A = 10.9$ eV, $U_{ss} = -8.9$ eV, $q_{Si} = 1.4$ eV, and $\Delta\mathscr{E} = 8.3$ eV. Though in this case the scheme was actually based on atomic parameters of Si, its systematic application has given results quite similar to those obtained in its previous version using the CNDO/BW technique, the difference being mainly caused by the distinctions in the two computational methods (CNDO/BW and MINDO/3) rather than by the different parametrization of the pseudo-atom A.

Let us return to the results obtained in the CNDO/BW realization of the CTP scheme. The calculations for the Si(OA)$_4$ cluster 1a give a quasi-band picture of electron levels with HOMO–LUMO splitting of ~ 12 eV that is somewhat (but quite reasonably) higher than the experimental estimate of the band gap in SiO$_2$. The HOMO is mainly composed of $2p$ AOs of O atoms, whereas the LUMO is constituted by $3s$ AOs of Si that are quite in agreement with the band structure of SiO$_2$.

When turning to a surface cluster, it is important to specify the type of surface group or surface fragment to be considered. Thus the surface hydroxyl group is simulated merely by the substitution of a single atom A by H, with $R_{OH} = 0.96$ Å, giving the HOSi(OA)$_3$ cluster (1b). The bridged oxygen is then modeled by the (AO)$_3$SiOSi(OA)$_3$ cluster (1c), etc.

The requirement of stability of the results upon enlarging the cluster, which is quite natural in any cluster computation of surface structures and chemisorption, presents a condition of logical consistency of the scheme of calculations. Such an examination of the stability of the charge distribution and of the picture of one-electron levels of bulk cluster (*32*) gave satisfactory results [the clusters were extended by successive substitution of atoms A by $Si(OA)_3$ fragments]. A similar test was also attemped for surface clusters of silica and aluminosilicates simulating the surface hydroxyl groups of these oxides (*33*). The energy of hydrogen ion abstraction, E_{H^+}, from a terminal surface hydroxyl group was also calculated in that case as a qualitative index of the acidity of the OH group. It was found that $E_{H^+} = 17.12 \pm 0.34$ eV and the charge of the hydrogen atom of a hydroxyl group $q_H = (0.312 \pm 0.005)|e|$. The uncertainty ranges of these values correspond both to uncertainties in the geometrical parameters of clusters (bond lengths, valence angles, etc.) modeling the silicas of different types and to computational errors. Note that similar calculations for the bridged OH groups of aluminosilicates give $E_{H^+} \simeq (14.82 \pm 0.19)$ eV, that is, the difference in E_{H^+} between neutral hydroxyl groups in silicates and acidic hydroxyl groups in aluminosilicates (~ 2.3 eV) is far larger than the uncertainty of computations.

Similar to the above pseudo-atom A, which may also be called "pseudo-silicon" (Si*), one can also introduce pseudo-atoms representing the tetrahedral aluminum in the silicate structure. Such an approach may be useful in the case of large clusters. For example, one could mention the parametrization by Mikheikin *et al.* (*34*) of pseudo-atom Al* modeling Al in its nearest oxygen environment in the aluminosilicate lattice (i.e., AlO_3 fragment) at sufficiently high Si/Al ratios. For this purpose the electronic structures of both extended cluster $(AOH)Al(OA)_2OSi(OA)_2OAl(OA)_2(HOA)$ (Fig. 1a) and effective cluster $Al*OSi(OA)_2OAl*$ (Fig. 1b) were calculated. The atomic orbital for Al* was taken to be the same as for atom A. $VOIP_{Al*}$ was adjusted so as to reproduce the charge distribution of the extended cluster in the effective cluster. This gave $VOIP_{Al*} = 7$ eV. The respective atomic charges are presented in Figs. 1a,b.

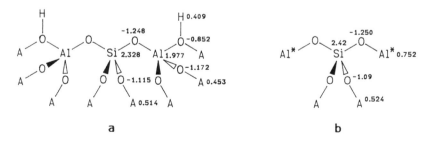

FIG. 1. (a) Extended and (b) minimal clusters used in choosing the parameters for pseudo-aluminum in the CTP scheme. The atomic charges are presented in the figure.

It is worth noting that the above CTP computational scheme opens a simple and convenient way for computer simulation of qualitative dependences. Thus, systematic variation of $VOIP_A$ in the $HOSi(OA)_3$ cluster results in the successive change of the acidity of OH groups (as represented by E_{H^+}), and thereby the qualitative influence of the acidity of OH group on its chemisorption properties and reactivity with respect to a particular reaction can easily be followed. Such an approach was rather widely used in our works (35, 36).

The considered CTP approach can be directly applied to covalent crystals. However, it meets with certain difficulties when used for closely packed lattices with high coordination numbers (Al_2O_3, MgO, ZnO, transition-metal oxides). It appears more convenient in this case to modify an ionic cluster corresponding to the ionic model of the lattice. Actually, just the excess (with respect to a neutral cluster) or lack of electrons serve here to saturate the broken valencies of a cluster. To obtain the neutral cluster, one may compensate the excess charge by properly fitting the core charges of terminal atoms, as suggested by Chuvyulkin et al. (37).

It seems of interest to compare the covalent and ionic cluster models as applied to the same system. Such a comparison was made within the scope of CNDO/2 calculations of some characteristics of surface hydroxyls of SiO_2 (38), magnetic resonance parameters of surface sites (39), and chemisorption (40). Though somewhat different, the results obtained within these models were quite close. Nevertheless, to make a more categorical conclusion, more extensive studies are required. It is likely that in several cases the close-packed oxide structures would be better represented by stoichiometric clusters (e.g., $Mg_{2n}O_{2n}$), although even in these cases it seems appropriate to take into account the Madelung field of the nearest part of the resting lattice.

One more type of cluster can be obtained by intracluster saturation of the broken bonds (41). This model was used in the comparative description of the surface centers of aluminophosphates and aluminosilicates (see Section V).

To conclude this section, let us note that it is quite reasonable to put forward the question of how sensitive are various physical and chemical characteristics of surface sites to the boundary conditions imposed on a cluster (40). This dependence proves to be considerably different for different characteristics, and they may therefore be separated into weakly and strongly dependent characteristics.

III. Silica Gel

Both crystalline and amorphous silicas of various types contain neither strong acid nor strong base sites (42). They are therefore of interest only as adsorbents and as supports of active components in supported catalysts. This is why the chemisorption properties of silicas comprised the main object of the quantum-chemical studies.

Silicon–oxygen tetrahedrons in variants of their interconnection in a silica framework (via either vertices or edges, etc.) form the principal structural bases of these systems. Essentially similar structure is characteristic of many other systems important in heterogeneous catalysis, for example, aluminosilicates and aluminophosphates. According to Harrison's classification, these are typical examples of so-called "open" solid systems where the structure of the nearest environment of a given atom in the lattice is determined by directed valence bonds (covalent crystals) (43). The fact that these interactions are decisive in determining the structure of silicates is evidenced by the similarity of the structural parameters (valence angles, bond lengths) of the disiloxane group SiOSi in a solid and in a $(H_3Si)_2O$ molecule, as emphasized by Gibbs (44). The package effects, long-range crystalline potential, etc. are, on the contrary, only of secondary importance.

Si–O bond length in silicates ranges from 1.55 to 1.76 Å, the mean value being 1.625 Å. The tetrahedral structure of SiO_4 is rigid enough, and the most labile part is the SiOSi fragment. Actually, the SiOSi bond angle ranges rather widely from 135 to 180° (44).

The question of the role of the $3d$ AO of Si in chemical bonding in silicates is the matter of discussions in the quantum-chemical treatment of these systems. Although a great number of works have dealt with this problem, there is still no common viewpoint. We do not have the opportunity to dwell upon numerous pro and con arguments presented in the literature, and therefore we will merely illustrate the situation by recent *ab initio* calculations of such model fragments of silica structure as $(SiH_3)_2O$ (45, 46) and $Si(OH)_4$ (47). The analysis of the comparative computations with and without regard to the d AO of Si has led Meier and Ha (45) to the conclusion that the calculated values of the geometrical parameters (SiO bond length and SiOSi angle) depend significantly on whether the d AO was considered or not. In contrast, Oberhammer and Boggs (46) and Sauer (47) have concluded that the role of the d AO in forming the equilibrium geometry is quite unessential. The problem drawing the main attention here is why the Si–O bond is so markedly short as compared to the sum of the covalent radii of Si and O. Pauling's well-known concept attributes this effect to $(d-p)\pi$ conjugation. However, it can be equally associated with a considerable ionicity of the bond (46). Our own experience comes out in favor of neglecting the d AO of Si due to its small contributions to the various calculated properties. Some examples of the comparative computations will be given below.

Formally, if various possible charged and radical defects are neglected, three types of adsorption sites on silicates may be specified. These are surface hydroxyl groups, bridged oxygen, and silicon atoms, and they are discussed below. We will not consider here a very interesting class of modified silicas with different surface functional groups, since there are still no systematical quantum-chemical calculations in this area.

A. SURFACE TERMINAL HYDROXYL GROUPS

As mentioned above, $HOSi(OA)_3$ may be taken as the simplest cluster model of the terminal hydroxyl group in silicas. Indeed, even with this cluster CNDO/BW provided a quite satisfactory description of the lower part of the curve representing potential energy as a function of the OH stretching vibration coordinate R_{OH} (Fig. 2) (48, 49). The respective experimental curve was plotted by Kazansky et al. (49) based on the analysis of the fundamental frequency ν_{OH} and the first overtone of the characteristic OH stretching vibration in terms of the Morse potential function. The frequencies of the second and third overtones were also determined in that work, and it was shown that the Morse potential reproduced well the potential curve within a rather wide range of R_{OH}.

One of the main characteristics of the hydroxyl groups on oxides is that their acidity governs their chemisorption and catalytic properties. A number

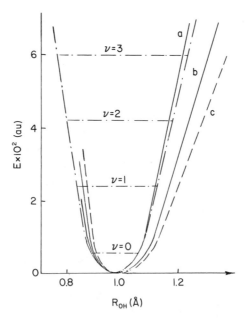

FIG. 2. Potential curves for the OH bond of the $HOSi(OA)_3$ cluster (a) in the free state ($VOIP_A = 10–16$ eV) and in the presence of an ethylene molecule at (b) $VOIP_A = 10$ eV and (c) 16 eV. The dashed lines represent the "experimental" Morse curve and first vibrational levels.

TABLE I

Cluster Modeling of Terminal Hydroxyl Groups of Different
Acidities [(AO)$_3$S:OH Cluster, CTP Scheme,
CNDO/BW Method]

	VOIP$_A$ (eV)				
	10.0	10.68	12.0	13.6	16.0
$(+) q_H|e|$	0.313	0.317	0.325	0.336	0.356
$(-) q_O$	0.836	0.835	0.832	0.827	0.819
$(-) q_{OH}$	0.523	0.518	0.507	0.491	0.463
E_{H^+} (eV)	17.0	16.8	16.3	15.5	14.8

of quantum-chemical works of different authors have been devoted to this problem, of which the monograph by Dunken and Lygin (50) seems of primary interest. Great attention was paid in these studies to the surface of silicas. Most computations were carried out within the frames of EHT and CNDO/2 using simple clusters involving one or more silicon–oxygen tetrahedra. Boundary conditions were taken into account by means of terminal H atoms. The main outcomes of these investigations are as follows. It was concluded that wide variation of the acidities of terminal hydroxyls did not practically influence such characteristics as bond order, homolytic diffraction energy, and Wiberg index. At the same time, the acidity could be characterized by such quantities as the charge on the H atom and E_{H^+} as well as the change in OH bond order due to its interaction with adsorbed molecules.

Close conclusions were also drawn in our studies using the CTP scheme and CNDO/BW method. Extensive variation of the acid–base properties of a terminal OH group by means of systematic change of VOIP$_A$ from 7 to 16 eV in the HOSi(OA)$_3$ cluster (Table I) did not noticeably affect the shape of the potential curve of the OH bond (Fig. 2). However, in the presence of an adsorbed ethylene molecule, the curve deviated considerably (48). This is in accordance with the known relatively slight differences in v_{OH} for hydroxyl groups of different acidity and the pronounced effect of the acidity on the shift, Δv_{OH}, caused by adsorption of some molecules (51).

1. Chemisorption Interaction with Simple Molecules

In this section we shall discuss the results of quantum-chemical calculations of the chemisorption interaction of the surface terminal hydroxyl groups in silicas with water and ammonia molecules. Two types of one-center coordination of H$_2$O molecules by a hydroxyl group of SiO$_2$ are presented in Fig. 3. CNDO/BW calculations using these clusters gave adsorption energies

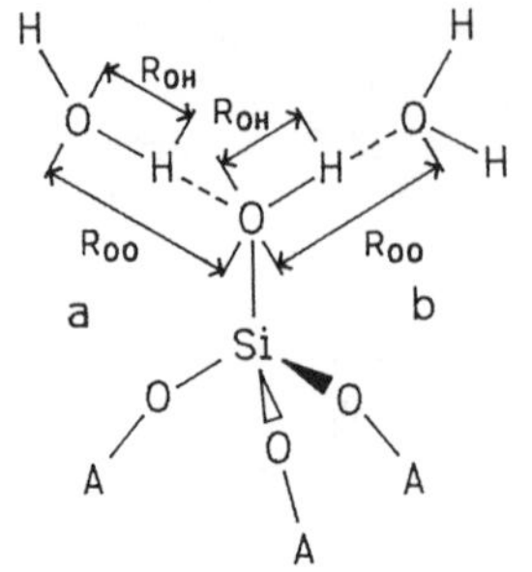

FIG. 3.   Two forms of water molecule adsorption on the terminal hydroxyl group.

of ~ 10 kcal/mol (a) and ~ 3 kcal/mol (b) (11, 52). This result was interpreted as an indication that the hydrogen bonding with water molecules is not characteristic of a surface hydroxyl group. This is in qualitative agreement with the relatively slight effect of the water adsorption on the frequency of the stretching vibrations of OH groups in silicates (53). In the above calculations, the distance R_{OO} was assumed to be 2.5 Å, that providing maximal hydrogen bonding in H_2O dimers as computed by CNDO/BW. R_{OH} was varied and was found to be ~ 1 Å.

The question of the preferential form of water adsorption on OH groups of silicates being of certain importance, Sauer et al. (54) and Senchenya et al. (52) have performed a similar comparative calculation at the nonempirical level using the STO-3G basis set. Terminal bonds of the cluster were saturated by hydrogen atoms. The results of our computations for different R_{OO} have shown that the energy minimum is attained at $R_{OO}^a = 2.77$ Å and $R_{OO}^b = 2.66$ Å. The corresponding interaction energies were $Q_a = 4.4$ and $Q_b = 6.7$ kcal/mol in a qualitative contradiction to the results of previous calculations. This allowed the authors (54) to call into question the validity of CNDO/BW as applied to such investigations. At the same time, a seemingly quite reasonable alternative belief was that the observed discrepancy should be mainly attributed to the difference between hydrogen atoms and the terminal pseudo-atoms A which increased the acidity of a hydroxyl group. Therefore Zelenkovskii (16) has repeated the calculation of this system, although with boundary pseudo-atoms A. The electronegativity of a pseudo-atom A was varied by adding the δ-like potential $K \exp(-\alpha r^2)$ to the core potential. The linear coefficient was adjusted to reproduce the stoichiometric charge distribution in the bulk $Si(OA)_4$ cluster. The calculations have shown that (a)-type water adsorption ($R_{OO}^a = 2.68$ Å, $Q_a = 7.75$ kcal/mol) was more favorable than (b)-type adsorption ($R_{OO}^b = 2.69$ Å, $Q_b = 5.4$ kcal/mol).

In the case of the strongly basic ammonia molecule, the computation (CTP, CNDO/BW) predicted considerable preference of (b)-type adsorption

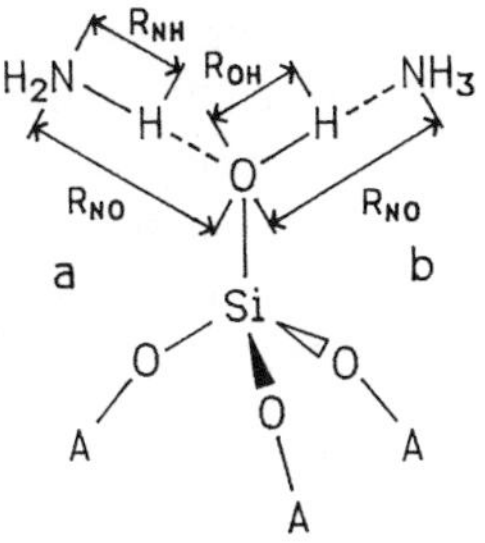

FIG. 4. Two forms of ammonia molecule adsorption on the terminal hydroxyl group.

(Fig. 4): $Q_a = 11.3$ kcal/mol, $Q_b = 26.5$ kcal/mol. Both R_{ON} and R_{OH} were varied (55). In the (b) structure the bond length R_{OH} was increased insignificantly. All these results are in qualitative agreement with IR data on ammonia adsorption on a hydroxylated SiO_2 surface. Strong perturbation of OH stretching vibrations can be observed there, but no adsorption band corresponding to ammonia ion.

2. H/D Isotopic Exchange with Terminal Hydroxyl Groups

One of the most direct ways to modify the surface of a silica is to substitute the hydrogen atoms of hydroxyl groups by cation-containing molecular groups. In particular, the transition-element complexes can be anchored in this way on the silica surface. Such a chemical modification would change the adsorption properties of a surface and can in principle give rise to catalytic activity. Therefore the mechanism of such exchange reactions is of great interest.

Because of the weakness of the Brönsted acidity of the surface hydroxyl groups, one can suppose that the synchronous mechanism is relatively efficient for these substitution reactions. The simplest reaction of this type is H/D isotopic heteroexchange with D-containing molecules. The heteroexchange with water is known to proceed rapidly at a temperature as low as 20°C. With molecular deuterium, such exchange occurs at 350–400°C with an activation energy of ~ 15–20 kcal/mol (55).

Application of the quantum-chemical methods to such surface reactions seems interesting and promising. One should, however, take into account the relative roughness of the reaction potential surface computations, certain ambiguity in active site structure and its modeling, practical inability to analyze all possible conformational rearrangements, etc. All this leads to the necessity of very clearly formulating the problem in such studies. Thus, for example, when considering the above problem of the H/D heteroexchange mechanism, it would be unjustified to try to prove numerically that the reaction follows a synchronous mechanism. This would require the comparison of the efficiencies of different reaction channels with the aim of finding the most optimal one. Such a comparison needs high accuracy of calculations and modeling of a reaction system which is now inaccessible in the surface reaction computations. A more realistic quantum-chemical investigation of this subject would likely give answers to the following questions.

(1)   Does the synchronous mechanism actually require relatively low activation energy?

(2)   How will one or another structural or chemical change in a system (surface site, reactants) affect the efficiency of the synchronous mechanism?

From general considerations, it should be anticipated that the effect of such changes would be essentially different for synchronous and stepwise mechanisms. Therefore theoretical studies of this difference may be useful in interpreting the experimental data and in planning the experiments. Thus, quantum-chemical calculations, some examples of which are given below, clearly indicate that the synchronous mechanism involving the surface hydroxyl groups only slightly depends on their acidity, in contrast to the stepwise protonation mechanism, where the acidity is a governing characteristic. Thus a wide field is opened here for comparative quantum-chemical studies.

Two simplest intermediate states of C_{2v} and C_{3v} symmetry with identical geometrical parameters for H and D in a reaction complex (Figs. 5a,b) can be suggested for the synchronous mechanism of H/D exchange between a water

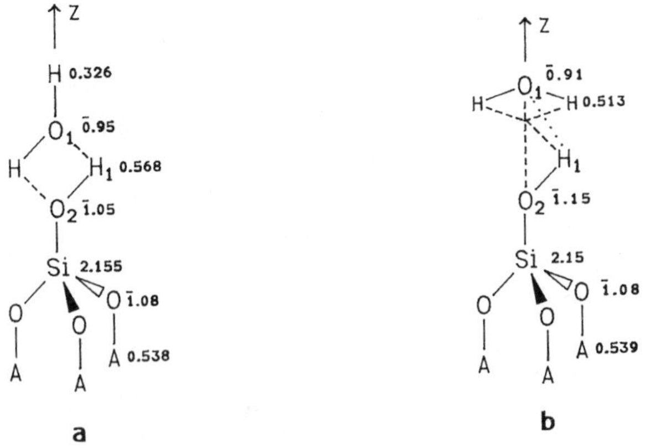

FIG. 5.   Models of the symmetric intermediate states of (a) C_{2v} ($E_a \simeq 0$ kcal/mol) and (b) C_{3v} ($E_a \simeq 30$ kcal/mol) symmetry for the reaction $Si(OA)_3OH + D_2O \rightarrow Si(OA)_3OD + HDO$. The atomic charges are given for the geometry corresponding to the minimal total energy.

molecule and a terminal hydroxyl group (11). A calculation (CTP, CNDO/ BW) of these states was carried out with $r_{O_1O_2} = 2$ Å, $r_{O_2H_1} = 0.9$ Å, and with optimization of hydrogen atom positions. The estimated energies of these structures with respect to noninteracting water molecules and clusters were ~ 0 (a) and 30 kcal/mol (b). An outcome of the computations was the conclusion that the synchronous mechanism is potentially feasible for this reaction and that the reaction path via the C_{2v} state should be preferred.

Such a calculation of a single intermediate structure provides only a tentative picture of the activation hindrances in the mechanism considered. To draw more quantitative conclusions, one should perform a complete calculation of the energy profiles along the reaction paths, discriminating between the reaction paths beginning from various adsorption states of water molecules and those corresponding to a free molecule coming in from the gas phase. For the latter case, the potential energy profile was calculated for the system with an oxygen atom of H_2O approaching along the z axis. The estimated activation energy for this path was ~ 6 kcal/mol (11). Note also that the "planar" C_{2v} structure was characterized by a more considerable electron density redistribution than the C_{3v} intermediate (Fig. 5).

To answer the second of the above questions, it seems expedient to make use of the CTP scheme, where progressive variation of pseudo-atom parameters permits the qualitative treatment of the influence of various characteristics of a molecule and active site on different reaction channels. Senchenya et al. (35) have calculated (CTP, CNDO/BW) the adsorption forms and intermediate C_{2v} state for a model hydroxyl-containing molecule A_1OH, with A_1 simulating the R fragment in an ROH molecule. Varying $VOIP_{A_1}$, one can reproduce a wide range of acidities of ROH molecules. The increase of $VOIP_{A_1}$ corresponds to the increase of the Brönsted acidity of a molecule, the base-to-acid transition occurring at $VOIP_{A_1} = VOIP_H = 13.6$ eV. It should naturally be anticipated that the influence of this parameter would be decisive in the energetics of both adsorption and proton exchange.

This is also confirmed by the direct quantum-chemical computations with substituent R taken in its explicit form (35). Figure 6 shows the structures corresponding to adsorption (a, b) and intermediate (c) states of the molecule A_1OH on a $HOSi(OA)_3$ cluster. The geometry of these structures was assumed to be similar to that in the case of water discussed previously. Special estimations involving the variation of $VOIP_{A_1}$ from 7 to 17 eV have actually demonstrated that optimal distances r_{OO} and r_{HH} ranged over only 0.02 Å, the respective energies ΔE ranging over ~ 2 kcal/mol. The results of the computations are listed in Table II. The main qualitative outcome of these calculations is that the acidity of the ROH molecule is not essential for the synchronous mechanism.

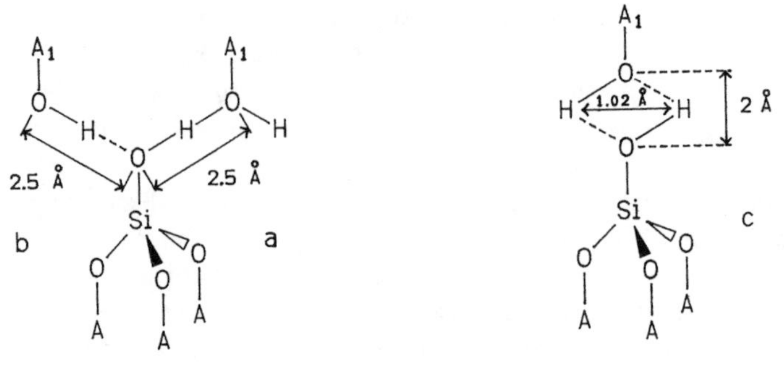

FIG. 6. Cluster models for (a, b) two types of adsorption of ROH molecules and for (c) the intermediate state (C_{2v}) of the H/D isotopic exchange reaction with terminal hydroxyl groups.

Thus, if the experiment shows that the exchange on silica gel can proceed under relatively soft conditions for a sufficiently great number of ROH-type molecules, the probability of the synchronous mechanism will be very high, since it seems unlikely that the reaction can proceed via an ionic structure (e.g., of ROH_2^+ type) with equally low activation energies for both hydroxyl-containing acids and bases.

It should be noted that the energetic parameters of adsorption and the intermediate structures presented in Table II characterize only a *relative*

TABLE II

Dependence of the Energies of the
Symmetric Transition State [ΔE
(kcal/mol)] and of Adsorption States ($Q_{a,b}$)
on the Type of ROH for Clusters in Fig. 6.

A_1OH	ΔE	Q_a	Q_b
$VOIP_{A_2}$ (eV)			
7	1.57	3.77	4.83
9	0.94	5.02	4.20
11	0.38	6.65	3.64
13	0.13	8.60	3.07
15	0.25	10.73	2.38
17	0.88	12.80	1.82
19	2.07	—	—
H_2O	0.00	9.9	2.83
CH_3OH	−1.32	—	—
C_2H_5OH	−1.63	8.35	4.96
$(CH_3)_3COH$	−1.69	—	—
HCOOH	5.7	13.1	4.58
CH_3COOH	0.82	11.92	6.27

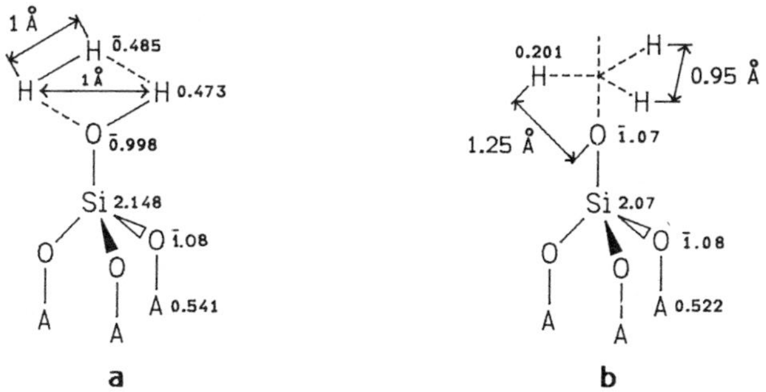

FIG. 7. Cluster models for symmetric intermediate states of (a) C_{2v} ($E = 8$ kcal/mol) and (b) C_{3v} ($E = 60$ kcal/mol) symmetry for the reaction $Si(OA)_3OH + D_2 \rightarrow Si(OA)_3OD + HD$.

energy scale of possible structural rearrangements, and therefore the positive values of ΔE should not be so surprising. The calculation did not involve full geometrical optimization of the transition state, and therefore it may deviate somewhat from the exact saddle point. Finally, the energy of noninteracting molecules was taken as a zero reference point for ΔE, whereas the real prereaction state may be represented by an adsorption state [e.g., like (a, b) in Fig. 6]. Thus it is only the orders of magnitudes and the directions and scales of variation that are essential in considering the data listed in Table II.

Similar investigations with varied acidity of a terminal hydroxyl group have led to essentially the same conclusion. An increased formation energy of the symmetric transition state should be expected only for sufficiently strong ROH acids and acidic surface OH groups. In all other cases the synchronous mechanism was predicted to be quite probable.

The calculation of the intermediate C_{2v} and C_{3v} structures for a synchronous mechanism of H/D exchange with molecular hydrogen (11) (Fig. 7a,b) foretells, as in the case of water molecules, the preference for the C_{2v} structure, with a formation energy of only ~ 8 kcal/mol. This correspondence led to the qualitative conclusion that C_{2v}-type structures are of primary importance for the efficiency of a sysnchronous mechanism. This conclusion permits a very simple qualitative test consisting of the comparison between the reaction paths of H/D isotopic exchange of terminal hydroxyl groups with ethylene and acetylene. Indeed, in the case of ethylene only the C_{3v} state is possible, whereas acetylene molecules can yield only the C_{2v} intermediate (Figs. 8a,b). From this it follows that the synchronous mechanism should be efficient only in the case of acetylene. Direct calculations (56) supported this conclusion: the estimated energies of the symmetric intermediate structures were 4.7 and 47.1 kcal/mol for acetylene and ethylene, respectively. It is worth noting that the experimental studies of Yates and Lucchesi (57) and Lucchesi et al. (58) indicate that H/D isotopic exchange between acetylene and surface hydroxyls

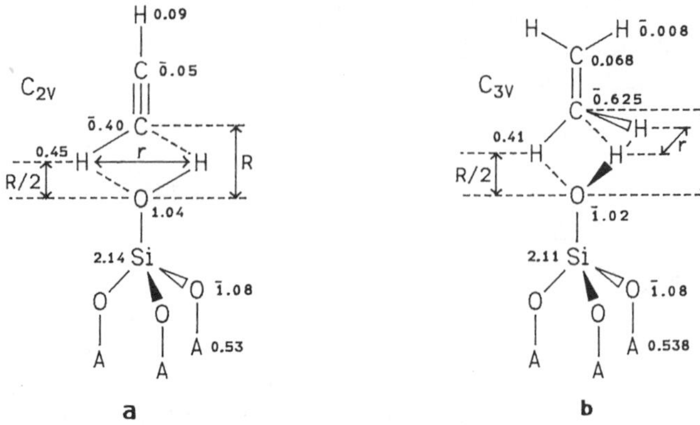

FIG. 8.   Possible models of symmetric intermediate states for the interaction of (a) acetylene (C_{2v}) and (b) ethylene (C_{3v}) with a surface OH group of silica gel. R and r correspond to the minimum of the total energy of the system.

of silica gel proceeds rather rapidly. This reaction was not observed for ethylene.

B.   BRIDGED OXYGEN ATOMS

Bridged oxygen atoms linking silicon–oxygen tetrahedra comprise possible chemisorption centers of a base nature. They may, for example, coordinate a water molecule, as represented by structure 2. Relatively few papers (59–62) have dealt with quantum-chemical studies of the chemisorption properties of such sites. The EHT calculations of Dunken and Hoffman (59) have led to assignment of a positive binding energy for structure 2. Ab initio computations of Sauer et al. (54) with a STO-3G basis and of Hobza et al. (61) with a 4-31G basis gave binding energies of 5.4 and 5.3 kcal/mol, respectively. However, the calculations were carried out for the $H_3SiOSiH_3$ cluster that seemed to be too crude a model of the site. Therefore, Zelenkovskii et al. (62) performed an ab initio computation with a STO-3G basis for adsorption

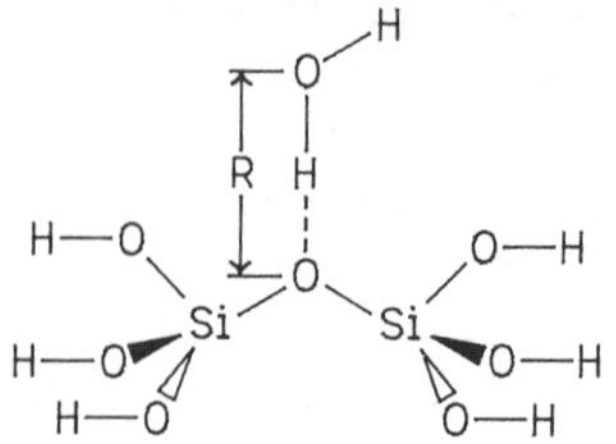

2

complex 2. The distances between the oxygen atoms of a water molecule and of a siloxane bond were optimized, giving $R_{OO} = 2.7$ Å and an adsorption energy of 4.0 kcal/mol. These results did not contradict the experimental data on the hydrophobic nature of a SiO_2 micropore (60).

Thus the calculation predicts the possibility of the chemisorption of H_2O on a bridged oxygen atom of SiO_2, though with a lower energy than in the case of adsorption on a terminal hydroxyl group. It should be noted in conclusion that more extensive quantum-chemical investigations of this chemisorption center are needed. First of all, the influence of the wide variation of the SiOSi angle on the chemisorption properties is of primary interest. Such variations, as mentioned above, actually occur in different forms of silicates. The lability of this structural element allows one to suppose that the geometry of this group on a real surface may vary considerably within the limits of the same structural form. Dehydroxylation under relatively soft conditions would likely favor such structural inhomogeneity, whereas annealing at elevated temperatures would, on the contrary, unify the surface structure.

C. On the Possibility of Coordinative Chemisorption on the Silicon Atom

The question of the possibility of coordinative chemisorption of molecules on a tetracoordinated silicon atom of SiO_2 is still extensively discussed in the literature. The assumption of the existence of such adsorption is sometimes used to explain various transformations of molecules on the surface of silica gel. Even this circumstance itself draws attention to this problem.

The hypothesis that molecules can be bound by a silicon atom to increase its coordination number (an idea originally formulated for water adsorption) was inspired by two experimental facts: high initial heats of adsorption on silicates (~ 20 kcal/mol) and only a weak perturbation of the stretching vibration bands of terminal OH groups with the inlet of water at early stages of the process (63). At the same time, an alternative explanation of these data is quite possible. High initial heats may be attributed to adsorption on structural defects, which is in accordance with the quantitative estimate of the amount of this anomalously fixed water ($\sim 10^{13}$ particles per cm^2). The absence of a perturbation of vibration bands of terminal OH groups may be merely caused by the preferential adsorption following the scheme (a) in Fig. 3.

Indirect evidence in favor of the coordinative binding is the existence of the hexacoordinated organosilicon compounds. However, a counterargument is that ammonia, possessing a greater ability for coordinative binding than water and having approximately equal molecular dimensions, has even a somewhat lower initial heat of adsorption on silica gel.

These examples of pro- and counterarguments may be supplemented by the results of quantum-chemical calculations, which are also far from always consistent with one another. The first computations of Pak et al. (64) and Dunken and Hoffman (59) were carried out using EHT and led to the conclusion that the coordinative binding between a water molecule and a silicon atom of SiO_2 should be strong enough, although particular values of the binding energies differed significantly (by more than factor of 10). However, taking into account known limitations of EHT in total energy calculations, one could hardly consider these results as conclusive in solving the problem under discussion.

A more rigorous approach was followed in the work of a group of Ukrainian authors (65–67), in which the computations of such complexes were performed using the Basilevsky–Berenfeld self-consistent perturbation theory as applied to CNDO/2 basis functions. According to Gorlov et al. (65), adsorption complexes of water with a silicon atom of a silanol groups appears to be the most favorable, with a bonding energy of 38–57 kcal/mol, depending on the structure of the complex. When discussing the conclusiveness of these calculations, however, some critical remarks should be made. The first one concerns the use of CNDO/2, which, generally speaking, is not adapted for the total energy computations. On the other hand, the question of whether the application of self-consistent perturbation theory for estimating the bonding energies would improve the computations should be carefully investigated.

The second point is the choice of the model clusters and their geometry. In (65), the broken boundary valencies were saturated by SiH_3^* fragments (H* means quasi-hydrogen atom). The chosen method of describing the clusters leads to a considerable overestimate of the SiO bond length (1.8 Å) as compared with experimental values (~ 1.63 Å). This would certainly facilitate coordinative bonding. In addition, the clusters had wrong charge stoichiometry in Si and O.

All these reasons have prompted us to perform nonempirical quantum-chemical calculations that could clarify the problem. Zelenkovskii (16) carried out such computations of adsorption complexes with H_2O and NH_3 molecules. Figure 9 shows two types of adsorption of water, where adsorption is considered either (a) on the face or (b) on the edge of a silicon–oxygen tetrahedron. The calculations involved a series of adsorption complexes with different distances R between the oxygen atom of water and a silicon atom and with various values of angle α describing the deviation from the tetrahedral configuration. STO-3G, STO-3G*, and 3-21G basis sets were used in these computations.

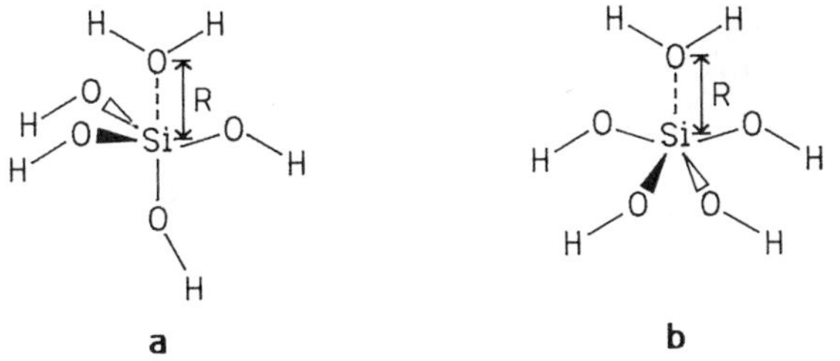

FIG. 9. Cluster model of the coordinative adsorption of a H_2O molecule on (a) the face and (b) the edge of a silicon–oxygen tetrahedron in SiO_2.

STO-3G calculations have revealed that the potential energy surface for (b)-type adsorption had no bonding region within the limits of $R = 2$–3 Å and $\alpha = 0$–30°. The calculated repulsion energy, for these ranges of R and α, varied from 7 to 86 kcal/mol. For (a)-type adsorption, the region of weak positive bonding was found at $R = 2.6$ Å and $\alpha = 0$. However, the maximal bonding energy was as low as 2.3 kcal/mol, and even a slight increase of α was accompanied by a transition to a repulsion region

STO-3G* computations were aimed at examining the role of the d atomic orbital of Si, and although they have led to somewhat different electron structures for adsorption complexes (e.g., the electron charge on Si decreased by $0.3e$), the binding energy remained practically the same. Thus, consideration of the d AO did not reveal any new qualitative effect.

As mentioned above, overestimated Si–O bond lengths used in the calculations might present a source of considerable increase in the energy of coordinative bonding of water by silicon atoms. In order to test this hypothesis, STO-3G calculation of the adsorption complex (a) in Fig. 9 was carried out, with the bond length being 1.84 Å, as in Gorlov et al. (65). The energy of the complex thus obtained actually increased from 2.3 kcal/mol ($R_{SiO} = 1.63$ Å) to 6.1 kcal/mol ($R = 1.84$ Å), that is, as much as by a factor of 3. Such a strong dependence implies that, to obtain a conclusive result, the geometry of the cluster should be chosen very carefully. It should be noted, however, that simultaneous elongation of four Si–O bonds from 1.63 to 1.84 Å would require about 90 kcal/mol, which exceeds many times the gain in the bonding energy.

To test the reliability of the results obtained, additional computations were performed on the adsorption complexes with extended $(HO)_3SiOSi(OH)_3$ and $Si(OH)_4 + H_2O$ complexes using the extended 3-21G basis set. The results appeared qualitatively quite similar. One could mention only some decrease (to 0.4 kcal/mol) of the energy of (a)-type coordination observed for the extended cluster. This may likely be ascribed to the increased repulsion from the "bridged" oxygen atom of the extended cluster caused by the

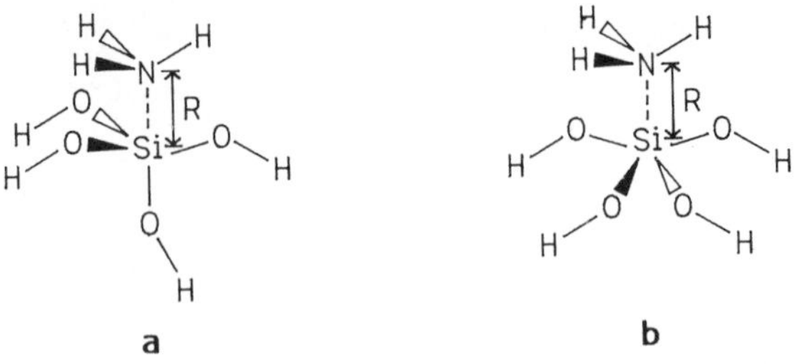

a   b

FIG. 10.   Cluster model of the coordinative chemisorption of an NH_3 molecule on (a) the face and (b) the edge of a silicon–oxygen tetrahedron in SiO_2.

increase of its negative charge by $0.1e$. This trend should certainly be retained when passing to larger (and more realistic) clusters.

In the calculations of the adsorption of NH_3, as in the case of H_2O, two types of adsorption complexes were considered. The computations were also carried out using STO-3G and STO-3G* basis sets for a range of values of R and α. The conclusions were essentially similar. Only for the (a) type structure was a region of weak positive bonding found, with $\Delta E = 4.3$ kcal/mol at $R = 2.2$ Å and $\alpha = 10°$. However, as in the case of water, the maximal energy of the coordinative bonding of NH_3 was four times lower than the energy of bonding with a terminal hydroxyl (Fig. 10).

Thus, ab initio calculations show that the coordinative binding by a silicon atom of a silanol group in the perfectly tetrahedral configuration is not characteristic of both water and ammonia adsorption. At least, such a bonding should be considerably weaker than the interaction with a terminal hydroxyl group. All this is in accordance with the results of ^{29}Si NMR studies of water adsorption on aerosols (68), where no increase of coordination number of Si was found. On the other hand, the above computations indicate that the cluster distortion can considerably increase the strength of the coordinative binding. Such distortions are most probable for amorphous silica; however, the number of such defects and the degree of distortion are unknown.

IV. Aluminosilicates

Aluminosilicates represent solid acids (42). In general, the nature of their acidity is quite clear and is associated with the chemical peculiarity of the substitution of a tetrahedral Si^{4+} ion by an Al^{3+} ion in the silicate structure. At the same time, the detailed structure of BASs and LASs in a particular system still remains the subject of experimental studies. Recently quantum-chemical methods have also come to find ever-increasing use in this field. To a greater degree they were applied to discuss the properties of crystalline aluminosilicates (zeolites) that were dictated by the extensive use of zerolites in catalysis. The other reason was that their structures were better investigated. Therefore, we shall initially consider the quantum-chemical simulations of active sites in zeolites and then turn to some properties of amorphous aluminosilicates.

A.   Crystalline Aluminosilicates

The original literature devoted to the structures and properties of zeolites was reviewed in Rabo (69). Therefore, in what follows we shall discuss only some aspects which are relevant to a quantum-chemical treatment.

The crystalline structure of the most intreresting forms of zeolites is now well investigated. It involves the primary structural units in the form of tetrahedra TO_4 occupied either by Si or Al atoms. The secondary units are constituted by the tetrahedra joined by their corners. Such units include 4, 5, 6, and 8-membered rings, hexagonal prisms, etc.

Brönsted acidity, characteristic of H forms of zeolites, is associated with bridged hydroxyl groups. Another type of center is represented by the cations in the cationic forms of zeolites. They are mainly located within 4- and 6-membered rings. Just these two types of sites have comprised the principal subject of quantum-chemical studies. They will be discussed below.

1.   Bridged Hydroxyl Groups

A simple cluster model of a bridged hydroxyl group in a zeolite is cluster **3**. Such a cluster with A = H was used by Chuvylkin et al. (70) as early as 1975 to discuss the properties of possible intermediate structures in the catalytic isomerization of butenes on aluminosilicate surfaces in terms of CNDO/2 approximation. Mikheikin et al. (34) have used a similar cluster with terminal pseudo-atoms A to study the Brönsted acidity of zeolites and its dependence on the Si/Al ratio.

3

a. Acidity of Bridged Hydroxyl Groups. Both CNDO/BW (34) and MINDO/3 (13) calculations for cluster **3** with pseudo-silicon atoms Si* as pseudo-atoms A have predicted a considerable increase in the acidity of a bridged OH groups as compared with a terminal OH group. The difference in E_{H^+} was about 2 eV, and the difference in the charges on protons was $\sim 0.1e$.

An important problem arising here is to realize the relative significance of different factors in affecting the acidity of such groups. These factors can be subdivided into structural ones and chemical ones. The structural factors represent various changes in the geometry of a bridged group, for example, variation of $Si_1-O(H)$ and $Al-O(H)$ bond lengths or angle α, as shown in structure **4**. The chemical factor consists of the change of the chemical composition of the nearest environment of the hydroxyl group, for example, the Si/Al ratio in the cluster. This could be done by the successive substitution of silicon atoms in the cluster by aluminum atoms.

4

In our view, the quantum-chemical cluster calculations unambiguously point to a relatively slight influence of the structural factors on the acidity of OH groups, whereas the chemical factor is of key importance. The results of the computations for cluster **4**, illustrating the effects of structural and chemical factors on the acidity of a bridged OH group, are listed in Table III. The range of structural changes ($\delta R \simeq 0.07$ Å, $\delta\alpha \simeq 5°$) approximately corresponds to the known scatter of structure parameters of the SiOHAl unit in faujasites. To obtain more conclusive results, the computations were carried out using both CNDO/BW and MINDO/3 techniques. Since vibrational infrared spectroscopy is the principal physical method of studying the hydroxyl groups, Table III also includes the values of the respective frequency shifts of the stretching (δv_{OH}) and bending (δv_{SiOH}) vibrations. It is essential that, contrary to such characterstics of acidity as E_{H^+}, vibrational frequencies are quite sensitive to the structural factors and almost independent of the chemical factor. This indicates that the infrared data on vibrations of free OH groups should be considered with extreme care when applied to the acidity of zeolites.

Based on the comparative CNDO/2 cluster calculations of six-membered rings in faujasites and five-membered rings in high silica containing ZSM-5

TABLE III
Influence of the Structural and Chemical Factors on the Properties of Bridged
Hydroxyl Groups (Cluster 4)

Factor	Method	ΔE_{H^+} (kcal/mol)	$\delta\nu_{OH}$ (cm^{-1})	$\delta\nu_{SiOH}$ (cm^{-1})
Structural				
$\delta R = 0.07$ Å	MINDO/3	4	41	-63
$\delta\alpha = 5°$	CNDO/BW	8	55	-77
Chemical	MINDO/3	26	1	11
Si → Al + H	CNDO/BW	43	15	22

zeolite, Beran (71) have suggested that the peculiar catalytic properties of H forms of ZSM-5 zeolites may be caused by OH groups of increased acidity. He has ascribed this increased acidity to the effects of the structural factor. Indeed, some structure units of ZSM zeolites are charaterized by the high values of the SiOAl angle (attaining 175°), whereas respective angles in faujasites fall within the considerably narrower range of 139–145°. However, if this hypothesis is valid, then, in accordance with the data presented in Table III, a strong shift of stretching vibration frequencies, ν_{OH}, would be expected for these OH groups. Meanwhile, the IR investigations of ZSM zeolites performed by Kazansky et al. (72) in the broad spectral range have shown that frequency characteristics of their OH groups were identical to those of bridged OH groups in mordenite. This seems to indicate that, in ZSM zeolites, the probability of the localization of a proton within the units with relatively large SiOAl angles is comparatively small. Moreover, the very hypothesis of specific acidic properties of bridged hydroxyl groups in ZSM zeolites seems questionable, since the comparison of the shifts of OH stretching vibration frequencies, $\delta\nu_{OH}$, caused, in ZSM zeolites and mordenite, by hydrogen binding with various base molecules, has shown that the acidities of OH groups in these zeolites are quite close (72).

In our opinion, the main factor which governs the acidity of bridged OH groups in zeolites is the chemical one. If the local nature of Brönsted sites is taken into account, the following rational classification can be proposed for the bridged OH groups of zeolites with regard to their acidity and Si/Al ration in the framework (34).

According to Loewenstein's rule (73), two aluminum–oxygen tetrahedra could never be in neighboring positions because of the strong Coulomb repulsion. Therefore, the following four possible types of hydroxyl groups can be specified for a hydrogen form of a zeolite, depending on the number of surrounding aluminum atoms:

OH-I   OH-II

OH-III   OH-IV

Different chemical environments will result in different acidities of these groups.

To estimate qualitatively the trend in the variation of the acidities in this series, it is convenient to make use of the CTP technique and to change successively the VOIPs of pseudo-atoms entering the left-hand side of the cluster 3 from 10.68 eV (pseudo-silicon) to 7 eV (pseudo-aluminum; cf. also Section II,B). The results of such CNDO/BW calculations are collected in Table IV (where q is the electron charge). In a qualitative agreement with experimental data, the calculations predict the decrease in the acidity of bridged OH groups with increasing number of Al atoms in the nearest environment of these OH groups (i.e., with decreasing the Si/Al ratio).

Each of the above four types of cluster representing the bridged OH groups has its specific chemical composition. Therefore, the ordered crystal lattice of a zeolite composed of clusters of a single type should have both uniform Brönsted acidity and a definite Si/Al ratio. For example, a perfect chain structure including only repeated clusters of the first type has a $[SiAlSi_3]_n$

TABLE IV

Results of Quantum-Chemical Calculations of the Parameters of Structural Hydroxyls in Zeolites (CTP, CNDO/BW)

Type of OH group	Number of surrounding Al atoms	Si/Al ratio	q_H	q_O	q_{OH}	E_{H^+} (eV)
OH-I	1	7	0.404	−0.870	−0.466	14.96
OH-II	2	3	0.392	−0.873	−0.481	15.32
OH-III	3	1.7	0.384	−0.884	−0.500	15.60
OH-IV	4	1	0.376	−0.883	−0.507	15.89

composition with Si/Al = 4. For a three-dimensional structure of similar type, Si/Al = 7. These figures are certainly approximate and could change somewhat depending on the configuration of the cluster connection in a real crystal lattice. The clusters could also be separated by additional tetrahedra occupied by Si. However, it is evident that these figures determine some critical Si/Al ratio beyond which **OH-I** groups should predominate. Therefore, one might expect that only **OH-I** groups would mainly exist in high-silica containing zeolites. It is also clear that the probability of finding only the structural hydroxyls of the first type increases with the Si/Al ratio.

In a similar way, the composition of structures constituted only of clusters of the second type is $[AlSi_3AlSi_3]_n$, with Si/Al = 3. Such a Si/Al ratio is characteristic of Y-type faujasites. Therefore, the predominance of **OH-II** may be anticipated for these structures.

For zeolites composed only of clusters of the third type, $[Al_2Si_2AlSi_3]_n$, the ratio Si/Al is ~ 1.7. This is just the border between Y- and X-type faujasites. Therefore, for low-silica-containing Y zeolites and X zeolites, one might expect the presence of both **OH-III** and **OH-II** groups. The former should predominate in X-type zeolites.

Finally, the structure composed of clusters of the fourth type, $[Al_3SiAlSi_3]_n$, has Si/Al = 1. Therefore, the H forms of A-type zeolites would contain only **OH-IV** groups. This is, however, only a hypothesis, since the H forms of A zeolites are unstable.

Hence, the hydrogen forms of zeolites should contain only several definite types of structural hydroxyls. In addition, each of them should predominate for a distinct crystal structure and chemical composition (Si/Al ratio). The exceptions include high-silica-containing zeolites, where the presence of only **OH-I** groups should be anticipated.

Consider now the stereoisomerism of structural hydroxyls. For this purpose it is more convenient to take a cluster containing a silicon atom at the central position. Then only one stereoisomer would exist for **OH-I**, with a proton migrating over three equivalent positions. For the cluster with two Al atoms in the first coordination sphere, two stereoisomers, **OH-II** and **OH-II'**, are already possible:

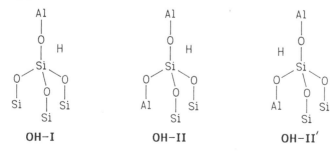

The proton of **OH-II'** is additionally hydrogen bonded to the neighboring basic oxygen. Therefore, its stretching vibration frequency, ν_{OH}, should be lower and its bending vibration frequency higher than those of **OH-II**. On the other hand, ν_{OH} for **OH-II** should be higher than that for **OH-I** (Table III). For the clusters containing three Al atoms, there are still two similar possibilities, **OH-III** and **OH-III'**, with the proton localized between Si and Al or betweeen two Al atoms. Thus, since the H forms of A zeolites are unstable, the existence of the following five types of structural hydroxyls may be expected, with acidity decreasing in the given sequence: **OH-I**, **OH-II**, **OH-III** and **OH-II'**, **OH-III'** for each subset.

In the papers by Kazansky (74–76), this classification was compared with IR spectra of hydrogen and cationic forms of different zeolites. The analysis of the experimental data revealed that for high-silica-containing zeolites with Si/Al ≥ 4 (mordenite, ferrierite, omega zeolite, clinoptilite, and ZSM zeolites), only one type of structural hydroxyl was actually observed, with the fundamental frequency of stretching vibrations being about 3610–3620 cm^{-1} (77–81). Kustov (72) has attempted special studies of high-silica-containing zeolites with widely varied Si/Al ratios (ranging from 4 to 125). The IR diffuse reflectance technique was used for this purpose. As compared to the traditional transmission technique, it possesses a greater sensitivity and also widens the range of spectral measurements that allow simultaneous recording of the fundamental, overtone, and compound frequencies of OH stretching plus bending vibrations. These investigations have also confirmed the identity of the structural hydroxyls in all high-silica-containing zeolites.

The hydroxyl groups of the first kind are already absent in faujasites, quite in accordance with the cluster model predictions. Instead, two other bands with fundamental stretching vibration frequencies of 3640 and 3650 cm^{-1} were observed in Y zeolites. They were assigned to **OH-II** groups located within the large cavities (3650 cm^{-1}) and to **OH-II'** groups positioned inside the hexagonal prisms (3555 cm^{-1}) (82).

It was also noted that the stretching vibration frequencies of these groups somewhat increased (by 20–30 cm^{-1}) with increasing content of aluminum in a faujasite lattice (83). From the cluster quantum-chemical viewpoint, this should be attributed to the effects of the aliminum atoms of the second and next coordination spheres of a cluster.

In accordance with the above discussion, **OH-III** and **OH-III'** groups should be expected to be present in X zeolites, along with **OH-II** and **OH-II'**. Their vibration bands, although unresolved in the fundamental region, appeared quite distinct in the region of compound stretching plus bending vibration frequencies. This allowed the assignment of the characteristic **OH-III** frequencies to be made as in Table V. Table V also lists the parameters of other kinds of hydroxyl groups which are characteristic of different types of

TABLE V
Positions of the Maxima of Bands in the IR Spectra Characteristic of Structural
Hydroxyls of Different Types

Type of OH group	Si/Al ratio	$\omega_{0\to1}$ (cm^{-1})	$\omega_{0\to2}$ (cm^{-1})	$\omega_{v+\delta}$ (cm^{-1})	Systems where the corresponding OH groups were observed
SiOH	—	3745	7325	4550	Silica gel, all zeolites ZSM,
OH-I	7	3610	7060	4660	mordenite, clinoptilolite, omega
OH-II	3	3640	7125	4660	Omega, erionite, Y zeolites
OH-II'		3555	6970	4610	
OH-III	1.7	3660	7155	4690	X zeolites
OH-III'		3580	7005	4580	

zeolites. In accordance with the cluster simulations, both **OH-I** and **OH-II** groups were observed in the transition region from the high-silica-containing zeolites to Y-type zeolites (erionite, dealuminated faujasites).

Recently Jacobs (*84, 85*) has proposed another system of structural hydroxyls in zeolites based on the mean electronegativities of zeolite lattices according to Sanderson. It was concluded as well that the Brönsted acidity of zeolites varies continuously with their Si/Al content. Let us compare these two approaches.

It is quite evident that they are based on the same physical model that an alumino–oxygen tetrahedron in a zeolite lattice comprises a strong donor of electron density, which results in a considerable decrease in the acidity of hydroxyl groups with increased aluminum content due to their negative charging. It should be noted, however, that both of the models represent extreme cases, since one overestimates the local properties and the other overestimates the cooperative electron properties of zeolites. Although the possibility of aluminum affecting the acidity of hydroxyl groups through change in the electronegativity of a lattice as a whole cannot be completely excluded, we believe that our local model is more realistic and provides a better description of the acidic properties of zeolites at the atomistic level. In our opinion, the macroscopic effects of the changes in the electronegativity would be mainly manifested in fine distinctions in the acidic properties of the principal kinds of OH groups and in relatively small shifts of the stretching vibration frequency. For instance, its value for **OH-II** groups increases from 3640 to 3660 cm^{-1} when passing from Y to X zeolites. Let us also emphasize that the average electronegativity of the entire zeolite lattice cannot account for the simultaneous existence of several types of hydroxyls of different acidities.

The distinction arises between the two models only in the case of high-silica-containing zeolites, whose Brönsted acid sites are identical in our model, whereas the electronegativity approach predicts that their strength in ZSM zeolites should be higher than in mordenite and then should increase upon further increasing the Si/Al ratio. This contradicts, however, the above-mentioned IR data and the results of the ammonia adsorption heat measurements on high-silica-containing zeolites (86). Moreover, the long-range effects seem rather unlikely in such insulating oxides as zeolites.

b. *Chemisorption of Simple Molecules.* As already mentioned above, the aim of quantum-chemical calculations of chemisorption is to estimate the relative priorities of the adsorption states, their geometry, and their energetics. The dependence of these characteristics on the properties of adsorption sites (e.g., their acidity) is also of interest. In addition, the changes in the electron structure of reactants caused by chemisorption could provide information on the mechanisms of the surface catalytic reactions.

The chemisorption calculations require rather improved computational techniques, especially if sufficiently accurate structure predictions are needed. In this respect the nonempirical approach is no doubt preferential; however, the accessible size of a model cluster is rather small in this case. Of the semiempirical methods, as mentioned above, CNDO/BW seems the most suitable (and quite widely used) for calculations of those adsorption states in which the hydrogen bond may be of essential importance. Similar in its capacities, the MINDO/3 method fails to describe the hydrogen bonding. Of certain advantage here is MNDO. Its parametrization, however, is far from being perfect and complete. CNDO/2 and EHT methods are rather inappropriate for the total energy calculations. Therefore, the most reliable information provided by these methods is that concerning the electron structure (charge distribution, qualitative characteristics of orbital bonding, etc.). Thus, any attempt at direct energy calculations should necessarily be supplemented by the analysis of the validity of the method applied to a particular problem or a specified system. Unfortunately, this is usually neglected.

Let us first consider the adsorption of water, which represents a classical chemisorption problem of catalysis. In the paper by Senchenya *et al.* (52), the CNDO/BW version of a CTP scheme was used to treat some forms of water nondissociative adsorption on a structure unit containing a bridged hydroxyl group. The effects of successive variation of the acidity of the hydroxyl group were studied at the model level by means of introducing, at the boundary of the cluster (Fig. 11), the pseudo-atoms B ($VOIP_B = 7\,eV$) modeling aluminum atoms in their oxygen environment in place of the pseudo-atoms A ($VOIP_A = 10.68\,eV$) modeling the silicon atoms in the corresponding environment. These calculations were inspired by the fact that a rather wide range of adsorption heats was observed for water chemisorption of zeolites (69).

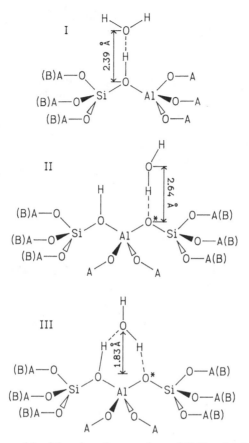

FIG. 11.   Cluster models of the adsorption complexes of H_2O molecules on the acid–base sites of a structure unit containing a bridged hydroxyl group, showing one-center (I, II) and two-center (III) adsorption forms.

The results of such calculations would be useful in judging both calorimetric and thermodesorption data for these systems.

Within the framework of the clusters proposed previously, one can visualize three types of water molecule adsorption, as shown in Fig. 11. In Mikheiklin *et al.* (*87*) we examined types I and III adsorption for the case of a cluster that does not contain pseudo-atoms B. It was found that the energy of two-point adsorption according to type III actually decreased sharply (from ∼25 to 9.4 kcal/mol) if the cluster contained the nonplanar fragment and Si(OA(B))$_3$ groups were turned in different directions around the AlO axes at an angle of about 21°, in accordance with the real structure of the zeolite. Therefore, all the data listed in Table VI for the cluster consisting of three tetrahedra were obtained using such "unwound" clusters.

In calculations of the interaction energy of the structures shown in Fig. 11, the parameter R was varied to minimize the total energy of a system. In models I and II the standard geometry of the water molecule was retained, whereas in structure III the two "lower" hydrogen atoms were placed in a position corresponding to the surface OH groups, and the location of the OH group of water was varied along the z axis at constant $R_{OH} = 0.96$ Å (the values of R corresponding to the minimal total energy of a structure with no pseudo-atoms B are given in Fig. 11).

Let us now return to the results of the computations.

Model I. The introduction of pseudo-atoms B into the left-hand part of the cluster leads to a decrease in the acidity of the OH group and to a gradual reduction of the positive charge on its proton. This, in turn, is accompanied by a decrease in the intraction energy of a water molecule with the cluster from ∼ 11 kcal/mol, when the cluster contains no pseudo-atoms B (clusters 1, 3), to 6.3 kcal/mol, when it includes three such pseudo-atoms (cluster 12 of Table VI).

Model II. Introduction of pseudo-atoms B into the left-hand part of the cluster lowers the electon charge on the O* atoms, whereas their introduction into the right-hand part increases the charge. The simultaneous introduction of pseudo-atoms B both into the left- and right-hand parts of the cluster also increases the charge on the O* atom. Sympatric with the electron charge on the O* atom is the interaction energy of a water molecule with the cluster that varies from 14.1 to 21.0 kcal/mol (cf. the sequence of clusters 10, 8, 1, 2, 4, and 13 in Table VI).

Model III. The interaction of this type appears to be extremely sensitive to changes in the acidity of the OH group as well as to changes in the charge state of the O* atom caused by the introduction of pseudo-atoms modeling aluminum–oxygen tetrahedra on the boundaries of the clusters. The data listed in Table VI enable one to draw the following conclusions. The energy of the two-point interaction of H$_2$O with a cluster in accordance with model III can vary within a wide range and may either increase or decrease with the change of the local content of aluminum. Calculations also showed that the basicity of the O* atom had a greater influence on the energetics than did the acidity of the OH group. Simultaneous weakening of the OH acidity and O* basicity resulted in a sharp decrease of the interction energy in two-point adsorption.

TABLE VI

Results of Calculations (CTP, CNDO/BW) of the Energetic Characteristics for Different Forms of Water Molecule Adsorption on Acid–Base Sites of H Forms of Zeolites as a Function of the Strength of the Acid site[a]

No.	Number of tetrahedra	Number of pseudo-atoms B in cluster Left-hand side	Number of pseudo-atoms B in cluster Right-hand side	E_{H^+} (eV)	q_H, $|e|$	q_O, $|e|$	Q (kcal/mol) according to type I	Q (kcal/mol) according to type II	Q (kcal/mol) according to type III
1	3	0	0	14.88	0.3996	−1.1782	11.81	16.7	9.44
2	3	0	1	15.01	0.4002	−1.1845	—	19.6	13.05
3	2	0	—	15.10	0.4040	—	10.17	—	—
4	3	0	2	15.15	0.4007	−1.1914	—	21.0	16.32
5	3	1	0	15.24	0.3967	−1.1756	—	15.8	7.66
6	3	1	1	15.38	0.3972	−1.1821	—	—	11.23
7	2	1	—	15.47	0.3959	—	8.41	—	—
8	3	2	0	15.60	0.3890	−1.1729	—	15.1	5.90
9	2	2	—	15.83	0.3883	—	7.34	—	—
10	3	3	0	15.96	0.3913	−1.1700	—	14.1	4.08
11	3	2	2	16.01	0.3901	−1.1868	—	—	12.74
12	2	3	—	16.20	0.3801	—	6.28	—	—
13	3	3	3	16.37	0.3826	−1.1913	—	20.89	—

[a] The clusters are presented in Fig. 11.

Thus, calculations demonstrate that water adsorption on the H forms of zeolites (the low degrees of surface coverage are certainly considered) should occur preferentially on the lattice oxygen (model II). Also, there may exist adsorption states corresponding to models I and III. The sum of the results presented indicates that, even for zeolites with uniform bulk distribution of aluminum, a rather wide spectrum of heats of water adsorption may occur. The conclusion about the preferential form of adsorption II is supported by the data of IR spectroscopic studies (69) indicating the formation of hydrogen bonds between the hydrogen atoms of a water molecule and the oxygen atoms of a zeolite framework.

The data presented in Table VI enable us to discuss in more detail the energetics and mechanism of H/D exchange in water involving the surface OH groups as well as to judge the influence of traces of water on the diffusion of hydroxyl-group protons of zeolites. Actually, they could be used to construct the energy diagram corresponding to these reactions on the assumption that they follow the synchronous mechanism. Moreover, if it is adopted that the initial and final state comprise the adsorption complexes of type II while the complexes of type III act as intermediates, it can be seen that both reactions (H/D heteroexchange in water and the diffusion of the acidic protons in a zeolite framework) proceed, as shown by computations, with activation energies not exceeding ~ 10 kcal/mol.

It could also be concluded that the protons of bridged OH groups should tend to occupy the positions corresponding to the maximal number of aluminum atoms surrounding the lattice oxygen atoms. In other words, less acidic OH groups are preferentially formed. The rate of H/D heteroexchange should likely be limited by the desorption process rather than directly by the proton transfer. As regards the proton migration over four lattice oxygen atoms surrounding Al atoms, the direct calculations have revealed that the activation energy of such a process in the absence of water molecules should be as high as 40–45 kcal/mol, that is, the traces of water and possibly ammonia in a sample (55) may substantially influence this reaction in a catalytic manner. This is in agreement with the results of NMR studies (88).

The optimization of the bond length R_{OH} in adsorption complex I gave no considerable effect. The increase of OH distance, ΔR_{OH}, was about 0.05 Å, with the corresponding change in the adsorption energy being only 0.5 kcal/ mol. Thus even rather acidic bridged OH groups show no marked trend toward proton transfer to a water molecule to form oxonium ion, H_3O^+. To check this conclusion, the water adsorption on the $(HO)_3SiOHAl(OH)_3$ cluster was recalculated nonempirically by Zelenkovskii et al. (62) using a STO-3G basis set. Initially the distance R_{OO} between the oxygen atoms of the bridged OH group and of a water molecule (adsorption form I, Fig. 11) was optimized, and then the distance R_{OH} was optimized at a fixed R_{OO} of 2.5 Å.

This value of R_{OH} proved to be unchanged within the limits of the computational errors. Thus the *ab initio* calculation leads to the same conclusion as the semiempirical one, although the nonempirical adsorption energy (20.4 kcal/mol) is substantially higher than the semiempirical one (cf. Table VI). In our view, this is connected with the fact that H atoms were taken as the terminal atoms in the cluster rather than with the roughness of the CNDO/BW method. This leads to a higher acidity for the OH group than would be expected in the system with terminal pseudo-silicon atoms A. In turn, this increases the adsorption energy. It is also significant that even in this case of a more acidic OH group, there is still no reason to consider adsorbed H_3O^+ ion formation.

In the context of discussing the problem of proton transfer to the adsorbed molecules, it seemed of interest to discuss the adsorption of a strongly basic molecule. Ammonia represented a convenient example. Its adsorption was investigated by Senchenya *et al.* (*55*) using the CTP scheme and the CNDO/BW method. The same cluster and the same two types of adsorption complexes I and II (Fig. 12) were considered as in the case of water adsorption. Structure I appeared more stable (contrary to the case of H_2O). Its energy was 37.9 kcal/mol at optimal bond lengths $R_{ON} = 2.29$ Å and $R_{OH} = 1.1$ Å. In this case, the trend toward proton transfer from the hydroxyl group is much more distinct, and essentially the proton can be certainly ascribed to neither the surface hydroxyl nor the ammonia molecule. Total positive charge on the NH_4 group is rather high ($\sim 0.7|e|$). This makes it similar to a free NH_4^+ ion; however, the question of how close are the vibrational characteristics of these ions needs special consideration. Note also that the calculated adsorption energy for state I is close to the experimental values estimated from the initial heats of NH_3 adsorption on HM zeolites (*89*) ($\gtrsim 30$ kcal/mol).

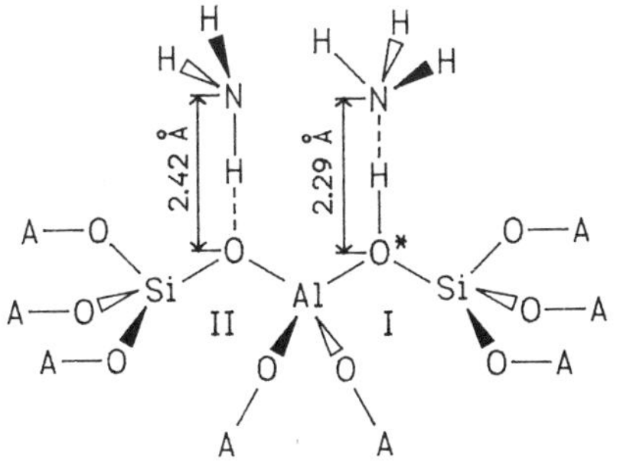

FIG. 12. Cluster models of one-center adsorption complexes of NH_3 molecules on the acid–base sites of a structure unit containing a bridged hydroxyl group. I, 37.9 kcal/mol; II, 19.9 kcal/mol.

As seen from the above examples, one should not overestimate the capacity of a bridged OH group to protonate an adsorbed molecule. Even in the case of such a strong base as ammonia, it would be better to talk about local surface structure with charge transfer than about relatively free ammonia ion. This is even more the case for the adsorption of organic molecules (olefines, alcohols) possessing lower basicities. Carbonium ion forms should likely be considered here only as excited structures.

Calculations of the adsorption of organic molecules on active sites of zeolites, including the structural hydroxyls, are still very limited. A wide field of applications opens here for the quantum-chemical methods.

Lumpov *et al.* (*90*) have calculated (CTP; CNDO/BW) the adsorption of propylene. Figure 13 shows the model of propylene adsorption on a cluster. The C–C–C plane was perpendicular to the z axis. The distances R, r_1, and r_2 were optimized. As expected, r_1 slightly decreased (from 1.54 to 1.52 Å), whereas r_2 somewhat increased (from 1.39 to 1.42 Å), as compared to the respective values in a free propylene molecule. This points to the trend to equalize the lengths of single and double bonds in such an adsorption complex. The adsorption energy amounted to 26.7 kcal/mol at $R = 2.4$ Å. Similarly, the carbocation and carbanion adsorption states were treated with complete proton transfer to propylene and cluster, respectively (Fig. 13). The corresponding binding energies were 14.1 and 15.8 kcal/mol. Thus the ionic states with complete proton transfer are activated with respect to the covalent adsorption state.

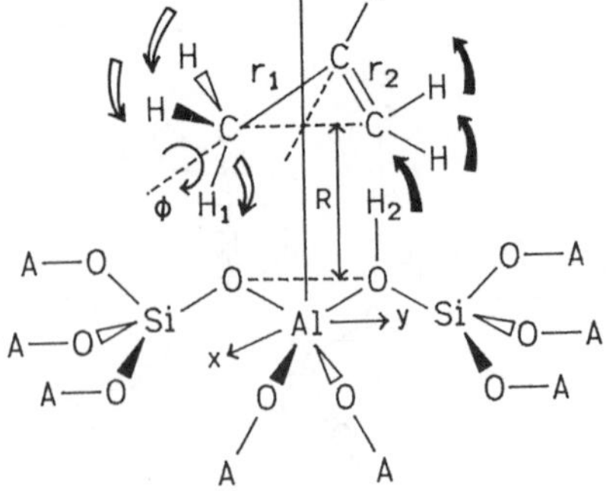

FIG. 13.   Model of propylene molecule adsorption on clusters containing a bridged hydroxyl group. The arrows indicate the atom motions corresponding to different mechanisms of the double-bond migration (black arrows, carbocation mechanism; white arrows, carbanion mechanism; in the case of the synchronous mechanism these motions are simultaneous).

Another interesting conclusion is that concerning the closeness of acidic and basic characteristics of the adsorption sites under discussion, if they are considered as acid-base pairs. The base is here represented by the lattice oxygen. This implies a possible efficiency of the synchronous mechanism for the double-bond migration reaction that is actually substantiated by direct computations (90). Experimentally, the synchronous mechanism can be recognized by its relatively weak dependence on the acidity of OH groups, distinguishing it from the carbocation mechanism of isomerization, which should be strongly affected by this factor.

On the contrary, similar EHT calculations of possible intermediate structures in the isomerization reaction (also exemplified by propylene) (91) predicted the stabilization of the carbocation on a surface that contradicted the experimental data (92). This was likely due to the limitations of EHT as applied to the total energy computations, especially of the charged forms. Essentially the same conclusion, that EHT overestimated the stability of the cationic form of propylene, was drawn by Schliebs et al. (93), who compared different mechanisms of the double-bond migration in zeolites using the EHT calculations for the cluster composed of four $Si(Al)O_4$ tetrahedra.

The interaction of ethylene and propylene molecules with the structural hydroxyl groups of zeolites was also computed by Beran et al. (94) using CNDO/2 for the closed $Si_3AlO_4H(OH)_8$ cluster modeling a four-membered window in faujasites. It was concluded that the adsorption form resulting from the interaction of the π bond of ethylene with a OH group was preferential. In essence, this adsorption form was the same as those considered in the works of Lumpov et al. (90, 91, 93) and Chuvylkin et al. (70, 95). The bonding energies for ethylene and propylene were predicted to be close and equal to ~ 13–15 kcal/mol, which was somewhat lower than the values obtained in CNDO/BW calculations (90).

Senchenya et al. (96) have treated the adsorption of ethanol on a structural hydroxyl group (Fig. 14) using a CTP scheme and the CNDO/BW method. The separation of a molecule and cluster with respect to the z axis was optimized, the optimal values being $r = 1.19$ Å and $R = 1.28$ Å The adsorption energy was 23.2 kcal/mol, which was close to the experimental value (97). Note that this was essentially the two-point adsorption involving both acid and base sites. This case is quite similar to the above propylene adsorption (90). There is also no definite trend toward proton transfer from the hydroxyl group of a zeolite to the alcohol molecule. The carbocation state is also predicted to be activated. This, in turn, increases relative efficiency of the synchronous mechanism (with the same recommendation for its experimental examination). The estimation (96) of the energetics of the intermediate structures of the synchronous mechanism showed that such a mechanism is quite realistic.

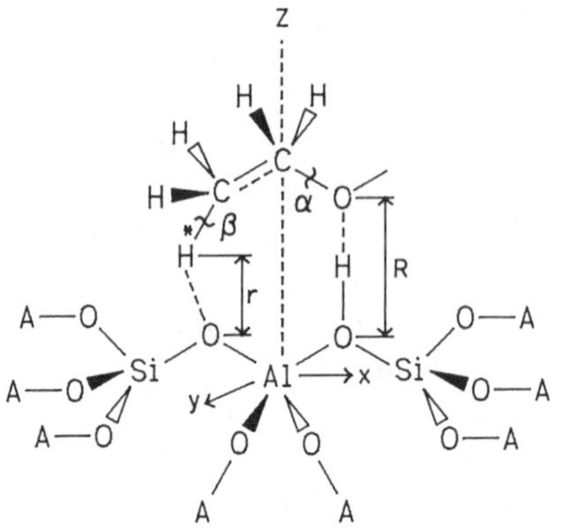

FIG. 14.   Cluster model of the adsorption complex of ethanol on a bridged hydroxyl group.

To conclude this section, it should be noted that the calculations of the potential energy surfaces for heterogeneous catalytic reactions, even by semiempirical methods, still remain a matter for the future. Insufficient accuracy of the semiempirical methods, the approximate nature of cluster modeling, the large volume of a configurational space, a variety of possible reaction paths, etc., considerably restrict the utility of quantum chemistry as applied to this field. There is, however, no doubt that these difficulties will be successfully overcome. The value of conclusive quantum-chemical calculations can hardly be overestimated. They are able to answer questions which the most sophisticated and refined experiments would fail to answer.

2. Cationic Forms of Zeolites

The cationic forms of zeolites are experimentally investigated rather well by various spectroscopic techniques, including IR, ESR, NMR, and UV–VIS spectroscopy. Structural interpretation of the spectroscopic data requires quantum-chemical computations of the optical and magnetic properties of $3d$-element ions in different environments. For this purpose, Mikheikin et al. (98, 99) have applied a simple crystal-field theory to discuss the optical and magnetic properties of ions with a d^n shell ($n = 1, \ldots, 9$) in the crystal field of low symmetry. The results could be used to discuss optical and ESR spectra of faujasites containing Cu^{2+}, Co^{2+}, Cr^{3+}, Fe^{3+}, Ni^{2+}, and Mn^{2+} ions and their complexes with various molecules (H_2O, NH_3, CH_3OH, C_2H_5OH, etc.).

The following questions are usually dealt with so far as the cationic forms of zeolites are concerned: the positions of the cations in the zeolite framework, the comparative energetics of the localization of different cations, the nature of bonding between a cation and a zeolite framework and its influence on the electron structure and properties of the framework, and chemisorption properties of low-coordinated cations and their influence on the stability of zeolites. To solve these problems, a number of approaches at quite different levels are used, Bosacek and Dubsky (100), for instance, have combined a simple atomic scheme with an electrostatic approach to calculate the adsorption of Ar, Kr, Xe, and N_2 on Na and K forms of Y zeolites. According to Proynov et al. (101), the electrostatic approach supplemented by consideration of some characteristics of the electron structure of transition-metal ions also provides a quite satisfactory interpretation of the peculiarities of the localization of Cr^{3+}, Ni^{2+}, Cu^{2+}, and Fe^{3+} ions within the Y-type zeolites. However, only quantum chemistry allows one to solve these problems properly.

The first direct quantum-chemical calculations used too simplified cluster models, sometimes as simple as a single metal ion. Such an approximation is obviously too poor for chemisorption computations. Thus, comparative calculations by Sauer et al. (54) have shown, for instance, that the neglect of the environment of a cation may result in overestimating the interaction energy by several times.

A series of papers by Beran et al. (102–112) was devoted to quantum-chemical studies of the cationic forms of zeolites using cluster models and the CNDO/2 technique. As a rule, cyclic clusters consisting of four or six TO_4 (T = Si, Al) tetrahedra were considered as models for four-membered (S_{III} sites) and six-membered (S_I, S_{II}, and $S_{II'}$ sites) oxygen windows in the zeolite structures.

The charge distribution in sodium forms of faujasites was investigated by Beran and Dubsky (102, 103). A fragment of the potential surface corresponding to the localization of Na^+ ion near two four-membered windows of X- and Y-type zeolites (S_{III} sites) was calculated, and the estimated activation energy for the ion migration between two equivalent positions was found to be ~ 5 kcal/mol.

One general remark, however, should be made concerning most of the comparative cluster computations of the cationic forms of zeolites. When considering clusters with different net charges, the charge distributions, bond orders, and total energies should be compared with extreme care, especially if the Madelung field is neglected in the calculations. For example, it was found in the above-mentioned works (102, 103) that the charge on a sodium ion located at S_{III} site ($0.56|e|$) was considerably higher than that on a sodium ion at a $S_{I'}$ or S_{II} site ($0.25|e|$). The authors attributed this unexpected result to the difference in the bonding of sodium with oxygen atoms of four- and six-membered rings. However, this is hardly the case. Rather, it originates from the incorrect comparison of the results obtained for the clusters with different net charges.

The cluster calculations for Li^+, Na^+, and K^+ ions in six-membered windows ($S_{I'}$ and S_{II} sites) were performed by Beran (104). It was concluded that in this series the properties of a zeolite framework (charge distribution, bond orders, Lewis acidity or basicity as characterized by LUMO and HOMO energies) only slightly depend on the type of cation. The decrease of water adsorption heats in this sequence was explained by the assumption that the strength of the water–cation interaction correlates with the strength of the interaction between a cation and lattice oxygen atoms.

Similar model clusters $T_6O_6(OH)_{12}$ were used in Beran (105) to compute atomic charges, bond orders, etc. of faujasites with different Si/Al ratios containing Ca^{2+} or $CaOH^+$ ions at S_{II} and $S_{I'}$ sites. Comparison of the results of calculations for clusters with different charges ranging from $+2e$ to $-2e$ was also carried out. The bonding energy between Ca^{2+} ion and zeolite framework was found to be of an essentially ionic nature. The authors concluded that a OH group bound with a Ca^{2+} ion should be more basic than the structural hydroxyls of zeolites.

Similar computations were carried out by Beran (106) for Mg^{2+}- and $MgOH^+$-containing zeolites. The results were compared with those obtained for Ca-containing zeolites. The distinctions were mainly ascribed to the stronger electron-acceptor properties of magnesium as well as to stronger electrostatic fields in the Mg-containing zeolites. In both cases the water–cation interaction was predicted to be rather weak.

Similar cluster calculations performed by Beran et al. (107, 108) for zeolites containing Al^{3+}, $AlOH^{2+}$, and $Al(OH)_2^+$ ions with different Si/Al ratios have shown that these cations located at S_{II} and $S_{II'}$ sites should possess quite strong Lewis acidity decreasing in the sequence $Al^{3+} > AlOH^{2+} > Al(OH)_2^+$. This was indicated by the LUMO energies of the calculated clusters. It was also concluded that the introduction of these cations in zeolites can substantially increase the stability of their structures. The bonding of OH groups with aluminum ions was found to be of a basic nature, in contrast to the case of structural hydroxyls.

Zeolites containing $3d$ transition-metal ions were considered in Beran et al. (109–112). The peculiarities of the donor–acceptor interactions of these cations located within six-membered rings with a zeolite lattice were discussed in terms of atomic charges, bond orders, and orbital energies. The redox properties of the cations, the acid–base properties of zeolites, and the dependence of these characteristics on the Si/Al ratio were discussed as well. The authors noted that the forms containing univalent copper and nickel ions should possess the highest electron-donor ability and consequently the highest oxidizability. All other forms of zeolites should possess extremely low electron-donor ability. The values of E_{LUMO} show that the system containing Fe^{2+}, Co^{2+}, Ni^{2+}, and Ni^+ ions should have the lowest Lewis acidity, whereas Al^{3+}- and Co^{3+}-containing zeolites should act as strong Lewis acids. These results are in good agreement with the common view of the electron-donor properties of the systems considered.

To conclude the discussion of the works of Beran et al. (102–112), let us note again that the comparison of the results obtained for clusters with different net charges seems rather questionable. In particular, it is not clear which real situation corresponds to the models where multicharged metal ions (Cu^{2+}, Ni^{2+}, Fe^{3+}, Fe^{2+}) are located near a six-membered oxygen window containing only one or no aluminum–oxygen tetrahedron. It is quite evident that such a situation should be extremely unfavorable energetically as compared with the case of sites with several neighboring aluminum–oxygen tetrahedra, whose negative charge would stabilize the cations. It is also not clear whether one can consider, say, the localization of a single cation with the charge of $+1e$ or $+2e$ near a ring containing three aluminum–oxygen tetrahedra. It seems likely that in a real situation the excess negative charge of such a system would be compensated by additional protons or other cations. The inclusion of such ions in the cluster calculations would significantly change their results and the conclusions drawn on their basis.

Simultaneously with the above-mentioned works, Mortier et al. (113, 114) have performed similar CNDO/2 calculations of clusters $Si_6O_{18}H_{12}$, $(Si_6O_{18}H_{12}Mg)^{2+}$, $(Si_3Al_3O_{18}H_{12})^{3-}$, and $(Si_3Al_3O_{18}H_{12}Mg)^-$ modeling S-I' sites in magnesium-containing faujasite-type zeolites with different Si/Al ratios. The charge distributions (calculated both in initial and deorthogonalized basis sets), MO energies, and bond orders were found. Then, the obtained charge distributions were used to compute, in CNDO/2 approximation, so-called molecular electrostatic potentials (MEPs), and the peculiarities of the adsorption of CO and CO_2 on magnesium-containing X- and Y-type zeolites were discussed in terms of these MEPs. The results obtained showed that isomorphic substitution of silicon by aluminum only slightly influences the charge state of the lattice oxygen. However, it affects considerably the charge state of the silicon. In the author's opinion, this accounts for the distinctions in the behavior of K_β emission bands for Si and Al observed for zeolites with different Si/Al ratios.

Similar computations were carried out by Mortier and Geerlings (114), both for zeolites containing Na^+ or Mg^+ ions at S_{II} and S_{III} sites and their H forms. It was concluded (see also Section IV,A,1a) that the effect of the geometry of the SiOHAl fragment on the acidity of structural hydroxyls is negligible as compared to the influence of the chemical composition and crystal environment.

3. *On the Structural Stability of Zeolites*

An important characteristic of zeolites, determining to a large extent their efficiency as catalysts, adsorbents, and ion exchangers, is their structural resistance to various factors, including thermal, chemical, and mechanical ones. Different theoretical aproaches are now being applied to this problem. Although quantum-chemical methods are only now coming into use in this field, certain results have already been obtained and some of them are summarized below.

The choice of a calculated parameter, to which the structural stability of a zeolite would be related, presents a key question in formulating the problem. Two approaches can be mentioned here: the "integral" and the "local" approach. The integral approach consists of calculating the dependence of a specified component of the total energy of a cluster on its structure and composition (usually, the dependence of the bonding or atomization energy). The model clusters are chosen so as to reproduce the main structural elements of a zeolite framework. The paper by Beran (*115*) presents a typical example. In that work two characteristic energies, E_D and E_S, were calculated for the clusters simulating six-membered windows of the large cavities and hexagonal prisms of faujasites. The destabilization energy E_D described the destabilization due to the substitution of one or several atoms of Si on a ring by Al atoms, while the stabilization energy E_S characterized the increased stability of a structure due to the coordinative binding of a cation (Na^+) at S_{II} and $S_{I'}$ sites by oxygen atoms of the zeolite ring. The CNDO/2 method was used in computations. Averaged geometrical parameters of clusters were taken equal for all the systems compared. The main conclusions were as follows.

(1) Substitution of Si by Al atoms results in progressive destabilization of a structure (it lowers the binding energy).

(2) The stabilization energy E_S decreases when passing from the conformation in which Al atoms are separated by two Si atoms to the structures in which they are separated by a single Si atom.

(3) The stabilization effect of sodium on a zeolite framework, which increases upon decreasing the Si/Al ratio, appears less pronounced than the destabilization due to the substitution of Si by Al. Thus, the bonding energy in the $Si_6O_6(OH)_{12}$ cluster is higher by 342 kcal/mol than that in the $Si_5AlO_6(OH)_{12}Na$ cluster. The stabilization and destabilization effects can also be followed by the variation of the HOMO energy. The changes in this energy, however, are less regular than those of E_S.

The above remark that the comparison of differently charged clusters should be made with care remains valid in this case as well. Meanwhile, such a comparison is implied in the very definition of E_D (115), which requires calculation of the difference in the bonding energies of clusters with different net charges, for example, between $Si_6O_6(OH)_{12}$ and $(Si_5AlO_6(OH)_{12})^-$, etc. Similarly, E_S is defined as the energy of withdrawing a Na^+ ion from a zeolite window, which also implies a change in the net charge of a zeolite structure. In our view, comparisons between the neutral structures would be physically more justified.

Let us also mention the work of Beran et al. (116) discussing the energetics of a zeolite stabilization due to the substitution of the skeleton Al atoms by Si atoms, with Al cations being localized at the cationic positions. The CNDO/2 calculation was performed for closed six-membered clusters with different Si/Al ratios. Unfortunately, the compared structures were again of different net charges. For a similar reason, care should be taken when considering the results of Hass et al. (117, 118), who carried out ab initio computations, in the minimal STO-3G basis, of open $[(HO)_3T_1OT_2(OH)_3]^{n-}$ (117) and closed four-membered $T_4O_4(OH)_8$ (118) clusters, with T = Si, Al and compensating charge n^-. Li^+ and Be^{2+} ions were used as the charge-compensating cations. A general conclusion was drawn that the bridged Al—O—Al structure is unstable but can be stabilized by interaction with a cation. These results were used to discuss the validity of the Lowenstein rule as applied to different systems.

Ab initio calculations of $[(HO_3TOT(OH)_3]^{n-}$ clusters were also performed by Sauer and Engelhardt (119), whose formulation of the problem was quite similar to that of the previous works. The only distinction in the scheme of computations was that the charge compensation was achieved by means of a crystal field simulated by six point charges q [$q = \frac{1}{6}e$ or $\frac{2}{6}e$ for (Si, Al)$^-$ and (Al, Al)$^{2-}$, respectively]. In addition, the alternative structures with point charges $q = e$, $2e$ located at the cation positions in a zeolite were also considered.

Calculations of the total binding energy in clusters, as applied to the analysis of the structural stability of zeolites, give results quite fitting the experiment. Some remarks, however, should be made on this subject. The main point is the rather obvious statement that the stability of a given structure is determined not only by the increase in the total binding energy but also by the properties of the weakest bonds whose rearrangement could govern the thermal decomposition of a zeolite. The donor–acceptor bond $Al \cdots O$ represents such an element in zeolites. In quantum-chemical computations, the difference in the strengths of Si–O and Al–O bonds is reflected by the different values of the Wiberg index p_{AlO} for different structural modifications of zeolites. It is hoped someday to provide a more proper description of the changes in the structural stability of zeolites.

Such a local approach is now widely used and works quite well. It provides a natural explanation of the trend toward increasing zeolite stability with Si/Al ratio. It also allows one to follow easily the influence of cations entering a zeolite structure on its stability. Quite a number of such comparative calculations can be found in the literature. We shall confine ourselves to only a few representative examples.

Based on the analysis of p_{AlO} in H forms of faujasites (computations were carried out for four-membered closed clusters $Si_{4-n}Al_nO_4H_n(OH)$ and for some of their anion analogs), Beran (104) has concluded that the formation of a skeleton (bridged) hydroxyl group results in considerably weakening the Al–O bond. Multivalent cations also, although notably less than protons, destabilize Al–O bonds, the effect being sensitive to their nature (104, 107, 109, 111). Alkali metal cations Li^+, Na^+, and K^+ affect p_{AlO} rather weakly (104). The influence of aluminum and transition-metal ions is more pronounced (107). The Al–O bond weakens in the following sequence: $Fe^{3+} > Fe^{2+} > Fe(OH)^+$, etc. (111). It seems likely that the effect of weakening the Al–O bond is a representative characteristic of the stability of a zeolite structure, although it is complicated in the same series by an opposite trend toward additional stabilization of a structure due to the donor–acceptor bonding of cations with the oxygen atoms of a zeolite framework.

Beran et al. (111) have also considered a $Si_5FeO_6(OH)_{12}Na$ cluster modeling a six-membered ring with Fe atoms included in a zeolite framework. It was concluded that Fe and Al affect the properties of a zeolite in a qualitative similar manner, although the destabilizing effects of Al are stronger.

The local approach to discussing the stability of a zeolite lattice seems quite justified but is substantially restricted by the use of a rather qualitative p_{AlO} index. In this respect, the analysis of the rearrangement of the weakest link of the structure deserves attention. With this aim in view, Pelmenshchikov et al. (120) have considered the energetics of the trigonal-to-tetrahedral aluminum transitions corresponding to the reversible formation and disruption of the Al $\cdots$ O donor–acceptor bond in a zeolite framework, as exemplified by the model $Al(OH)_3 + H_2O$ system. The MINDO/3 calculations for trigonal aluminum with no interaction between $Al(OH)_3$ and H_2O have shown that the planar geometry of structure 5a is preferential for the AlO_3 fragment since it provides the minimal repulsion between oxygen atoms. The transition to the pyramidal state 5b, comprising a step of the structural rearrangement to the tetrahedral aluminum, requires $E_r = 17.9$ kcal/mol. The energy released due to the formation of a new bond between the pyramidal $Al(OH)_3$ and a water molecule is $E_c = 30.1$ kcal/mol. Thus, the tetrahedral state of aluminum is energetically favorable in the $Al(OH)_3 + H_2O$ system structure 5c). However, a major part of the energy of the coordination bonding is expended

5a

5b

5c

to compensate the energy consumed in the structural rearrangement of the AlO_3 fragment. The increase in the electron–donor ability of the oxygen atom involved in $Al \cdots O$ bonding would increase the stability of the tetrahedral aluminum as compared with its trigonal state. Indeed, the energy gain in the $Al(OH)_3 + OH^-$ system is $E_c = 98.4$ kcal/mol.

It seems quite natural to suppose that the structural stability of the TO_4 fragments in a zeolite framework is also mainly determined by the interfragmental interactions. This allows one to explain the increase in the thermal stability of zeolites with the same Si/Al ratio observed in the following sequence of charge-compensating cations: $H^+ < Li^+ < Na^+ < K^+ < Rb^+$ (121). Actually, the electron-acceptor properties of cations decrease in this row. Hence the electron-donor properties of the bridged oxygen atoms forming the donor-acceptor bonds with aluminum atoms are increased in this sequence.

It also seems of interest to compare the stabilities of tetrahedral states of aluminum and boron in terms of this simple qualitative picture. The calculations have revealed that the planar structure is preferential for $B(OH)_3$ as well. The value of E_r for $B(OH)_3$ is 30.1 kcal/mol, that is, it exceeds substantially the value for $Al(OH)_3$. This is a manifestation of the difference in the covalent radii of B and Al. The value of E_c for the $B(OH)_3 + H_2O$ system is no longer sufficient to compensate E_r, that is, in this system the tetrahedral coordination of the boron atom is unfavorable. However, in the $B(OH)_3 + OH^-$ system, $E_c = 88.4$ kcal/mol, that is, the presence of the highly electron-donating oxygen atom of OH^- should make the tetrahedral state of boron preferential. Actually, the tetrahedral boron is known to exist in a number of mixed oxides.

Of particular interest are the zeolites with aluminum atoms partially substituted in their framework by boron. Calculations with optimization of the geometry of clusters have shown that the energy gains in the trigonal-to-tetrahedral state transformations in the H forms of such zeolites were 1.9 and 14.1 kcal/mol for boron and aluminum, respectively. It may therefore be concluded that H forms of boralites, if they exist, should be considerably less stable than those of zeolites.

B. AMORPHOUS ALUMINOSILICATES

Amorphous aluminosilicates represent a wide variety of systems with surfaces whose states are strongly dependent on the "biography" of a system. The very specificity of the amorphous structures and the more limited possibility of application of physical methods to their study results in much poorer knowledge about the structures of the amorphous aluminosilicate surfaces than in the case of crystalline systems. This makes quantum-chemical treatment considerably more qualitative in this case. Cluster models of active sites appear here mainly *a priori* and experimentally independent and provide, to some extent, an additional way of studying such systems.

The main fraction of the hydroxyl groups of aluminosilicate surfaces are the terminal hydroxyls, which exhibit no notable Brönsted acidity. At the same time, sufficiently strong BASs were quite unambiguously detected on these surfaces by various methods, including kinetic investigations of catalytic reactions, indicator techniques, IR studies of the adsorption of strong organic bases, etc. The nature of these sites is still discussed in the literature. A probable candidate is a bridged hydroxyl group, in close analogy with zeolites. This viewpoint was strongly supported by Borovkov *et al.* (*122*), who were able to observe directly the bridged hydroxyls in the IR spectra of amorphous aluminosilicates, although in much lower concentration than in zeolites. The general conclusion of their studies was that the acidic hydroxyl groups of amorphous and crystalline aluminosilicates are quite similar both in their spectral and acidic properties. As applied to quantum-chemical modeling of the active sites, this gives the direct recommendation to use for these BASs the same clusters as in the case of zeolites.

Another possible type of BAS in these systems might be represented by a water molecule bonded with a trigonal aluminum. This is quite unambiguously indicated by the results of quantum-chemical calculations which will be discussed below.

Of great interest is the question of the role of trigonal aluminum, which is usually assumed to act as a LAS. Such a center should be quite typical of Al_2O_3, where it may appear as a result of surface oxygen vacancy formation. These vacancies may either develop due to dehydroxylation or be of a biographical nature. A similar situation may take place in the case of such mixed oxides as amorphous aluminosilicates. Uytterhoeven, Cristner, and Hall (*123*) have concluded that trigonal aluminum could also appear as a LAS upon dehydroxylation of H forms of zeolites. Their scheme was criticized, however, by Kühl (*124*), who has undertaken X-ray fluorescence studies of the dehydroxylated forms of faujasites and found that the dehydroxylation was accompanied by dealumination of a zeolite framework with formation of extralattice aluminum which could also exhibit the Lewis acidity.

At the same time, specific properties (primarily the Si/Al ratio) of a zeolite should be taken into account when discussing the mechanism of its dehydroxylation. It is quite possible that the mechanism typical of H forms of faujasites would be completely improper for high-silica-containing zeolites. Thus, in their studies of dehydroxylated forms of ZSM-5 zeolite by means of IR spectroscopy of molecular hydrogen adsorbed at low temperatures, Kazansky et al. (72, 76) have demonstrated that the Uytterhoeven–Cristner–Hall scheme seems valid in this case.

1. *On Some Capacities of Cluster Quantum-Chemical Calculations in Comparing Specific Catalytic Activities of Different BASs of Aluminosilicates*

There are three traditional structures usually adopted as probable BASs in amorphous aluminosilicates: a water molecule coordinated by an electron-acceptor center (I), a bridged OH group (II), and a surface H_3O^+ ion (III) (125, 126). The catalytic activity of these sites is obviously determined by their properties and surface concentrations. Pelmenshchikov et al. (127) have attempted to compare these characteristics for the above types of BAS in aluminosilicates in terms of the cluster approach. For this purpose they considered a sequence of states of the model fragment of a dehydroxylated surface plus two water molecules (Fig. 15). State S_0 corresponds to a dehydroxylated surface, states S_I, S_{II}, and S_{III} represent the sites of the I, II, and III types and states S_{Ia} and S_{IIa} correspond to centers I and II at a higher coverage. The relative energies of these structures obtained using the CTP scheme and the CNDO/BW technique are presented in Fig. 15. The relative surface density of the sites, $\log(n_i/n_j)$, was estimated as the relative probability of their occurrence:

$$\log(n_i/n_j) = 0.43(E_j - E_i)/RT$$

For $T = 573$ K, $\log(n_I/n_{II}) \simeq \log(n_{Ia}/n_{IIa}) \simeq 0$, that is, the surface densities of BASs of the I and II types are approximately equal and independent of coverage. Also, $\log(n_{Ia}/n_{III}) \simeq \log(n_{IIa}/n_{III}) = 12$, that is, the surface density of BASs of the III type is too low to be of any practical interest in catalysis. Consideration of the extended fragment of a dehydroxylated surface plus five water molecules, where a surface H_3O^+ ion receives additional possibilities for solvation, did not qualitatively change the conclusion.

The orders of the relative atomic catalytic activities of the sites, $\log(k_i/k_j)$, in the alcohol dehydration reactions were estimated from the data of Malysheva et al. (128) by means of the empirical relationship

$$\ln k_i = -\alpha AP_i/RT + \text{const.}$$

similar to the Brönsted equation in the homogeneous acid catalysis. Here $\alpha = 0.44$ at $T = 573$ K, the constant is determined by the basicity of the

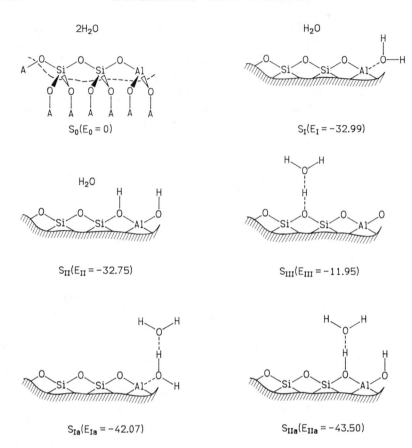

FIG. 15. Cluster models of a series of S_i states of a model fragment of dehydroxylated aluminosilicate surface with two water molecules. The calculated energies of these states (CTP, CNDO/BW) are as follows: S_0, $E_0 = 0$; S_I, $E_I = -32.99$; S_{II}, $E_{II} = -32.75$; S_{III}, $E_{III} = -11.95$; S_{Ia}, $E_{Ia} = -42.07$; S_{IIa}, $E_{IIa} = -43.50$.

molecule of a particular alcohol, and AP_i is the proton affinity of the ith structure. Bearing in mind that $AP_I - AP_{II} = E_{II} - E_I = -0.24$ kcal/mol, one can find that $\log(k_I/k_{II}) \simeq 0$ and estimate the relative specific activities of these sites as

$$\log(n_I k_I / n_{II} k_{II}) \simeq 0$$

Thus, the following conclusions concerning BASs of aluminosilicates can be drawn from the results of the calculations:

(1)   The catalytic activity of aluminosilicates in typical reactions of acid catalysis is determined by two types of BASs: bridged hydroxyl groups and

water molecules coordinated on a trigonal aluminum atom. These centers are characterized by approximately equal surface densities and atomic catalytic activities.

(2) The surface density of H_3O^+ ions appears to be too low to influence the catalytic activity of aluminosilicates.

2. Trigonal Aluminum

As noted above, it is widely adopted that trigonal aluminum is one of the most important chemisorption and catalytic sites in aluminosilicates. Formation of these centers is usually associated with dehydroxylation. In the preceding section this concept was used to discuss different types of BASs formed as a result of water adsorption on the dehydroxylated surface of a model aluminosilicate fragment. The activity of the trigonal aluminum atoms was particularly manifested in the strong activation of the coordinatively bonded water molecules. In the chemical sense, such a site comprises a typical Lewis acid, which is also confirmed by quantum-chemical calculations.

a. Trigonal Aluminum as a Lewis Acid Site. Senchenya *et al.* (*129*) have performed a quantum-chemical consideration of trigonal aluminum as exemplified by the clusters presented in Fig. 16. Recall that pseudo-atoms A

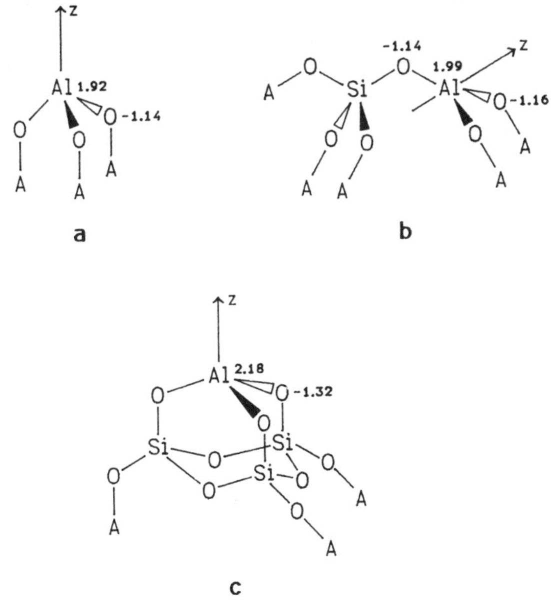

FIG. 16. Cluster models of trigonal aluminum LASs in aluminosilicates.

186   G. M. ZHIDOMIROV AND V. B. KAZANSKY

may represent both lattice Si and Al atoms in their oxygen environment (in the CNDO/BW method their VOIPs are 10.68 and 7 eV for Si* and Al*, respectively). Varying $VOIP_A$ within reasonable limits, one can model different crystalochemical environments of the center. The computations were carried out for two positions of aluminum atom in the clusters.

The calculated one-electron energy levels of clusters (a)–(c) are presented in Fig. 17. In all the clusters, LUMO is mainly represented by the sp_z^3 hybrid AO. The picture obtained actually corresponds to centers accepting an electron pair. HOMO–LUMO splitting is about 12 eV. HOMO is composed mainly of AOs of oxygen atoms. Substitution of Si* by Al* results in

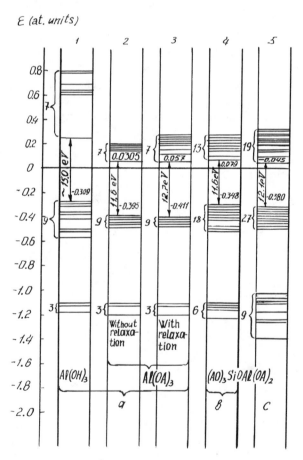

FIG. 17.   Scheme of one-electron energy levels of clusters (a)–(c) (Fig. 16) as obtained by the STO-3G (1) and CNDO/BW (2–5) methods. Scheme 3 was obtained with allowance made for the shift of the Al atom along the z axis.

diminishing the gap between HOMO and LUMO while simultaneous increasing the HOMO energy by ~ 1 eV, which reflects the decrease in the acceptor ability of the site.

Some shift Δ of the aluminum atom along the z axis from the perfectly tetrahedral position toward the oxygen atom plane was required to minimize the total energy. For cluster (a), the respective energy gain was ~ 24 kcal/mol for $\Delta \simeq 0.3$ Å. In this case, the LUMO energy increased (i.e., the acceptor ability decreased) and HOMO–LUMO splitting also increased. The Lewis acidity of a trigonal aluminum site can be clearly seen if one considers the adsorption of a set of model bases $A'OH$ with extensively varied $VOIP_{A'}$ and, hence, basicities. The calculations showed that the increase in $VOIP_{A'}$ causes a systematic decrease both in the energy of the interaction with a trigonal aluminum and in the degree of the charge transfer from the testing molecule to the cluster. Thus, the computations indicate that all three parameters considered (the relative position of the LUMO, the interaction energy between the center and a testing molecule, and the degree of the charge transfer from the testing molecule to the site) behave in accordance with the common view of the properties of LASs and can serve as measures of the acidity of such systems in their quantum-chemical simulations.

The above qualitative conclusion obtained by the CNDO/BW method was also confirmed by MINDO/3 and *ab initio* computations. The latter were carried out using both minimal STO-3G (*129*) and extended STO-3G*, 3-21G, and 3-21G* basis sets (*16*). The simplest cluster of type (a) was considered with hydrogens as terminal atoms. As in the semiempirical case, the *ab initio* calculations predicted certain, although less in absolute value, shifts of Al atom is along the z-axis. This shift decreased with extension of the basis set. However, the introduction of polarization functions (3d AO of Al) in a basis somewhat increased the effect.

b. Chemisorption Interaction with Simple Molecules. Quantum-chemical calculations have demonstrated that the trigonal aluminum sites possessed quite strong adsorbing properties (*129*). The structures of molecular adsorption considered are presented in Fig. 18. The CNDO/BW method was used in the computations. Some structures were calculated *ab initio* in a STO-3G basis. The results of the computations are listed in Table VII. There, Q is the adsorption energy, q is the charge of an adsorbed molecule, and the R and r are the same as in Fig. 18. The values of Q were calculated for cluster (a) with an Al atom in the optimal position. Although the calculated interaction energies Q appear somewhat overestimated, both the trend toward the charge transfer from an adsorbed molecule to the cluster and the values of R and r, as found by optimization procedures, seem quite reasonable. It is also worth noting that the charge transfer to the cluster increases with the basicity of an adsorbed molecule. An unexpected result that $Q_{H_2O} > Q_{NH_3}$ should be

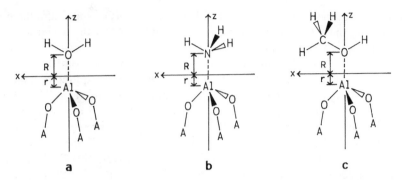

FIG. 18.   Cluster models of the molecular adsorption forms of (a) H_2O, (b) NH_3, and (c) CH_3OH on trigonal aluminum.

considered as a defect of the CNDO/BW scheme. It is eliminated in a STO-3G calculation (Table VII).

The calculated heats of dissociative adsorption of water, ammonia, and methyl alcohol (Fig. 19) are also presented in Table VII. The heat of water adsorption was 61 kcal/mol (CNDO/BW) or 43.7 kcal/mol (STO-3G), and the respective value for ammonia was 38.7 kcal/mol (CNDO/BW). If the pseudo-atoms A model the lattice silicon atoms, structure (d) is somewhat

TABLE VII

Results of Calculations of Nondissociative Adsorption of Simple Molecules on Trigonal Aluminum LASs[a]

Adsorbed molecule	Parameters	CNDO/BW	STO-3G
H_2O	Q (kcal/mol)	71.7	61.3
	q (e)	0.171	0.212
	R (Å)	1.43	1.84
	r (Å)	0.21	0
CH_3OH	Q	87.6	—
	q	0.197	—
	R	1.35	—
	r	0.15	—
NH_3	Q	65.9	76.7
	q	0.222	0.231
	R	1.55	1.94
	r	0.15	0

[a] Cluster presented in Fig. 18.

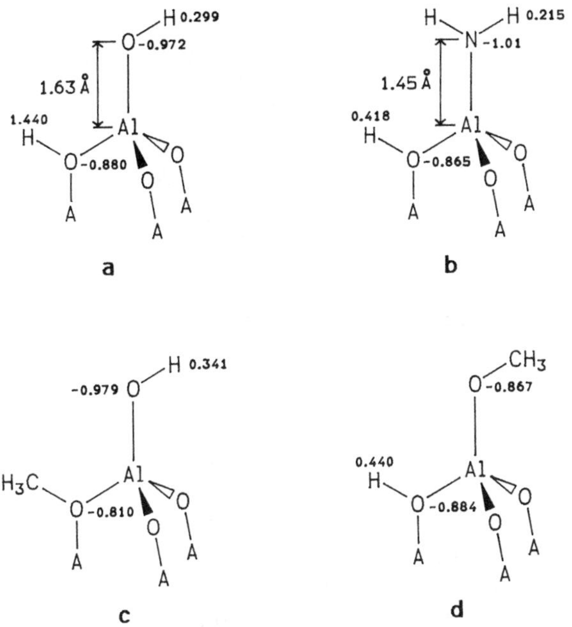

Fig. 19. Cluster models of the dissociative forms of adsorption of (a; c) H_2O, (b) NH_3, and (d) CH_3OH on trigonal aluminum.

more preferential for methyl alcohol adsorption than structure (c). However, the energies of these structures are equalized as $VOIP_A$ (and hence the strength of the LAS) increases.

As noted above, the calculations of molecular and dissociative adsorptions gave unexpectedly high adsorption heats. This result can be partly attributed to the imperfections of the computational methods (both CNDO/BW and STO-3G). Actually, the *ab initio* calculation of the energy of water adsorption on cluster (a) in 3-21G* basis (*16*) gave a lower value of 39.7 kcal/mol, in better agreement with some averaged experimental estimates. At the same time, the results of the computations suggest that "pure" trigonal aluminum sites, as presented in Fig. 16, do not occur in real systems due to their screening by surrounding oxygens, which lowers their chemisorption ability. Possible variants of such a screening and its consequences will be considered in the next section. If, however, some centers remain unscreened by virtue of local steric hindrances, they would possess anomalously strong chemisorption properties.

To conclude this section, let us consider the results of *ab initio* calculations of the chemisorption of a H_2 molecule on a trigonal aluminum site (*16*). Two types of shift of the stretching vibration frequencies of H_2 were observed in the studies of the low-temperature adsorption of H_2 on a silica gel, amorphous aluminosilicate, γ- and η-Al_2O_3 (*130, 131*). The weaker shifts of 30 to 60 cm^{-1} were ascribed to H_2 interacting with surface hydroxyls, while the stronger shifts of ~ 140 to 170 cm^{-1} observed for η-Al_2O_3 and aluminosilicate were assigned to H_2 adsorbed on a LAS or acid–base pair M^+O^-. Dissociative adsorption of H_2 on η-Al_2O_3 occurs only at room temperature, giving rise to surface Al–H and hydroxyl group formation.

A set of different forms of adsorption of H_2 on as trigonal aluminum site was calculated in 3-21G basis. These were one-point (**6a, 6b**), two-point (**6c**), and dissociative (**6d**) forms. In the case of one-point adsorption, the center of gravity of H_2 was fixed at the z axis. When this fixation was removed, it appeared that the one-point form of adsorption did not correspond to a minimum of the potential energy surface, that is, it was unstable. Three parameters were varied in the calculations of the two-point adsorption form **6c**: the distance R from the center of gravity of H_2 to the Al–O axis, the distance from the projection of this center on the Al–O axis to the Al atom, and the angle α between the Al–O and H–H directions. The optimal values of these parameters were $R = 2.4$ Å, $\alpha = 20°$, and $r = 0.25$ Å. The corresponding adsorption energy was 4.8 kcal/mol, which seemed quite reasonable. Of certain interest is that the Al–H separation in the adsorption complex is rather large, and that the H–H bond is only slightly stretched, from 0.74 to 0.75 Å. At the same time, the H_2 molecule is markedly polarized, which provides a qualitative explanation of the occurrence of a H_2 stretching vibration band in the IR spectra. The dissociative structure **6d** appeared by

2.4 kcal/mol to be more favorable than the molecular adsorption form **6c**. The activation barrier of the respective transition was not estimated.

c. Effect of Structural Screening on the Chemisorption Properties. Activated Chemisorption. In real systems one should expect a fairly strong screening of the low-coordinated cations by oxygen anions of the oxide lattice. In general, a cation in a lower coordination state should be considered as a site with an excess energy. In our view, this is especially essential for transition-metal cations, whose trend toward the higher coordination is the most pronounced. Pelmenshchikov *et al.* (*132*), however, have shown that a similar tendency is also characteristic of non-transition-metal ions.

Somewhat simplifying the situation, one can specify two types of the trigonal surface aluminum sites in silica aluminas, depending on the type of coordination by surrounding oxygen atoms. The first type (LAS-I) represents the case where the screening is due to the interaction with the nearest SiO_3 fragments of the surface layer, as shown in Fig. 20a. The second type (LAS-II) deals with coordination by interlattice oxygen ions (Fig. 20b, position 1).

These two types of LASs differ in their chemisorption properties. The LAS-I are more accessible for gas-phase agents. However, in this case, as compared with a free trigonal aluminum, the adsorption is less energetically favorable or sometimes even completely impossible, depending on the geometry of the AlO_3 fragment. Thus, the relative energies of LAS-I increase in the sequence $\delta E_{struct} = 0, 3.1, 5.0, 9.1, 12.0$ kcal/mol with increasing angle $\alpha = 10, 20, 30, 40, 50°$. These values were estimated as the differences of the total energies of the respective clusters calculated using a CTP scheme and the MINDO/3 method.

The corresponding seqence of the water adsorption energies was $\delta E_{ads} = 11.0, 9.1, 7.9, 6.9, 3.1$ kcal/mol. Thus, the LAS-I model provides a qualitative

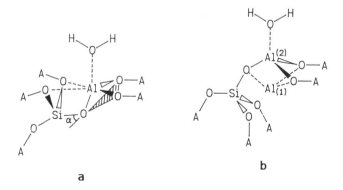

FIG. 20. Cluster models of two structural types of trigonal aluminum LASs in aluminosilicates. State 1 in structure (b) corresponds to the structural coordination, whereas state 2 represents the adsorption coordination.

explanation of the possible scatter of the adsorption heats that is in agreement with the experimental observations.

In the case of LAS-II, trigonal aluminum may be in the state of either "structural" (position 1) or "adsorption" (position 2) coordination. These two types of geometry alternate, and the transition between them requires some activation energy. The relative stability of the "adsorption" state and the probability of "extracting" such Al atoms to the surface layer should increase with the basicity of an adsorbate. In the case of water molecules, the result of the competition between structural and adsorption coordinations also depends, for LAS-II, on the AlOSi angle. Its increase makes the adsorption coordination preferential. Rough estimations showed that the transition to the exothermic adsorption should occur at angles of about 140–170°. The conclusion concerning the activated adsorption on LAS-II might be a subject of experimental studies.

V. Aluminophosphates

The catalytic activity of aluminophosphates has been discussed by Tada *et al.* (*133–135*). It seems also of interest to perform a theoretical investigation of possible active centers (both BAS and LAS) in these systems and to compare them with the respective centers in aluminosilicates. Such a comparison implies certain requirements both to the scheme of computations and to the choice of the cluster models. Most important is that the procedure of saturating the dangling bonds of a cluster should affect the results to a minimal extent. A simple way of attaining this aim is to construct closed clusters with terminal bonds mutually saturating each other (*41*).

Catalytic activity of aluminophosphates depends on the method of their preparation, the P/Al ratio, and the thermal pretreatment (*136*), Pelmensh-chikov *et al.* (*41*) have examined the acid sites that could occur on the surface of aluminum orthophosphate, which has the same P/Al ratio as the real acid aluminophosphate catalysts. A fragment of the hydrated surface of aluminum orthophosphate is shown in Fig. 21a. It provides a basis for constructing a model cluster [(b) of Fig. 21]. The other possible BASs in this system were represented by clusters (c) and (d) (Fig. 21), which were obtained from (b) by transferring a hydrogen atom to other positions. The respective clusters for an aluminosilicate are presented in Figs. 21e and 21f. The MINDO/3 method was used in the calculations. The lengths of bonds marked in Fig. 21 by crossing were optimized; the other were fixed at their experimental values (*136*). In all the clusters, both tetrahedral coordination of cations and linking of the tetrahedra through their corners were retained.

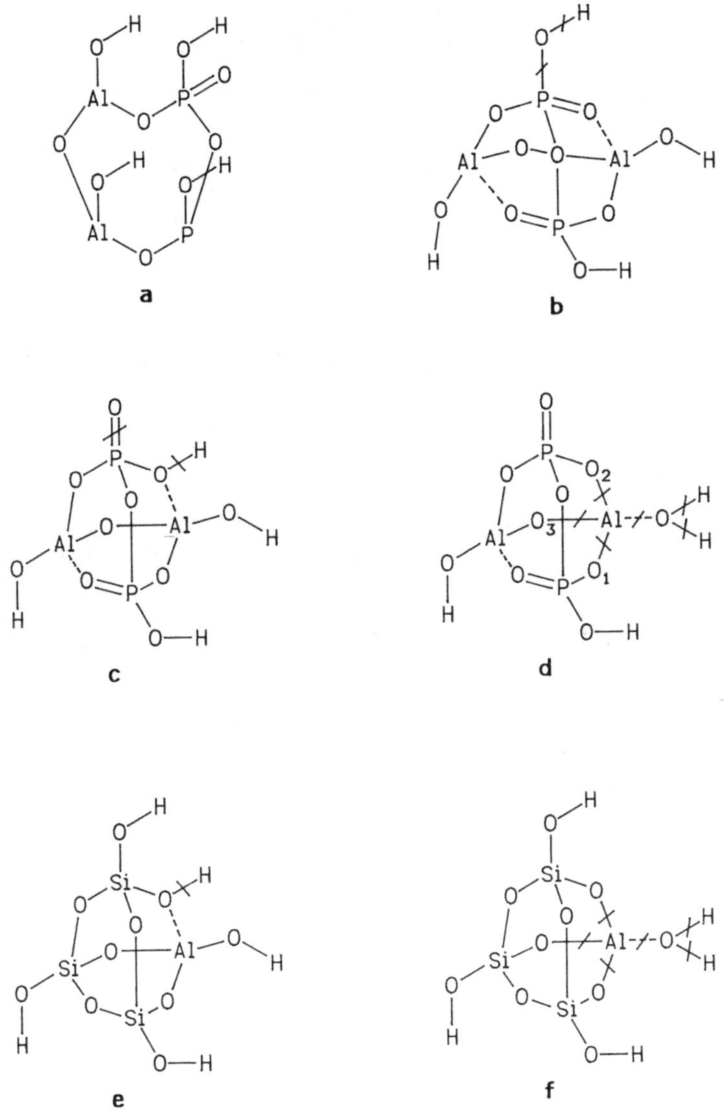

FIG. 21.   Model clusters of aluminosilicates and aluminophosphates. The lengths of bonds marked by crossing were varied to minimize the energy of the system. Optimization of the bond lengths in structure (d) gave rise to activationless migration of the Al atom through the $O_1O_2O_3$ plane inside the cluster, with simultaneous loss of the water molecule.

TABLE VIII
Relative Binding ($\mathscr{E}$) and Proton Abstraction (E_{H^+})
Energies[a]

Cluster	Site	$\mathscr{E}$ (eV)	E_{H^+} (eV)
d	P—OH	0	12.64
c	P—OH—Al	0.54	12.10
e	Si—OH—Al	0	12.34
f	Al···OH$_2$	−0.17	12.54

[a] Clusters presented in Fig. 21.

Relative binding ($\mathscr{E}$) and proton abstraction (E_{H^+}) energies for the clusters considered are listed in Table VIII. Of particular interest is site (d) (Fig. 21), representing a water molecule coordinated by a LAS of an aluminophosphate. As mentioned above, in the surface–gas system, two different ways of coordination are possible for trigonal aluminum on the surface: the adsorption one, with a water molecule held by the center, and the structural one, representing linking with internal oxygen. The computations have shown that, unlike aluminosilicates (Fig. 21f), structure (d) appeared unfavorable in the case of aluminophosphates, since the Al atom migration inside the cluster with simultaneous loss of the water molecule showed no activation when the lengths of Al–O$_i$ ($i = 1, 2, 3$: Fig. 21d) bonds were varied.

Naturally, structures (d) and (f) do not exhaust all possible states of low-coordinated Al atoms on the surface of the oxides considered. The calculations, however, seem quite sufficient to suggest that water molecule coordination by a LAS is energetically less favorable for aluminophosphate than for aluminosilicate surfaces. This conclusion is also in accordance with IR data, which indicate that LASs of the both oxides quite similarly interact with pyridine, whereas the LASs of aluminophosphates do not coordinate CO_2 molecules (136). Indeed, in the case of a sufficiently strong base (pyridine), adsorption interaction appears stronger than the structural coordination and therefore stabilizes the Al atom in the adsorption state. On the contrary, for CO_2, which is certainly a very weak base, the interaction is strong enough in the case of aluminosilicates but is insufficient for the adsorption stabilization of aluminum in aluminophosphates.

According to Table VIII, P–OH groups (Fig. 21b) represent the most stable Brönsted acid sites on the aluminophosphate surfaces. At the same time, P–OH–Al sites (bridged hydroxyl groups) exhibit the strongest acidity. Their relative surface concentration with respect to P–OH could be estimated as

$$\frac{n(\text{P-OH-Al})}{n(\text{P-OH})} = \exp\left(-\frac{\mathscr{E}_c - \mathscr{E}_b}{kT}\right) \simeq 10^{-5}$$

However, if one bears in mind that n(P-OH) $= 10^{14}$ cm^{-2} at $T = 600°$C (137), it becomes evident that the bridged groups could hardly be detected by the IR technique. Therefore, the observed Brönsted acidity of aluminophosphate should likely be determined by P-OH sites, whereas in the case of aluminosilicates it is simultaneously determined by both Si-OH-Al and Al$\cdots$OH$_2$ sites, which are very close in their stability and acidity. It should also be stressed that $E_{H^+} = 16.34$ and 16.02 eV values for Al-OH groups in structures (b) and (c) (Fig. 21) are considerably higher than that for the bridged SiOHAl group of structure (e) ($E_{H^+} = 13.91$ eV). This indicates that the former centers cannot be considered as strong Brönsted acid sites.

VI. Aluminum Oxide

It was mentioned above that the cluster modeling of the surface sites of highly coordinated oxide lattices faces certain difficulties. This is probably the reason why only a few computations were performed for such systems. The aluminum oxide structure is just of this type.

The theoretical works on Al$_2$O$_3$ have dealt mainly with studies of the properties of the surface hydroxyl groups (primarily of their vibration freqencies) as well as with LASs corresponding to low-coordinated aluminum. We shall not consider here various peculiarities of the structure of different modifications of Al$_2$O$_3$, since, on the one hand, the catalytic aspects of the problem have already been reviewed by Knözinger and Ratnasamy (138), and, on the other hand, the quantum-chemical investigations were aimed only at qualitative description of principal structural units and did not yet concern the detailed structure of the surfaces. The principal units were mainly considered to be the polyhedra containing Al in the tetrahedral and octahedral coordinations [in γ-Al$_2$O$_3$, in particular, 84% of Al atoms are located in the positions of T_d symmetry, while the rest (16%) occupy the positions of O_h symmetry)]. The more extended structural units are modeled by doubled or tripled tetrahedra or octahedra.

A. Lewis Acid Sites

Various surface defect sites in Al$_2$O$_3$ were discussed by Gagarin et al. (139) in terms of simple ionic cluster modeling of the surface fragments with a single hydroxyl group and Al in either tetrahedral or octahedral coordination: [O$_3$AlOH]$^{4-}$ and [O$_5$AlOH]$^{8-}$. They have also considered the defect structures formed by withdrawal of H$^+$, OH$\cdot$, or OH$^-$ from these clusters as well as a number of pair defects, as exemplified by doubled tetrahedral clusters. EHT was used in calculations. One-center parameters H_{3s}, H_{3p}, and H_{3d} for aluminum were adjusted so as to reproduce the experimental charge distribution in the bulk AlO_6^{9-} cluster and the width of its band gap. Special attention was paid to the new states occurring in the energy spectrum in the presence of surface defects. It was found that the oxygen vacancies result in the appearance of the local states on the Al atom. These vacancies gave rise to the Lewis acidity, which was higher in the case of trigonal aluminum. This was substantiated by direct computations of the chemisorption of H_2O, CO, C_2H_4, and NH_3 molecules. Qualitatively similar results were also obtained by the X_α-SW method (140) and using the scheme of so-called orbital-stoichiometric clusters (141) suggested by Litinskii (142). Surface clusters composed of two polyhedra (141, 143) open the way for considering such questions as the influence of the chemical modification of a surface (141) or of a degree of dehydroxylation (143) on the Lewis acidity. These calculations, quite in accordance with experiment, predict that fluorination would increase, whereas the alkalization and ammonation would decrease, the Lewis acidity (141).

B. SURFACE HYDROXYL GROUPS

The aluminum oxide surfaces exhibit only a weak Brönsted acidity. Therefore, studies of its surface hydroxyls have usually dealt with their structure. IR spectroscopy appears the most informative and appropriate method for these purposes. The IR spectra cover a wide variety of hydroxyl stretching vibrations, and their assignment is frequently quite a problem. Primarily, it was based on electrostatic concepts and on the electrostatic theory of valency (138).

The cluster approach opens the way for a direct quantum-chemical calculation of the influence of the nearest environment on OH vibration frequencies (12). In such a computation, the scheme of the neutral ionic cluster (12, 144) is used that allows one to construct easily a consistent set of the required cluster structures. Two examples are clusters 7a and 7b. Cluster 7a models the hydroxyl group bonded with the Al atom in tetrahedral coordination. Cluster 7b simulates the bridged hydroxyl group linking two

7a   7b

aluminum atoms in tetrahedral and octahedral coordination. They will be designated in what follows as OH^t and OH^{to}, respectively. Clusters OH^o, OH^{oo}, and OH^{oot} were constructed in a similar manner. The MINDO/3 method was used in computations.

The effective electron charge of the clusters was determined starting with the formal charges -2 for oxygen and $+3$ for aluminum. Total negative charge was compensated by introduction of the additional positive core charge on the terminal oxygen atoms O^* in order to make the whole cluster neutral. In addition, the correction δU was also introduced for diagonal matrix elements of the terminal O^* atoms (as compared to their conventional value accepted in MINDO/3 parametrization). It was chosen so as to maximize the vibration frequency ν_{OH^+}, in accordance with the results of Baumgarten and Weinstrach (145). Such an approach is quite similar to the above scheme of the cluster with terminal pseudo-atoms. The results of the calculations are listed in Table IX. They allow one to draw the following conclusions.

(1) The acidity (as estimated both by the deprotonation energies and by the positive charges on H atoms) increases in the sequence $OH^o < OH^t < OH^{oo} < OH^{to} < OH^{oot}$. This is in agreement with the acidities predicted by the electrostatic model of Knözinger (138).

(2) Frequencies ν_{OH} appear separated into three groups depending on the coordination number of oxygen, that is, on the structure of the first coordination sphere of hydroxyls. Each group is then subdivided depending on the coordination number of neighboring aluminum atoms.

(3) The calculated OH stretching vibration frequencies decrease in the following order: $\nu_{OH^t} > \nu_{OH^o} > \nu_{OH^{oo}} > \nu_{OH^{to}} > \nu_{OH^{oot}}$. This completely agrees

TABLE IX

Results of Calculations of Differently Coordinated OH Groups of Aluminum Oxide[a]

	q_H (e)	q_O (e)	ν_{OH} (cm^{-1})
OH^o	0.264	-0.842	3950
OH^t	0.302	-0.729	4100
OH^{oo}	0.305	-0.735	3700
OH^{to}	0.308	-0.613	3620
OH^{oot}	0.380	-0.616	3460

[a] q_H and q_O are the charges on H and O atoms of a hydroxyl group, respectively; ν_{OH} are the stretching vibration frequencies.

with the assignment suggested in Mardilovich (*146*) on the basis of the analysis of experimental data.

The problem of cluster modeling the aluminum oxide surfaces will no doubt be the subject of numerous future investigations. The quantum-chemical treatment of catalytic transformations of organic molecules on Al_2O_3 will require the consideration of larger fragments of the surfaces. However, until now only very simple model approaches, similar to those discussed above, were used. An example of this kind is the calculation of ethyl alcohol adsorption (*144*). The surface site was modeled by a $Al_2O_2O_2^*$ cluster with terminal pseudo-atoms O*. The computations were carried out using the CNDO/2 method. One molecular and two dissociative ($C_2H_5O^- + H^+$ and $C_2H_5^+ + OH^-$) adsorption forms were considered. The conclusion was drawn that the first of these dissociative mechanisms was preferential.

VII. Conclusion

Even now, quantum-chemical calculations make a considerable contribution to the theory of heterogeneous catalysis. In the present review, we have confined ourselves to treating a relatively narrow class of adsorbents and catalysts. Many more works, which remain outside the scope of this survey, are devoted to studies of metal surfaces. One can mention as well works dealing with transition-metal oxides and supported metals. It may be expected that the number of these works will soon rapidly increase. Such a widening of the range of investigated systems would require some methodological problems to be solved. These involve improvement of the cluster approach itself and increasing the efficiency of computer programs (first of all, for *ab initio* and X_α-DV calculations). Of key importance is also the proper selection of optimal subjects and directions of quantum-chemical studies. It is closely connected with the progress in the experimental knowledge of the mechanisms of catalytic reactions. There is no doubt that the most interesting problems and the main advances in the application of quantum-chemical methods to studies of chemisorption and catalysis are matters of the near future.

ACKNOWLEDGMENTS

The authors would like to express their profound gratitude to Drs. I. D. Mikheikin, A. G. Pelmenshchikov, and I. N. Senchenya for their invaluable assistance in preparing this review and to Dr. N. N. Weinberg for translating the manuscript.

REFERENCES

1. Dowden, D. A., *J. Chem. Soc.* p. 242 (1950).
2. Vol'kenshtein, F. F., "The Electronic Theory of Catalysis on Semiconductors." Pergamon, Oxford, 1963.
3. Morrison, S. R. "The Chemical Physics of Surface." Plenum, New York, 1977.
4. Ertl, G. V., *in* "Electron and Ion Spectroscopy of Solids" (L. Seerman, ed.). Plenum, New York, 1978.
5. McWeeney, R., and Sutcliffe, B. T., "Methods of Molecular Quantum Mechanics." Academic Press, New York, 1969.
6. Kunz, A. B., *in* "Theory of Chemisorption" (J. Smith, ed.). Springer-Verlag, Berlin and New York, 1980.
7. Hoffman, R., *J. Chem. Phys.* **39**, 1397 (1963).
8. Anderson, A. B., *J. Chem. Phys.* **62**, 1187 (1975).
9. Segal, G. A., ed., "Semiempirical Methods of Electronic Structure Calculation," Part A. Plenum, New York, 1977.
10. Baba-Ahmed, A., and Gaueso, J., *Theor. Chim. Acta* **62**, 507 (1983).
11. Mikheikin, I. D., Abronin, I. A., Zhidomirov, G. M., and Kazansky, V. B., *J. Mol. Catal.* **3**, 435 (1977-1978).
12. Pelmenshchikov, A. G., Ph.D. Thesis, Novosibirsk, Institute of Catalysis (1984).
13. Pelmenshchikov, A. G., Mikheikin, I. D., and Zhidomirov, G. M., *Kinet. Katal.* **22**, 1427 (1981).
14. Blyholder, G., Head, J., and Ruette, F., *Surf. Sci.* **131**, 403 (1983).
15. Ruette, F., Blyholder, G., and Head, J., *Surf. Sci.*, **137**, 491 (1984).
16. Zelenkovskii, V. M., Ph.D. Thesis, Institute of Organic Chemistry, Moscow (1984).
17. Gaddard, W. A., III, *Acc. Chem. Res.* **6**, 383 (1973).
18. Bagus, P. S., Schaefer, H. F., and Bauschlicher, C. W., Jr., *J. Chem. Phys.* **78**, 1390 (1983).
19. Bennett, A., McCarrol, B., and Messmer, R. P., *Phys. Rev. B: Solid State* **3**, 1397 (1971).
20. Evarestov, R. A., "Quantum-chemical Methods in the Theory of Solids." Izd. Leningr. Gos. Univ., Leningrad, (1982)
21. Evarestov, R. A., Sokolov, A. R., and Ermoshkin, A. N., *Khim. Fiz.* p. 299 (1982).
22. Pisani, C., *Phys. Rev. B: Solid State* **17**, 3143 (1978).
23. Pisani, C., and Ricca, F., *Surf. Sci.* **92**, 481 (1980).
24. Muscat, J. P., *Surf. Sci.* **110**, 389 (1981).
25. Whitten, J. L., and Pakkanen, T. A., *Phys. Rev. B: Candens. Matter* [3] **21**, 4357 (1980).
26. Fink, W. H., Butkus, A. M., and Lopez, J. P., *Int. J. Quantum Chem. Symp.* p. 331 (1979).
27. Sawada, S., *J. Phys. C: Solid State Phys.* **13**, 4823 (1980).
28. Minot, C., Van Hove, M. A., and Samorjai, G. A., *Surf. Sci.* **127**, 441 (1982).
29. Hayns, M. R., *Theor. Chim. Acta* **39**, 61 (1975)
30. Chen, Z., Weng, Z., Hong. R., and Zhang, Y., *J. Catal.* **79**, 271 (1983).
31. Krylov, O. V., "Catalysis by Nonmetal," p. 236. Khimia, Leningrad, 1967.
32. Mikheikin, I. D., Abronin, I. A., Lumpov, A. I., and Zhidomirov, G. M., *Kinet. Katal.* **19**, 1050 (1978).
33. Mikheikin, I. D., Lumpov, A. I., and Zhidomirov, G. M., *Kinet. Katal.* **20**, 501 (1979).
34. Mikheikin, I. D., Lumpov, A. I., and Zhidomirov, G. M., and Kazansky, V. B., *Kinet. Katal.* **19**, 1053 (1978).
35. Senchenya, I. N., Mikheikin, I. D., Zhidomirov, G. M., and Kazansky, V. B., *Kinet. Katal.* **21**, 785 (1980).
36. Mikheikin, I. D. Lumpov, A. I., Zhidomirov, G. M., and Kazansky, V. B., *Kinet. Katal.* **20**, 499 (1979).
37. Korsunov, V. A., Chuvylkin, N. D., Zhidomirov, G. M., and Kazansky, V. B., *Kinet. Katal.* **19**, 1152 (1978).

200   G. M. ZHIDOMIROV AND V. B. KAZANSKY

38. Korsunov, V. A., Chuvylkin, N. D., Zhidomirov, G. M., and Kazansky, V. B., *Kinet. Katal.* **21**, 402 (1980).

39. Chuvylkin, N. D., Korsunov, V. A., and Kazansky, V. B., *Kinet. Katal.* **24**, 832 (1983).

40. Pelmenshchikov, A. G., Senchenya, I. N., and Zhidomirov, G. M., *Kinet. Katal.* **24**, 827 (1983).

41. Pelmenshchikov, A. G., and Zhidomirov, G. M., *React. Kinet. Catal. Lett* **23**, 295 (1983).

42. Tanabe, K., "Solid Acids and Bases." Academic Press, New York, 1970.

43. Harrison, W. A., "Electronic Structure and Properties of Solids." Freeman, San Francisco, California, 1979.

44. Gibbs, G. V., *Am. Mineral.* **67**, 421 (1982).

45. Meier, R., and Ha, T.-K. *Phys. Chem. Miner.* **6**, 37 (1980).

46. Oberhammer, H., and Boggs, J. E., *J. Am. Chem. Soc.* **102**, 7241 (1980).

47. Sauer, J., *Chem. Phys. Lett.* **97**, 275 (1983).

48. Mikheikin, I. D., Abronin, I. A., Zhidomirov, G. M., and Kazansky, V. B., *Kinet. Katal.* **18**, 1580 (1977).

49. Kazansky, V. B., Gritskov, A. M., Andreev, V. M., and Zhidomirov, G. M., *J. Mole. Catal.* **4**, 135 (1978).

50. Dunken, H., Lygin, V. I., "Quantenchemie der Adsorbtion an Festkörperoberflächen." VEB Dtsch. Verlag für Grundstoff-industrie, Leipzig, 1979.

51. Little, L. H., "Infrared Spectra of Adsorbed Species." Academic Press, New York, 1966.

52. Senchenya, I. N., Mikheikin, I. D., Zhidomirov, G. M., and Kazansky, V. B., *Kinet Katal.* **23**, 591 (1982).

53. Kisel'ov, A. V., and Lygin, V. I., "Infrared Spectra of Surface and Adsorbed Species." Nauka, Moscow, 1972.

54. Sauer, J., Fiedler, K., Schirmer, W., and Zahradnik, R., *Proc. Int. Congr. Zeolites, 5th, 1980*, p. 501.

55. Senchenya, I. N., Mikheikin, I. D., Zhidomirov, G. M., and Kazansky, V. B., *Kinet. Katal.* **22**, 1174 (1981).

56. Senchenya, I. N., Mikheikin, I. D., Lumpov, A. I., Zhidomirov, G. M., and Kazansky, V. B., *Kinet. Katal.* **20**, 495 (1979).

57. Yates, D. J., and Lucchesi, P. J., *J. Chem. Phys.* **35**, 243 (1961).

58. Lucchesi, P. J., Carter, J. L., and Yates, D. J., *J. Chem. Phys.* **66**, 1451 (1962).

59. Dunken, H., and Hoffman, R., *Z. Phys. Chem.* [N. S.] **125**, 207 (1981).

60. Lagaly, P., *Adv. Colloid Interface Sci.* **11**, 105 (1979).

61. Hobza, P., Sauer, J., Morgeneyer, C., Hurych, J., and Zahradnik, R., *J. Phys. Chem.* **85**, 4061 (1981).

62. Zelenkovskii, G. M., Zhidomirov, G. M., and Kazansky, V. B., *React. Kinet. Catal. Lett.* **24**, 15 (1984).

63. Kiselev, V. F., and Krylov, O. V., "Adsorption Processes on the Surface of Semiconductors and dielectrics." Nauka, Moscow, 1978.

64. Pak, V. N., Kol'tsov, S. I., and Aleskovskii, V. B., *Kinet. Katal.* **14**, 1577 (1973).

65. Gorlov, Yu. I., Tkachenko, K. I., Konoplya, M. M., Tertykh, V. A., and Chuyko, A. A., *Adsorbtsiya Adsorbenty* **6**, 50 (1978).

66. Gorlov, Yu. I., Konoplya, M. M., Furman, V. I., and Chuyko, A. A., *Teor. Ekspe. Khim.* **15**, 446 (1979).

67. Konoplya, M. M., and Gorlov, Yu. I., *Teor. Eksp. Khim.* **16**, 166 (1980).

68. Lipmaa, E. T., Samoson, A. V., Brey, V. V., and Gorlov, Yu. I., *Dokl. Akad. Nauk SSSR* **259**, 403 (1981).

69. Rabo, J., ed. "Zeolite Chemistry and Catalysis," ACS Monogr. Vol. 171., Am. Chem. Soc., Washington, D. C., 1976.

70. Chuvylkin, N. D., Zhidomirov, G. M., and Kazansky, V. B., *J. Catal.* **38**, 214 (1975).
71. Beran, S., Z. *Phys. Chem.* [N. S.] **137**, 89 (1983).
72. Kazansky, V. B., Kustov, L. M., and Borovkov, V. Yu., *Zeolites* **3**, 77 (1983).
73. Loewenstein, W., *Am. Mineral.* **35**, 92 (1954).
74. Kazansky, V. B., *Proc. Nat. Symp. Catal, Ind. Inst. Techn., 4th, 1978* p. 14 (1978).
75. Kazansky, V. B., *Kinet. Katal.* **23**, 1334 (1982).
76. Kazansky, V. B., in "Structure and Reactivity of Modified Zeolites" (Jacobs et al., eds.), p. 61. Elsevier, Amsterdam, 1984.
77. Karge, H., and Klose, K., Z. *Phys. Chem.* [N. S.] **83**, 100 (1973).
78. Gianetti, J. P., and Perrota, A. J., *Ind. Eng. Chem. Process Des. Dev.* **14**, 86 (1975).
79. Weeks, T. J., Kimak, D. G., Bujalsky, B. J., and Bolton, A. P., *J. Chem. Soc., Faraday Trans. I* **72**, 575 (1976).
80. Detrekoy, E. G., Jacobs, P. A., Kallo, D., and Uytterhoeven, J. B., *J. Catal.* **32**, 442 (1979).
81. Vedrine, J. C., Auroux, A., Bolis, V., Desaifve, P., Naccache, C., et al., *J. Catal.* **59**, 248 (1979).
82. Ward, W., *ACS Monogr.* **171**, 118 (1976).
83. Jacobs, P. A., and Uytterhoeven, J. B., *J. Chem. Soc., Faraday Trans. I* **69**, 359 (1973).
84. Jacobs, P. A., and Mortier, W. J., *Zeolites* **2**, 226 (1982).
85. Jacobs, P. A., *Catal. Rev.* **24**, 415 (1982).
86. Kapustin, G. I., Kustov, L. M., Glonti, G. O., Bruyeva, T. R., Borovkov, V. Yu., Klyachko, A. L., Rubinshtein, A. N., and Kazansky, V. B., *Kinet. Katal* **25**, 1129 (1984).
87. Mikheikin, I. D., Senchenya, I. N., Lumpov, A. I., Zhidomirov, G. M., and Kazansky, V. B., *Kinet. Katal.* **20**, 496 (1979).
88. Freude, D., Oheme, W., Schmidel, H., and Staudte, B., *J. Catal.* **47**, 123 (1977).
89. Klyachko, A. L., Bruyeva, T. R., Mishin, I. V., Kapustin, G. I., and Rubinstein, A. M., *Acta Phys. Chem.* **24**, 183 (1978).
90. Lumpov, A. I., Mikheikin, I. D., Zhidomirov, G. M., and Kazansky, V. B., *Kinet. Katal.* **20**, 811 (1979).
91. Gagarin, S. G., Gati D., Zakharyan, R. Z., Zhidomirov, G. M., and Margolis, L. Ya., *Izv. Akad. Mauk SSSR, Ser. Khim.* **3**, 527 (1975); **8**, 1707 (1976).
92. Hightower, J. W., and Hall, W. K., *Chem. Eng. Prog.* **63**, 122 (1967).
93. Schliebs, R., Heidrich, D., Barth, A., and Hoffmann, J., *React. Kinet. Catal. Lett* **10**, 83 (1979).
94. Beran, S., Jiru, P., and Kubelkova, L., *J. Mol. Catal.* **12**, 341 (1981).
95. Chuvylkin, N. D., Zhidomirov, G. M., and Kazansky, V. B., *J. Catal* **44**, 76 (1976).
96. Senchenya, I. N., Mikheikin, I. D., Zhidomirov, G. M., and Kazansky, V. B., *Kinet. Katal.* **21**, 1184 (1980).
97. Kisel'ev, A. V., *Proc. Int. Congre. Catal., 4th, 1968*, Vol. 2, p. 194 (1971).
98. Mikheikin, I. D., Zhidomirov, G. M., and Kazansky, V. B., *Usp. Khim.* **41**, 909 (1972).
99. Mikheikin, I. D., Zhidomirov, G. M., and Kazansky, V. B., *Zh. Fiz. Khim.* **46**, 534 (1971).
100. Bosacek, V., and Dubsky, J., *Collect. Czech. Chem. Commun.* **40**, 3281 (1975).
101. Proynov, E., Halachev, T., Neshev, N., Andreev, A, and Shopov, D., *Commun. Dep. Chem. (Bulg. Acad. Sci.)* **13**, 157 (1980).
102. Beran, S., and Dubsky, J., *J. Phys. Chem.* **83**, 2538 (1979).
103. Beran, S., and Dubsky, J., *Chem. Phys. Lett.* **71**, 300 (1980).
104. Beran, S., *J. Phys. Chem. Solids* **43**, 221 (1982).
105. Beran, S., *J. Phys. Chem.* **86**, 111 (1982).
106. Beran, S., Z. *Phys. Chem.* [N. S.] **130**, 81 (1982).
107. Beran, S., Jiru, P., and Wichterlova, B., *J. Phys. Chem.* **85**, 1951 (1981).
108. Beran, S., *J. Phys. Chem.* **85**, 1956 (1981).

109. Beran, S., *Collect. Czech. Chem. Commun.* **47**, 1282 (1982).

110. Beran, S., *Chem. Phys. Lett.* **84**, 111 (1981).

111. Beran, S., Jiru, P., and Wichterlova, B., *Zeolites* **2**, 252 (1982).

112. Beran, S., Jiru, P., and Wichterlova, B., *J. Chem. Soc. Faraday Trans. I* **79**, 1585 (1983).

113. Mortier, W. J., Geerlings, P., Van Alsenoy, C., and Figeus, H. P., *J. Phys. Chem* **83**, 855 (1979).

114. Mortier, W. J., and Geerlings, P., *J. Phys. Chem.* **84**, 1982 (1980).

115. Beran, S., Z. Phys. Chem. [N. S.] **123**, 129 (1980).

116. Beran, S., Jiru, P., and Wichterlova, B., *React. Kinet. Catal, Lett.* **18**, 51 (1981).

117. Hass, E. C., Mezey, P. G., and Plath, P. J., *J. Mol. Struct.* **76**, 389 (1981).

118. Hass, E. C., Mezey, P. G., and Plath, P. J., *J. Mol. Struct.* **87**, 241 (1982).

119. Sauer, J., and Engelhardt, G., *Z. Naturforsch.* **37a**, 277 (1982).

120. Pelmenshchikov, A. G., Zhidomirov, G. M., Khuroshvili, D. M., and Tzitzischvili, G. V., in *"Structure and Reactivity of Modified Zeolites"* (Jacobs et al., eds.), p. 85. Elsevier, Amsterdam, 1984.

121. Berger, A. S., and Yakovlev, A. K., *Zh. Prikl. Khim.* **38**, 1240 (1965).

122. Borovkov, V. Yu., Alexeev, A. A., and Kazansky V. B., *J. Catal.* **80**, 462 (1983).

123. Uytterhoeven, J. B., Cristner, L. G., and Hall, W. K., *J. Phys. Chem.* **69**, 2117 (1965).

124. Kühl, G. H., *J. Phys. Chem. Solids*, **38**, 1259 (1977).

125. Kiselev, V. F., and Zarifiants, Yu. A., *Probl. Kinet. Katal.* **16**, 221 (1975).

126. Fripiat, J. J., Leonard, A., and Uytterhoeven, J. B., *J. Phys. Chem.* **69**, 3274 (1965).

127. Pelmenshchikov, A. G., Zhidomirov, G. M., and Zamaraev, K. I., *React. Kinet. Catal. Lett.* **21**, 115 (1982).

128. Malysheva, L. V., Kotsarenko, N. S., and Paukshtis, E. A., *React. Kinet. Catal. Lett.* **16**, 365 (1982).

129. Senchenya, I. N., Mikheikin, I. D., Zhidomirov, G. M., and Trokhimez, A. I., *Kinet. Katal.* **24**, 35 (1983).

130. Kustov, L. M., Alekseev, A. A., Borovkov, V. Yu., and Kazansky, V. B., *Dokl. Akad. Nauk SSSR* **262**, 1374 (1981).

131. Borovkov, V. Yu., Muzyka, I. S., and Kazansky V. B., *Dokl. Akad. Nauk SSSR* **265**, 109 (1982).

132. Pelmenshchikov, A. G., Senchenya, I. N., Zhidomirov, G. M., and Kazansky, V. B., *Kinet. Katal.* **24**, 233 (1983).

133. Tada, A., Yamomoto, Y., Ito, M., and Suzuki, S., *J. Chem. Soc. Jpn., Chem. Ind. Sect.* **73**, 1886 (1970).

134. Itoh, H., and Tada, A., *Nippon Kagaku Kaishi* p. 698 (1976).

135. Wendt, G., and Lindstrom, C. F., *Z. Chem.* **12**, 500 (1976).

136. Moffat, J. B., *Catal. Rev.—Sci. Eng.* **18**, 199 (1978).

137. Peri, J. B., *Discuss. Faraday Soc.* **52**, 55 (1971).

138. Knözinger, H., and Ratnasamy, P., *Catal. Rev.—Sci. Eng.* **17**, 31 (1978).

139. Gagarin, S. G., Kolbanovskii, Yu. A., and Plekhanov, Yu. V., *Kinet. Katal.* **21**, 919 (1980).

140. Gagarin, S. G., Teterin, Yu. A., Gubskii, A. L., Kovtun, A. P., and Bayev, A. S., *Teore. Eksp. Khim.* **17**, 507 (1981).

141. Gokhberg, P. Ya., Litinskii, A. O., Khardin, A. P., and Berjunas, A. V., *Kinet. Katal.* **22**, 1169 (1981).

142. Litinskii, A. O., *Zh, Strukt. Khim.* **23**, 40, 48 (1982).

143. Litinskii, A. O., Gokhberg, P. Ya., Khardin, A. P., and Lazauskas, V. M., *Kinet. Katal.* **24**, 230 (1983).

144. Korsunov, V. A., Chuvylkin, N. D., Zhidomirov, G. M., and Kazansky, V. B., *Kinet. Katal.* **22**, 930 (1981).

145. Baumgarten, E., and Weinstrach, F., *Spectrochim. Acta, Part A* **35A**, 1315 (1979).

146. Mardilovich, P. P., Ph. D. Thesis, Byelorussian State University, Minsk (1981).

ADVANCES IN CATALYSIS, VOLUME 34

Near-Edge X-Ray Absorption Spectroscopy in Catalysis

JAN C. J. BART*

*Department of Structural Chemistry
The Weizmann Institute of Science
Rehovot, 76100 Israel*

I. Introductory Statement

A. OBJECTIVES

Knowledge of the spatial arrangement of the atoms composing a catalytic solid (in particular the local geometry of the active sites) in relation to the electronic properties of the ensemble is a key ingredient toward the understanding of the catalytic phenomenon. Whereas electronic properties are generally best inferred from various spectroscopic studies, structural information (of ordered arrays of atoms) is usually obtained by means of some form of diffraction technique. Catalysts are often composed of solid multiphase chemical aggregates (not necessarily crystalline) in which the different phases are either closely associated or mutually interacting, and structural characterization of such entities is particularly difficult. Moreover, the catalytic experimentalist is usually faced with even more severe conditions. Namely, the object of his studies—the catalyst—cannot be considered as a static entity, which can be properly regarded for *ex situ* analytical studies. By its nature the catalyst interacts with the environment of reactants and products in a dynamic equilibrium. Consequently, *in situ* studies of working catalyst samples are actually highly desirable, a feature which severely limits the applicability of many spectroscopic tools. Since, moreover, catalysts are often composed of highly dispersed phases in low concentrations without much structural order, methods of physicochemical characterization of catalysts need usually be pushed to their limits. Not surprisingly, no single means of characterization of catalysts can normally provide insight into all aspects of

* Permanent address: DSM Research B.V., 6160MD Geleen, The Netherlands.

the catalyst structure, and consequently a multitechnique approach of catalyst characterization is an absolute need.

In this article we are concerned with the study of catalysts and related materials by means of X-ray absorption, a form of spectroscopy which simultaneously provides both structural and electronic information in a unique fashion.

The relatively slow development of the old X-ray absorption (XAS) method until its recent renaissance in the form of a variety of techniques [X-ray absorption edge spectroscopy (XAES) and extended X-ray absorption fine structure (EXAFS)] is closely related to the development of synchrotron radiation (SR) sources. The availability of intense, tunable, high-collimated X-ray sources permits rapid and simple measurement of high-quality spectra over a wide range of energies. Many catalytically interesting core states are now readily accessible, and sufficient intensity is available to measure spectra far beyond the edge. However, catalysis, and even more semiconductor technology, is one of the fields where the need for increasing photon flux is most dramatic. Due to the generally high dilution of the absorbers, this need appears even in the favorable 10-keV range; some catalytically important elements (such as Pd) are still difficult to study. Further instrumental improvements, such as dispersive X-ray spectroscopy, are expected to enhance catalytic process knowledge, particularly through *in situ* kinetic experiments.

As the region near an X-ray absorption edge is scanned in energy, the ejected photoelectron sequentially probes the empty electronic levels of the material. Theoretically, interest in core-state excitation has developed considerably since the work of Mahan (*179*) and Nozières and De Dominicis (*219*) on the singular response of the conduction electrons (in metals) to the sudden potential created by removal of a core electron. The resulting electron–hole pair excitations lead to a threshold edge asymmetry.

High-energy excitations in solids, which involve the core state of an atom, are of increasing practical interest, mainly because of the chemical specificity of such core-state energy levels. It is for this reason that such excitations are important for the study of chemically complex systems, surface properties, heterogeneous systems, etc. Another aspect of high-energy excitations is of even greater practical interest. This is the effect of local structure on the shape of photoelectric X-ray absorption edges in solids and molecules.

The X-ray absorption spectrum is conventionally divided into near-edge and extended fine structure. The so-called X-ray absorption near-edge structure (XANES) and extended fine structure (EXAFS) features are weak modulations of the ordinary X-ray absorption cross section; the largest and most detailed structures are in the near-edge region (Fig. 1). In this range, which extends up to about 50 eV off threshold, one can often distinguish pre-edge features and other shape resonances. The EXAFS region covers an energy range from about 50 eV to over 1000 eV past the absorption edge.

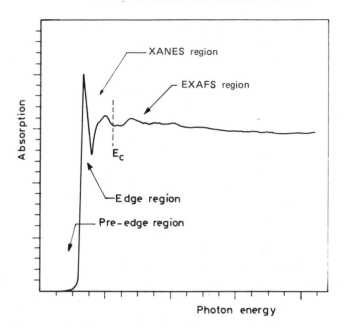

FIG. 1. Phenomenological energy ranges in an X-ray absorption spectrum; E_c is the energy where the wavelength of the initially excited photoelectron conforms to the interatomic distance.

In X-ray absorption the initial state is a dispersionless deep core level, which simplifies data analysis. This is particularly true for EXAFS. The relevant theory is understood in general terms because it can be treated approximately by perturbation theory. EXAFS features are mostly determined by the local geometry of an excited atom (161). However, EXAFS possesses a disadvantage in that structural information may be limited to the sharp distribution functions characteristic of first neighbors of the X-ray absorbing atom. Any information on higher neighbors is principally limited to the low-energy region (below about 50 eV from the edge), where the single-scattering approximations may be inadequate.

X-Ray absorption edges are usually the easiest part of the absorption spectrum to measure in complex chemical systems because of high intensities of the relevant spectral features and absence of overlap of fine structures (as may occur more readily in EXAFS). The near-edge structure probes more directly the angular momentum content of the unoccupied electronic states (83, 205, 208), but is yet generally rich in other chemical and structural information (270). XANES theory, however, is more demanding theoretically than EXAFS because the potential must be treated to all orders. Not surprisingly, therefore, the low-energy range is still troubled by several difficulties in theoretical understanding, as will be evident from this critical review. Yet, steady progress has been made since the time the fine structure of near inner-shell absorption thresholds was detected (92), as can be seen by comparing recent evaluations (7, 34, 43, 125, 237b). A more detailed theoretical appreciation of absorption-edge phenomena would provide the experimentalist with an even more powerful tool in the analysis of homogeneous and/or heterogeneous catalyst structures and processes.

X-Ray spectroscopic methods are often traditionally (erroneously) considered only as probes of the electronic structure of matter. The oldest and most heavily used methods, valence-band emission and absorption spectroscopy, probe the partial (s-, p-, or d-like) densities of states (DOS) because of core-level angular momentum selection rules. In combination, absorption and emission X-ray spectra provide a method for studying the electronic structure by probing many aspects of the unoccupied and occupied conduction-band states, respectively, of materials. Analysis of XAES data may lead from the simple confirmation that a given chemical element is present in the sample to a description of the X-ray absorber oxidation state, its site symmetry and local geometry, and electronic structure of the ligands bound to the metal (57, 64, 261). Values for the crystal field interaction, the involved Coulomb interactions, the hybridization, and the charge transfer can be obtained. Information about interaction between components of the catalytic material (e.g., supported matter and carrier) is readily provided as well as details on the interaction between the catalyst and reactant gases (chemi- or physisorption phenomena). Although the main results in the application of edge studies have been obtained for metal catalysts, oxide catalysts and aspects of chemisorption and catalyst reactivity have also been considered.

The organization of this review is as follows: after recalling early X-ray absorption catalyst studies and briefly describing modern experimental modes, it is stressed in Section III that the X-ray photoabsorption spectral range is highly structured and consists of a pre-edge region, an edge section, an immediate edge extension (~ 10 eV), XANES (up to about 50 eV), and EXAFS toward higher energies. Whereas the simple single-particle single-scattering approach is suitable for ab initio EXAFS data analysis and its application in catalysis has been reviewed (18), the physics of the low-energy features is more complicated. Pitfalls are numerous and correct interpretation often requires consideration of multiple-scattering and many-body effects. The review then continues by analyzing X-ray spectral features related to bonding and local structure determination and compares the results of various computational schemes of near-edge absorption structure with experiment to focus on certain systematic features. X-Ray absorption of matter related to catalytic interest (metals, insulators, and molecular complexes) is then considered in Section IV. Finally, Section V reviews the applications of near-edge X-ray absorption spectroscopy in a wide variety of catalytic problems. The power and present limitations of near-edge studies are illustrated on the basis of work published up to the end of 1984.

B. Near-Edge X-Ray Absorption Catalyst Studies: Past and Present

Near-edge structures in X-ray absorption spectra were originally interpreted by Kossel (146) in terms of assignment to excited electronic states, but chemical information using the near-edge structure was obtained only sporadically (199, 217).

X-Ray absorption edges of metals and oxides have been measured repeatedly in the past because these materials are of practical importance to solid-state sciences with important technological applications, including catalysis. Starting with Beeman and Friedman (24) the observed fine structures have been correlated to the energy-band structures of the metals, specifically, to the density of unoccupied states above the Fermi level. Hanson and Knight (118) have paid attention to the K edges of oxides and sulfides of Mn through Zn, following up previous work by Sanner (247). The virtual absence of low-lying absorption in these spectra is in line with the expectations for absorbers without $3d$ vacancies. Namely, low-lying absorption is understood to be independent of the symmetry and occurs when the $3d$ levels have a vacancy.

A first introduction to the use of X-ray K absorption edges in the study of catalytically active solids was presented by Van Nordstrand (217) at a time when there was little understanding of the various phenomena and when little experimental work had actually been carried out. The variation in activity of some supported chromia catalysts upon oxidative pretreatments was found to correlate with changes in absorption-edge spectra. Other observations related to the support effect for permanganate catalysts, which remain as permanganate on silica and are found as MnO_2 on activated carbon. Near-edge spectra were used to discriminate between inactive Co^0/C and active CoO/C catalysts. Van Nordstrand also describes that Co K-edge spectra of cobalt–alumina and $(Co, Mo)O/Al_2O_3$ catalysts contain cobalt in a state comparable to that in the spinel $CoMn_2O_4$, with cobalt being associated with alumina rather than molybdena. Comparison of a hydrous titania catalyst precursor, some coprecipitated mixed metal hydrous oxides, and rutile was finally used to gain information about the valence state and coordination of titanium (217).

In early work Boehm et al. (39) concluded the presence of Co^{3+} in the vitamin B_{12} catalyst, $C_{61-64}H_{86-92}O_{14}N_{14}PCo$. At the same time, K-edge spectra of supposed higher oxides of nickel, of catalytic interest, were examined by Hanson and Milligan (119). The Ni K absorption edge in supported nickel oxide catalysts was subsequently studied by Keeling (139). The same author also investigated the Co K edge position and shape of supported cobalt catalysts and has postulated two cobalt-containing species, namely, a dispersed (δ) phase at low concentrations, consisting of octahedrally coordinated Co^{2+} associated with the Al_2O_3 and $SiO_2 \cdot Al_2O_3$ supports (in accordance with magnetic susceptibility measurements) and a non-dispersed Co_3O_4 phase predominant at higher concentrations (*140, 141*). The fraction of cobalt present in the dispersed phase (N_δ) was found to decrease in the order $Co/SiO_2 \cdot Al_2O_3 > Co/Al_2O_3 > Co/SiO_2$ ($N_\delta = 0.5$ for loadings of 3.0, 1.5, and <0.5 wt.% Co on these supports). The variations suggest a connection between acidity and ability for dispersed phase formation. Since the dispersed phase in Co/Al_2O_3 is more difficult to reduce than Co_3O_4, this indicates an intimate association of the δ-phase cobalt with the structure of the support, while the Co_3O_4 particles do not interact with the support.

In an early series of papers, P. H. Lewis at Texaco helped pioneer the use of XAS to study samples of catalytic interest, in particular supported nickel and platinum catalysts (*165–168*). The reasoning behind his efforts was that important physical properties of transition metals along with catalytic activity are governed by the electrons that are in or near the electronic levels that form the valence bands. Just above the valence bands are several bands which are partially or completely electron free or empty. Lewis has applied XAS in order to observe how the threshold region is affected by (i) the size of the metal crystallites, (ii) interaction of the metal with adsorbed gas, and (iii) the effect of the support materials. In order to meet these objectives $Ni/\gamma\text{-}Al_2O_3$ (*165*), Ni/SiO_2 (*166*), and $Pt/\eta\text{-}Al_2O_3$ (*167*) catalysts were examined as well as the chemical state of platinum in Ca-exchanged zeolite *Y*, before and after reduction and outgassing (*168*).

In 1969 Monsanto's Wolberg and Roth (*310*) used X-ray *K*-absorption edge spectroscopy to study Cu^{2+} supported on $\gamma\text{-}Al_2O_3$ and distinguished three phases: isolated cupric ions, a copper aluminate surface phase, and cupric oxide. At low Cu^{2+} ion concentrations for given surface area conditions, a surface phase is present with a $CuAl_2O_4$-like structure. When samples containing the aluminate surface phase are reduced and reoxidized in the absence of moisture, the aluminate is converted to cupric oxide. Friedman *et al.* (*93*) later extended the investigation, leading to a more complete understanding of this important catalyst, which was among the first for which XAS has generated practical new information about the chemical nature of a highly dispersed species.

Although originally mainly the more pronounced near-edge features were observed and compared or interpreted (usually empirically), it is evident from a recent review (*18*) that today considerably greater emphasis is given to the study of the extended fine structure. This stands in relation to the improved experimental methods for detection of weak modulations and to the currently more advanced theoretical description of the EXAFS part of the spectrum. Complete understanding of the Kossel structure at the threshold part of an element's inner-shell spectrum, which contains among others valence orbital, ionization, and chemical shift information, is relatively slow due to the complex physical phenomena near the photoionization threshold and a sensitivity to the state of aggregation. Roughly speaking, the 0–10-eV range beyond threshold can most appropriately be considered on the basis of the local electronic structure, whereas the 10–50-eV range comprises multiple-scattering resonances of the excited photoelectron within the neighboring shells (*47, 230*) and can be applied for local structure determination in the form of a technique renamed XANES or NEXAFS (near-edge X-ray absorption fine structure), which is complementary to EXAFS. The importance of the near-edge region is slowly becoming better appreciated by catalytic chemists under the stimulus of advances in simulation of this part of the spectrum (*84, 155, 210*).

II. Description of Experimental Requirements

Briefly, the total linear absorption coefficient μ (cm^{-1}) varies as a function of the wavelength and the nature of the absorber as the photon energy is varied across and beyond the absorption edge. The logical setup for an absorption experiment in transmission mode therefore consists of three primary components (Fig. 2a): (i) an X-ray source, (ii) a monochromator (and collimator), and (iii) a detector. In this case Beer's law,

$$I(E) = I_0(E) \exp(-\mu(E)x) \qquad (1)$$

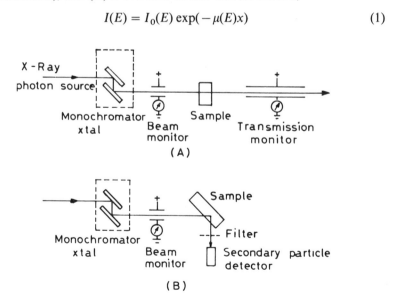

FIG. 2.   Schematic diagram of X-ray absorption measurements by (A) transmission and (B) secondary particle (photons, electrons, ions) detectors. From Brown and Doniach (*48a*).

applies, where I_0 and I represent the photon intensities before and after passage of the beam through the absorber of thickness x.

For details concerning catalyst sample preparation, photon sources, and element range for XAS studies, spectrometer design, and detection schemes, the reader is referred to Bart and Vlaic (18). XANES measurements require a higher resolution mode than EXAFS but a lower flux. The resolving power of laboratory XAS spectrometers is presently good enough for the study of near-edge structures.

III. Fundamentals of Near-Edge X-Ray Absorption Spectroscopy

The contribution to the X-ray absorption coefficient due to the excitation of a deep core level may be expressed as $\mu_c = n_c\sigma_c$, where n_c is the density of atoms with the core level of concern and σ_c is the absorption cross section for this level on a single atom. Assuming the X-ray field to be a small perturbation, the latter can be evaluated from the golden rule transition rate per unit photon flux. The general X-ray absorption cross section is given by

$$\sigma_c = \frac{4}{3} \pi^2 \alpha h v \sum_f |\langle f|\mathbf{r}|i\rangle|^2 \, \delta(E_i - E_f + hv) \tag{2}$$

where α is the fine-structure constant, hv is the energy of the absorbed photon, and $\mathbf{r}$ is the interaction operator. In a single-particle model, the initial state $|i\rangle$ is the core state from which the photoelectron is emitted, and $\langle f|$ represents the final state of the photoelectron, which must be above the Fermi level for any transition to occur. In order to solve the problem, we need a model to describe the effective potential seen by the final-state photoelectron.

Interpretation of X-ray absorption spectra (and most other types of core-electron spectra) is complicated by the creation of a core hole in one of the atoms in the solid. In many cases (e.g., for transition and rare-earth metals) the magnitude of this effect is not known as yet. Further, these spectra depend on the excited states of the electronic system, which are less well understood than the corresponding ground-state properties (202).

A.   THE EVENT

Absorption-edge spectroscopy deals with electronic transitions from a core atomic level to unoccupied conduction states above the Fermi level. In this process X-ray photons promote a bound electron from an inner level to the first empty energy state of the absorbing atom, thereby increasing the X-ray absorption coefficient. An incident photon is completely absorbed by an atom and used for ejection of a lower level electron when the energy hv exceeds the core-state binding energy E_B. The kinetic energy of the ejected core electron is $E = hv - E_B$ (relative to the Fermi level of the sample). An absorption spectrum thus reflects the transitions of the ejected photoelectron to empty states, bound or unbound, discrete or broad, atomic or molecular, of proper symmetry. This spectrum is superimposed on an absorption background due to transitions to high-energy continuum states from other occupied levels of the same atom and of other atoms. The probability of an X-ray photon to be absorbed by a core electron depends on both the initial and final states of the excited photoelectron [Eq. (2)].

Dipole selection rules ($\Delta l = \pm 1$) apply in the absorption process (see however refs. *135*, *138*). The fine structures of the K edge (s-to-p transition) and L_{iii} edge (p absorption) of an element are not identical. The structure of L_i (s absorption) is also quite different from the $L_{ii,iii}$ structures, which are alike, and correspond to processes in which an X-ray photon is absorbed by promoting an electron from $2p$ core states to d states (with a small and calculable p-to-s contribution). The L_{ii} X-ray absorption edge arises from $2p_{1/2}$ core states, while the L_{iii} X-ray absorption edge arises from the $2p_{3/2}$ core states and is at a lower energy. The d states are usually by far the most strongly peaked and most favored in the transition (*88*).

For X-rays well above the absorption edge ($\Delta E > 30$ eV), the final electron is unbound, that is, has a continuous distribution of allowed energies, and the density of allowed states $\rho(E_f)$ at the final-state energy E_f is then a smooth function which may be approximated by the density of states of a free electron of energy

$$E = E_0 + \hbar^2 k^2 / 2m \qquad (3)$$

The threshold energy E_0 is the energy of a free electron with momentum $k = 0$. The position of the main absorption edge itself varies with chemical bonding. At the high-energy side of each main absorption edge (within a range of a few tens of electron-volts), a fine structure reflects jumps to higher unoccupied levels, converging to the continuum of free-electron levels. The edge range is a fingerprint of the electronic structure of the absorber and provides information about binding energies, quantum numbers, and multiplicities of low-lying bound electronic excited states of the ionized absorbing atom and of low-lying resonant electronic states in the continuum of the absorbing atom. The qualitative features of the absorption peaks in the near-edge region are also sensitive to the chemical valency of the absorbing atom and the symmetry of the surrounding near-neighboring atoms. The electronic processes which determine the X-ray absorption features can be divided into atomic-like effects and solid-state effects. In the threshold spectra the main solid-state effect is the excitation of core electrons to the conduction bands.

Most XAS studies begin with general scans in transmission mode and expand to more detailed experimentation in the near-edge region. Whereas the high-energy side of the edge spectrum provides a wealth of quantitative structural information by examination of the EXAFS modulations (fluctuations of a few percent in μ), subtle electronic changes of the system are the features of principle interest in the low-energy range. Consequently, low-kinetic-energy XAS spectra probe the electronic state of the absorber and the local structure in matter through the electron scattering (back scattering) by neighbor atoms of the emitted electron at the central atom (absorber). This electronic and structural information is much desired for the description of catalytically active sites.

B. SIMPLIFIED SINGLE-PARTICLE MODEL

One basic assumption almost universally made in interpreting XANES is the validity of the single-particle picture, which assumes that a single-particle density of states can explain the various spectral features. This assumption has been strengthened by recent calculations (83, 126, 155, 205, 206). The appealing result of 15 years of both theoretical and experimental work is that a one-electron approach is usually really relevant. However, in some instances this simple single-particle picture for XANES should be used with caution (13, 273).

In the one-electron transition model it is assumed that only one core electron is excited to an unfilled state present in the initial, unperturbed solid. The remaining electrons are assumed to be unaffected, remaining frozen in their original states. In essence the one-electron model describes the potential seen by the final-state electrons as nonoverlapping spherically symmetric spin-independent potentials (muffin-tin scatterers) centered around an atom from which the X-ray cross section from a deep core level of an atom to final states above the Fermi level can in principle be calculated for any energy above threshold.

Failure of one-electron models to provide more than a qualitative understanding of the optical properties of solids in the X-ray region has long been recognized (48, 230). This failure is particularly evident near thresholds marking the onset of transitions from a given core level, where usually sharp structures, either in the form of line resonances or steep, atomic-like thresholds, are observed. In fact, very close to the edge (< 10 eV) many-body effects, such as singular behavior at threshold (219), multiplet structure (230), or plasma resonances (*121*), are to be considered. Also (relatively weak) shake-up processes have been observed in the X-ray absorption spectrum of atoms (*53, 143*), molecules (*42, 53*), and solids (*273*). A shake-up transition is a multielectron excitation produced by the sudden creation of the core-hole potential induced by the absorption of an X-ray photon. This sudden perturbation may excite a passive electron (not directly excited by the X-ray photon) into a discrete bound state (shake-up) or into the continuum (shake-off). The shortcomings of the single-particle (one-electron) model are less evident far from thresholds.

Within the single-particle picture some disagreement exists in whether the density of states of the initial solid (*83, 155, 205, 206*) should be employed or the density of states of the final system with the excited core hole (*126, 180, 302, 303*). Determination of the effective single-particle potential seen by the final-state electron is difficult and involves the screening reaction of the electrons to the developing presence of a core hole. In case of metals it has been assumed that the effective single-particle potential is that of a screened hole (*252*). Data presented by Materlik *et al.* (*188*) may be viewed as evidence for the final-state rule which claims that single-particle calculations can give realistic X-ray absorption spectra provided final-state energies and wave functions are calculated in the presence of the core-hole potential.

Within the single-particle picture the photoabsorption cross section μ due to core excitations factorizes into atomic and solid-state terms: $\mu = \mu_{\mathrm{at}}\chi$, where χ is the modulation due to the neighbors. This leads to the following simple understanding of X-ray spectra. The effect of the central potential is expressed by the atomic transition rate or the dipole transition strength connecting a core state to a muffin-tin orbital in a free-electron metal, which determines the overall magnitude and shape of a particular solid-state spectrum. It has a rather smooth energy dependence and varies rather weakly with the atomic number Z. Many-body effects appear in the atomic-like factor whenever the spin–orbit splitting becomes comparable to the electron–hole exchange interaction producing a shift of oscillator strength toward the higher energy line (*308*).

For extended systems the solid-state term, χ, due to the scattering of the ejected electron on the surrounding atoms can be described using either scattering or band-structure theory. The solid-state factor determines the fine structure of the spectrum. It has a rapidly varying energy dependence which is characteristic of the band structure of the system and shows up as oscillations about the atomic term. Features in the solid-state term lying at higher energies cannot be examined simply in these terms. The solid-state factor χ exhibits deviations from the single-particle picture due to broadening of the spectra (lifetime effects) and energy shifts (due to exchange-correlation effects) (*188*).

1. Many-Body Effects and Threshold Singularity

X-Ray spectral studies of solids became important in the 1930s as a means of confirming the simple concepts concerning the band structure of solids. Prior to 1967 virtually all X-ray absorption data were interpreted solely in terms of one-electron theories (87), but these are unable to explain certain spectral features, in particular anomalous $L_{ii,iii}$ X-ray edge spikes of some simple metals. Solid-state many-body effects, which are so evident in XPS, are usually greatly suppressed in XAS (250), but are sometimes dominant. The one-electron formalism is especially inadequate very near the edge. Although of little direct importance for catalysis studies, a discussion of the threshold singularity problem (as expressed by unusually peaked or rounded X-ray edge shapes) is appropriate in order to set the scene for the various factors influencing X-ray absorption spectra.

Absorption of the X-ray makes two particles in the solid: the hole in the core level and the extra electron in the conduction band. After they are created, the hole and the electron can interact with each other, which is an exciton process. Many-body corrections to the one-electron picture, including relaxation of the valence electrons in response to the core-hole and excited-electron–core-hole interaction, alter the one-electron picture and play a role in some parts of the absorption spectrum. Mahan (179–181) has predicted enhanced absorption to occur over and above that of the one-electron theory near an edge on the basis of core-hole–electron interaction. Contributions of many-body effects are usually invoked in case unambiguous discrepancies between experiment and the one-electron model theory cannot be explained otherwise. Final-state effects may considerably alter the position and strength of features associated with the band structure.

A theory predicting the structure very close to threshold (peaking and rounding-off effects) was originally given by Nozières and De Dominicis (219) and was given a great deal of attention by Mahan for simple metals (179, 181). In these solids the core hole charge is screened by conduction electrons. As a result, the X-ray absorption coefficient shows a singular behavior right at the threshold arising from strong interaction between the excited electrons above the Fermi edge and the deep core hole.

The X-ray singularity problem was originally solved in the asymptotic limit and the complicated many-body problem was turned into an effective one-particle problem (219). For the X-ray photon frequency ω very close to the threshold frequency ω_0, the absorption spectrum $\mu(\omega)$ for the process in which a deep, structureless core electron is excited to the conduction band by the absorption of an X-ray of frequency ω is expressed by the power law

$$\mu(\omega) = \mu_0 \left[\frac{D}{\omega - \omega_0} \right]^\alpha, \qquad (\omega - \omega_0) \ll D \qquad (4)$$

where the dimensionless coupling constant α describes the strength of the attractive interaction between conduction electrons and the deep core hole left behind. It follows that the transition rate at threshold is either infinite ($\alpha > 0$) or zero ($\alpha < 0$), as observed. Within the range of validity of the equation [about $0.03E_F$ of threshold (285)] experimental data are in accordance with the Mahan–Nozières–De Dominicis (MND) theory.

Subsequently the adequacy of the MND theory has been challenged and the influence of more conventional effects has been stressed (81, 82). Finally, Citrin et al. (61) have carried out a complete analysis of one- and many-body electron effects on X-ray absorption edge shapes (K and $L_{ii,iii}$ thresholds) in simple metals (Li, Na, Mg, and Al). These include the transition density of states (TDOS) (109), the core-hole lifetime (211) and phonon broadening, the MND many body response of the conduction electrons, the spin–orbit exchange, and the instrumental response function. It appears that all these phenomena contribute to the shape of the edge, though at times to different extents: the $L_{ii,iii}$ edges of Na, Mg, and Al are dominantly peaked by the many-body interactions (as described quantitatively by the MND theory, confirming the essential validity of the theory), whereas the K edges are variously affected by the other factors mentioned above. Therefore, in a particular case either one- or many-body phenomena may dominate. The influence of many-body effects on deep core-level spectra of metals has recently been reviewed by Hedin (120). Fujikawa (95) has successfully interpreted many-body effects observed in XANES in terms of Dyson orbitals (fully correlated one-electron orbitals in a many-electron system).

2. Extended Theory of the X-Ray Edge Singularity

The MND theory is rigorously valid only asymptotically close to threshold. Experimental resolution and core-hole lifetime broadening make this region inaccessible. Since the edge singularities, even when they exist, occupy only a narrow energy range of 0.5 to 1.0 eV near threshold, another theory is needed for the other parts of the spectra, that is, in the region between the pure many-particle edge region (~ 0.5 eV from the edge) and the pure one-particle regime, in which the transition density of states (TDOS) provides a good basis for calculating $I(\omega)$. There has been considerable recent theoretical progress on extending the solution away from threshold either in numerical (72, 181, 222, 285, 302) or analytical (114, 234) form. According to Hänsch et al., the MND model is conceptually inadequate for describing X-ray spectra for simple metals in the off-edge region because the dynamical aspects of the electron–electron interaction are neglected by construction of the model (115). Yet, the method is sufficiently simple and accurate. Using both multiple-scattering and determinant techniques, it is often assumed or asserted that in the remaining parts of the spectra the many-body effects can accurately be ascribed within a single-particle (one-electron) transition-rate amplitude expression. The question then arises *which* one-electron theory is suitable. According to Mahan (*181*), the one-electron theory should use the final-state potential of the central atomic core which has the screened core hole in absorption (but not in emission). It should be noted that the final-state rule, applied in numerical studies of the X-ray absorption problem (*72, 181, 303*), is still debated. Clearly, after removing an electron from the core state of an atom the potential of the central atom is different from the others since it appears as an ion of charge $Z + 1$.

Hänsch and Ekardt (*114*) have recently solved the ND equations governing the singular behavior of the X-ray response function for the more realistic Thomas–Fermi potential $-e^{-\lambda r}/r$ describing the attractive interaction between the deep hole and the conduction electrons. Since the solutions to the ND equations are valid not just for $t \rightarrow \infty$ but over the whole region of time, this leads to a response function in frequency space away from the edge. In fact, in time rather than energy theory, the probability amplitude for a hole lifetime of exactly t is the Fourier transform of that hole's energy spectrum. Numerical results are given for the Na $L_{ii,iii}$ edge.

C. SINGLE-SCATTERING THEORY OF NEAR-EDGE X-RAY ABSORPTION

Within the single-particle treatment of the final electron state, EXAFS is usually considered to be due to short-range order and XANES to long-range order (*8*). Correspondingly, the theoretical approach to EXAFS has been via a few single-scattering events from near neighbors (nn) for the outgoing final electron (*162*) and via band-structure calculations for XANES (*205*). [The single-scattering approximation for EXAFS is most likely successful for relatively loosely packed structures and breaks down for closely packed structures, such as most metals, in energy ranges where the scattering due to a single atom is large (*6*).] However, according to Schaich (*252*), a complete treatment from either point of view should lead to the same result. This has been substantiated by Durham *et al.* (*83*), who extended the short-range approach beyond the standard EXAFS formula by including multiple-scattering (MS) events. The scheme has an advantage over the band-structure method in that it can be applied to disordered materials.

Müller and Schaich (*207, 253*) have recently presented a simple single-scattering theory of X-ray absorption by calculating the efficiency of single-electron excitation from deep core levels to the continuum which reproduces well the results of more involved single-particle calculations. In this approach the scattering of the excited electron from the absorbing atom is treated exactly, while the influence of neighboring atoms is approximated by retaining only single backscattering events. The theory yields tractable equations that reduce to the conventional EXAFS version (*162*) for energies well above threshold and provides an appreciable improvement at low energies with small computational burden. The only difference between the exact single-scattering theory and the conventional EXAFS theory rests in the former's proper treatment of scattered wavelets. Similar single-scattering formulas for X-ray absorption have recently been derived independently by Gurman and Pettifer (*111*). The single-scattering approach not only yields accurate results over the entire energy range, and is reasonably successful close to the edge (the supposed multiple-scattering regime), but also offers a different conceptual view of the source of the near-edge structure.

Extra scattering terms are in principle important near threshold, where one observes good agreement between the generalized single-scattering expression of Müller and Schaich (*207*) and the exact MS calculation, and in case of shadowing (*162*). However, MS effects may be corrected for even when the scattering is strong (*254*) since both long MS paths and distant scattering events produce no structure that survives Lorentzian energy averaging in any single-particle calculation (a necessary procedure in order to approximately account for various decay processes, such as inelastic collisions, core-hole lifetime, and experimental resolution). In other words, with sufficient energy broadening only the effect of a single backscattering from the nearest neighbors is felt and any discrepancies between band-structure and single-scattering results disappear. It has been pointed out by Bunker and Stern (*52*) that except for very close to the absorption edge and/or for very short bond lengths, the shadowing effect is the main multiple-scattering effect missing from the single-scattering approach. Even this shadowing modification is easily included in the Müller and Schaich formalism (*207, 254*) via renormalization factors: relatively simple expressions may also be derived for other multiple-scattering paths (*41, 293*).

1. Single- versus Multiple-Scattering Theory

Until recently it has been claimed that many-body effects and higher order multiple-scattering processes set the scene near threshold. Consequently, XANES and EXAFS were originally distinguished on the basis of the scattering phenomena (*271*). Whereas single scattering (SS) is adequate in the EXAFS region, multiple-scattering (MS) corrections have explicitly been introduced in techniques developed for calculating XANES (*84, 155, 210*). If we consider that the photoelectrons are in the same energy range as LEED electrons, but in case of XAS consist of outgoing spherical waves from a central atom rather than plane waves as in LEED, it is at least intuitive that multiple-scattering effects from near-neighbor shells cannot be ignored. They appear to be especially important in the low-energy spectral region for closed-packed structures. Multiple-scattering expansion formulas with curved propagation waves have been used up to very near the rising edge of an X-ray absorption spectrum from deep core states. The expansion procedure breaks down at the lower energy bound, below which resonances in the near-edge region are present (in some cases within 1 Ry of the rising edge). Such a lower limit is material and polarization dependent. The energy range of full MS resonances (i.e., of the XANES region of XAS spectra) appears to be very sensitive to the shortest interatomic distance R of the local atomic cluster and may vary from 10 to 100 eV in different systems, depending upon the geometrical arrangement of the neighboring atoms. The general framework of full MS theory was until recently held to account for the XANES spectra of molecules and more complex systems (ordered and disordered solids).

The idea that MS dominates XANES has now been challenged (*207*), and a SS XANES theory (like EXAFS) has been put forward in which the planewave and short-wavelength approximations are no longer valid. This contrasts to the fact that according to Norman *et al.* (*218*) the SS theory is incapable of explaining the oxygen K-edge XANES spectrum of NiO. Müller *et al.* (*205, 206, 208*) have shown that a single-particle approach is adequate for the description of X-ray absorption in crystalline transition metals. Comparing various single-particle theories, it appears that a single-scattering approach offers considerable computational advantages in the near-edge region (*207*). Figure 3 compares the calculated single-scattering near-K-edge absorption using the generalized scheme of Müller and Schaich (*207*) with that derived from a band-structure approach (*205, 206, 208*) and according to the conventional EXAFS formula for the same structure (copper), scattering potential (periodic muffin-tin), and broadening function excluding manybody effects. These three approaches make essentially different approximations to χ_l, where

$$\chi_l = 1 + (-)^l \sum_j N_j \, \mathrm{Im}\left[e^{2i\delta_l} \frac{e^{2ikR_j}}{kR_j^2} f(\pi) \right] \tag{5}$$

where Im denotes imaginary part and the sum is taken over j shells of N_j neighboring atoms at a distance R_j; δ_l' are the scattering phase shifts of the absorbing atom and $f(\pi)$ is the backscattering amplitude of a single neighboring atom. All scattering properties are evaluated at energy E. It is important

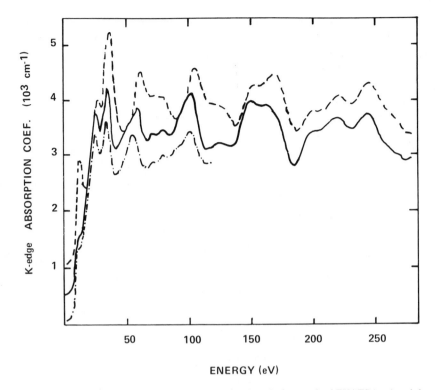

FIG. 3.   Comparison between the single-scattering (——), the standard EXAFS (---) and the band-structure K-edge ($\cdot$—$\cdot$) absorption spectra of copper. The energies are measured with respect to the muffin-tin zero of the potential, whereby the threshold energy is $E_F = 8.96$ eV. From Müller and Schaich (207).

to notice that Eq. (5) treats the effect of neighboring atoms only via single backscattering events and that except for the case of absorption of isolated atoms χ_l differs from unity. The band-structure scheme represents an exact solution for the given model potential since it contains all MS contributions and all shells of neighbors. In both the single-scattering and standard EXAFS scheme the contribution of the first 11 shells of neighbors has been included with partial waves up to $l_{max} = 10$. As may be seen from Fig. 3, the exact band structure and single-scattering schemes agree well over the first 100 eV above the edge both in energy position and in amplitude; discrepancies with the standard EXAFS results grow larger closer to threshold. As expected, the single-scattering and EXAFS schemes are identical beyond 200 eV from the edge. Similar behavior has been noted for other edges and transition metals. The agreement between single scattering and the exact results improves substantially with increasing spectral broadening. This indicates that MS events give rise mostly to rapid oscillations and that *all* major structural features of the spectra arise from SS events off the first few neighboring shells of atoms. Clearly, in the case of copper, XANES can be explained without invoking multiple scattering, as suggested earlier (*105*).

The cited controversy is of considerable interest. Namely, if SS dominates XANES its analysis would allow the same kind of information as EXAFS to be obtained. On the other hand, if MS dominates XANES bond-angle information can be extracted, which is inaccessible to EXAFS (*83, 293*). As noticed by Bunker and Stern (*52*), two types of MS events exist, namely, for approximately collinear central and scattering atoms (MS-1) and for the case far from collinearity (MS-2). The first case has similar importance in the XANES and EXAFS regions. Here the intervening atom focuses the photo-electron onto the backscattering atom and enhances its scattering. This "focusing effect" is large throughout the whole absorption fine-structure range. Consequently, it is possible to approximate the absorption spectrum by a relatively simple SS calculation when MS-2 is small [as suggested by Müller and Schaich (*207*)] modified by addition of a MS-1 term. Figure 4 illustrates a mechanism whereby multiple scattering gives higher sensitivity. In the metal carbonyl $X(CO)_6$ oxygen scattering is intensified due to carbon in a linear XCO arrangement. The large-angle MS-2 effect is negligible in the EXAFS region because large-angle scattering is weak here and is generally important within a few eV of the edge for bond lengths less than about 1.6 Å and lacking inversion symmetry (*52, 277*). This limits the usefulness of XANES as a supplement to EXAFS. Only the MS-2 type contains new bond-angle information, which can be extracted when its magnitude is substantial

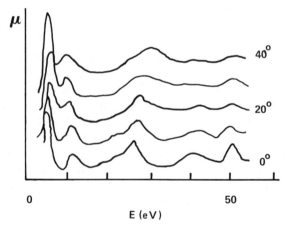

FIG. 4.   *K*-edge XANES calculated for the model carbonyl molecule $X(CO)_6$ for various angles of rotation, θ, of the oxygen shell. From Pendry (*233*).

(292). The general MS contribution could be a mixture of the two types, but XANES and EXAFS are physically distinct only when MS-2 is large.

2. White Lines

In extreme cases a multiple-scattering, sharp resonant structure can result in which the electron is in a quasi-bound state (155). One example is the "white line," which is among the most spectacular features in X-ray absorption and is seen in spectra of covalently bonded materials as sharp (~ 2 eV wide) peaks in absorption immediately above threshold (i.e., the near continuum). The cause of white lines has qualitatively been understood as being due to a high density of final states or due to exciton effects (56, 203). Their description depends upon the physical approach to the problem: for example, the $L_{ii,iii}$ white lines of the transition metals are interpreted as a density-of-states effect in band-structure calculations but as a matrix-element effect in scattering language.

White lines in the K- and L_i-edge spectra must be caused by p-symmetric final states. They occur when the positively charged hole remains sufficiently unshielded (as in a semiconductor or insulator) so as to bind the p state and produce exciton levels, thus giving the narrow energy width and high density of states to produce a strong transition. Although white lines are mainly atomic effects, orbitals of p symmetry hybridize with those on neighboring atoms to give an unoccupied antibonding level into which an electron can be excited from the K shell. Therefore, their characteristics are also determined by the first neighbor coordination shell. Its sharpness is a measure of how complete the shell is and thus how effective at preventing escape of the excited electron.

The shape of the white line is affected by multiple-scattering excitations and by the fact that the L-shell transition is degenerate in energy with a continuum of transitions. The continuum may cause the white line to become skewed at the high-energy side, as in the case of Pt, where a tailing of up to 30 eV past the edge is observed.

The L_i X-ray absorption edge of transition and rare earth metals does not usually have a white line, as opposed to the $L_{ii,iii}$ edges. This difference stands in connection to the fact that L_i initiates from the $2s$ state as opposed to the $2p_{1/2}$ (L_{ii}) and $2p_{3/2}$ (L_{iii}) states for $L_{ii,iii}$. While the latter probe the s- and d-symmetric portions, the L_i and K edges probe the p-symmetric portion of the density of final states.

Leapman et al. (159, 160) have examined the general features (threshold energies, widths, and intensity ratios) of $L_{ii,iii}$ white lines in $3d$ transition-metal oxides (TiO_2, Cr_2O_3, FeO, NiO, CuO), which arise from transition into a partially filled, large d density of states; the relative white-line intensities remain as yet unexplained.

Müller *et al.* (*206*) have been concerned with the $L_{ii,iii}$ white lines for Ca, Ti, Cr, Co, and Cu and L_{iii} white lines for Sr, Zr, Nb, Ru, Rh, and Pd. In the latter series the shape of the white lines with increasing atomic number is determined by (i) the narrowing and increase of the $4d$ density of states and (ii) progressive filling of the $4d$ band. Crystal structure effects are much less conspicuous in the L-edge spectra than in the K-edge spectra. The shapes of the individual $L_{ii,iii}$ white lines of the $3d$ elements Ca, Ti, Cr, Co, Ni, and Cu are similar to those encountered in the $4d$ series, but the $2p_{1/2}-2p_{3/2}$ spin–orbit splitting here is so small (from 6 eV for Ca to 20 eV for Cu) that the L_{ii} and L_{iii} spectra are superimposed. The relative size of the L_{iii} and L_{ii} white lines follows approximately the multiplicity ratio 2 to 1 of the $2p_{3/2}$ and $2p_{1/2}$ core states.

Also, molybdenum X-ray absorption $L_{ii,iii}$ edges of a variety of Mo compounds having varying oxidation states and coordination environments contain intense white lines. For high oxidation states the lines are split, reflecting the ligand-field splitting of the d orbitals. The splitting is smaller for tetrahedrally than for octahedrally coordinated molybdenum, which is consistent with predictions from ligand-field theory.

Our present understanding of the white-line phenomenon is incomplete. Without going into details, this is shown by the following examples. The L_{iii} X-ray absorption edge of Ti shows no white line (*90*) even though Ti is a transition metal with a high density of unoccupied d states. Also, the experimental results for Ni do not fit into the present theoretical framework. In metallic Pt a white line is observed at the L_{iii} X-ray absorption edge but is almost absent at the L_{ii} edge, whereas for α-PtO_2 a white line is observed at both the $L_{ii,iii}$ thresholds.

An example of an exciton effect is found in Cu_2O. Hulbert *et al.* (*130*) have observed that the prominent white lines at the Cu $L_{ii,iii}$ thresholds of Cu_2O are like those observed in CuO even though the Cu d states are filled in Cu_2O. It is known (*160*) that the ratio of the white-line peak height to the L_{iii} absorption step in CuO (d^9 configuration) is the smallest of the third-row transition-metal oxide series (TiO_2, Cr_2O_3, FeO, NiO, and CuO) due to the nearly filled Cu $3d$ shell in CuO. On the basis of this inverse relation between white-line strength and Cu $3d$ shell filling—d-like conduction bands are the usual criterion for appearance of a white line—the filled Cu $3d^{10}$ configuration in Cu_2O should preclude the existence of a white line such as that observed for CuO. Actually the two lines are about equally strong, though qualitatively different (the CuO white line is symmetric, whereas the Cu_2O edge feature is narrow and asymmetrically broadened on the high-energy side). The strong $L_{ii,iii}$ white lines of Cu_2O are interpreted as relatively local core excitons. Cu_2O has a very open crystal structure with a strong redistribution of charge which might tend to favor local atomic-like excitations. The strong edge features can be understood in part as transitions to antibonding MO states of the linear $O-Cu-O$ molecular cluster peculiar to copper in the Cu_2O crystal structure. Similarly, for the linear SO_2 molecule the $L_{ii,iii}$ threshold spectrum is also quite singular (316).

Finally, it is noticed that white lines are accompanied by exceptionally large changes in anomalous scattering.

D. NEAR-EDGE X-RAY ABSORPTION SPECTROSCOPY AND BONDING

1. Pre-Edge to Near Continuum

The edge region at the absorption threshold is sensitive to the electron structure near the Fermi level. The profile of an edge at the low-energy side (low-energy Kossel structure), which usually exhibits relatively weak absorption, reflects discrete and allowed electronic transitions to unoccupied high-energy bound quantum states just above the Fermi level but below the dissociation limit (146) and has been assigned to the partial local density of states of the conduction band, complicated by many-body effects. In the past, the threshold (or preionization) fine structure, which extends over a few tens of electron-volts, has provided basic information on the electronic structure of atoms and has been the experimental support for the quantum theory. Edge literature and absorption-edge spectra up to 1952 are collected in Landolt-Börnstein (157).

Comparison between the core-level X-ray absorption spectroscopy (XAS), emission (XES), and X-ray photoemission spectroscopies (XPS) usually shows that the spectral edges rarely coincide with each other and with the Fermi level. It is common practice, however, to place E_F at the emission threshold which corresponds to a fully relaxed ion core (16). For defining the structure of the edge, an energy resolution of at least 1–2 eV is required in the range of 5–20-keV X-ray photons. This can be achieved with Bonse–Hart channel-cut silicon monochromator crystals.

The Kossel model (146) of single-electron transitions to unoccupied states has been applied to the interpretation of the absorption-edge structure of isolated atoms (inert gases) as well as to molecules and solids, in which case use is made of band-model calculations, including the possible existence of quasi-stationary bound states as exciton states. Parratt (229), who has carried out the first careful analysis of the absorption spectrum of an inert gas, assumed that dipole selection rules govern the transition possibilities, with allowed transitions being $1s \rightarrow np$.

Interpretation of the observed preionization shifts and splittings in XAES spectra for covalently bonded molecular complexes has been attempted by molecular-orbital calculations (65, 66, 133, 256), which for large molecules are forbiddingly tedious. The electronic transitions associated with X-ray absorption are allowed by the selection rule $\Delta l = \pm 1$ (229). The transitions are decided by a number of considerations, including the symmetry and overlap of wave functions and the covalent mixing parameters in occupied and unoccupied orbitals.

The K absorption-edge maximum corresponds to the allowed $1s \rightarrow np$ transitions, which merge into the continuum at higher energy. Below the maximum are subsidiary peaks and shoulders reflecting transitions to empty orbitals, for example, $1s \rightarrow 3d$ and $1s \rightarrow 4s$. Their intensities are expected to be governed by dipole selection rules, $1s \rightarrow 3d$ and $1s \rightarrow 4s$ being weaker than $1s \rightarrow 4p$. The spacings between these levels can be perturbed from those obtained in the analogous free-ion states since the vacant orbitals are hybridized with the filled orbitals of ligand atoms, leading to antibonding orbitals of elevated energy. This hybridization is, of course, quite different for the d, s, and p orbitals since they are of quite different sizes and symmetries. The small maxima at the low-energy side of the edge are of considerable importance in the assignment of bonding orbitals (199). Obviously, the atomic description in terms of the allowed $1s \rightarrow np$ transitions does not explain the sensitivity of these transitions to the chemical state of the absorber.

Since XAES spectra correspond to electronic transitions from inner electronic levels to outer unoccupied levels, which are essentially the regions of interatomic orbital overlap, the energies and transition probabilities depend upon the formal charge of the absorbing atom, the actual charge, the site symmetry, the degree of covalency, and the nature of the ligands. In favorable cases (e.g., by systematic variation of the ligand set, coordination type, valency state, etc. of the absorber) the technique can be exploited to throw light on many stereochemical features of complexes, such as the bonding configuration, the ligand-field symmetry, the valency of the central metal ion, etc. Böke (40) has made an extensive edge study of complexes of transition elements (Cr–Zn) in order to evaluate the ligand influence on the shape of the absorption curves. It was shown that this shape is greatly dependent upon the ligands.

Certain types of symmetry are known to produce particular features in the near-edge spectrum. Atoms in tetrahedral sites often exhibit a peak just prior to the edge (though not always, cf. Ref. 199). In the case of sp^3 tetrahedral bonding low-energy absorption can still be observed, even when the $4p$ level is completely filled, as in $KMnO_4$. In this case assignment as a $1s \rightarrow 3d$ quadrupole transition ($\Delta l = \pm 2$) has been advanced (118). Similarly, in K_2CrO_4 a normally disallowed $1s \rightarrow 4s$ transition is supposed, after mixing with p orbitals (65). Also $1s \rightarrow 4s^*$ (antibonding) transitions have been

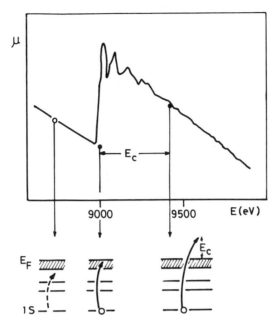

FIG. 5. Schematic of X-ray absorption cross section as a function of the photon energy showing the threshold region (including pre-edge and edge regions), the EXAFS region, and relevant electron processes: excitation of a core electron to a higher unoccupied atomic level, to the Fermi level (at absorption edge), and to the continuum (atom ionization).

described (*214*). Clearly, the selection rules for the lowest available "bound–bound" transitions (as in Fig. 5) of the type $1s \rightarrow 4s$, $1s \rightarrow 4p$, $1s \rightarrow 3d$ are dependent upon the symmetry group for the clusters. Kutzler *et al.* (*155*) have used one-electron multiple-scattered wave SCF X_α calculations for the interpretation of the bound transitions in the *K*-shell XAS spectra of ionic transition-metal complexes, such as MoO_4^{2-}, CrO_4^{2-}, and MoS_4^{2-}, and show that the first fairly intense peak on the low-energy side of the rising edge for molybdate and chromate (Fig. 6) is due to a dipole-allowed transition to a bound antibonding state of mainly metal *nd* character on the metal ion; this transition is possible because of mixing with the ligand *p* orbitals having the proper T_2 symmetry induced by the tetrahedral molecular potential. The results show that a cage which breaks inversion symmetry at the metal atom site can enable dipolar $1s$-to-*nd* transitions, normally dipole forbidden.

For $\bar{1}$ symmetry (e.g., square planar or octahedral geometries), the $1s \rightarrow 4s$ and $1s \rightarrow 3d$ transitions are forbidden, whereas for 1 symmetry (e.g., tetrahedral geometry) the $1s \rightarrow 4s$ and $1s \rightarrow 4p$ transitions are allowed. The origin of the oscillator strengths for these transitions is not always known, but their

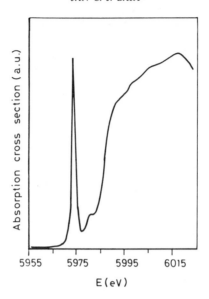

FIG. 6.   Absorption spectrum of a CrO_4^{2-} ion at the Cr K edge. From Kutzler *et al.* (*155*).

dependence on molecular symmetry is clearly illustrated in Fig. 7, where the $1s \rightarrow 3d$ transition is forbidden for the octahedral coordinations of vanado-cytes but allowed for less symmetrical arrangements (*291*). In general, different pre-edge peak [$1s \rightarrow 3d$ transition (*261*)] intensities are observed in the spectra of octahedral and tetrahedral transition-metal complexes with incompletely filled d shells; this feature allows site symmetry determination (*270*).

Pre-edge absorption is an outstanding feature in the K-edge XAES spectra of vanadium compounds. Empirically, the strength of this transition is found to be dependent on the size of the "molecular cage" defined by the nn ligands coordinating vanadium (*155*). The smaller the cage, defined by the parameter $\bar{R} = (1/n)\sum_i^n R_i$, where n is the number of nn bonds, the higher is the intensity of the pre-edge absorption. In vanadium oxides the K edge is described by a strong peak at ~ 20 eV [assigned as the dipole-allowed $1s \rightarrow 4p$ transition (*261*)], a lower energy shoulder [a $1s \rightarrow 4p$ shakedown transition (*13*)], and the pre-edge feature [a forbidden $1s \rightarrow 3d$ transition (*261*)]. At energies equal to and above the $1s \rightarrow 4p$ transition, absorption features may arise from a transition to higher np states, shape resonances (*74, 76, 78*), and/or multiple scattering (*83, 105*). The latter two effects are much more complicated to analyze. The oscillator strength of the pre-peak increases with progressive relaxation from perfect octahedral symmetry (as in VO) to distorted octa-

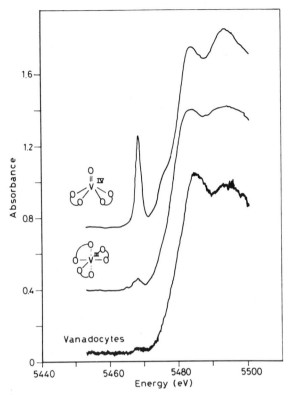

FIG. 7. Vanadium K-edge absorption spectra for compounds with vanadium in various oxygen coordinations; ÔÔ, acetylacetonate. From Doniach *et al.* (*80*).

hedral VO_6 groups, as in V_2O_3, V_4O_7, and V_2O_4, and to a lower coordination with a short V–O bond in square-pyramidal symmetry (as in V_2O_5).

Roe *et al.* (*243*) have observed that the intensity of the pre-edge features at about 10 eV below the edge (assigned to a $1s \rightarrow 3d$ transition) in an extensive series of Fe(III) complexes with (O, N) ligand spheres and a broad sampling of coordination environments varies inversely with coordination number. An extended Hückel molecular-orbital calculation shows a good correlation of the pre-edge peak areas with the total number of Fe $4p$ atomic orbitals mixed into the predominantly Fe $3d$ molecular orbitals. This correlation is expected for a dipole transition. Pre-edge features provide insight into the coordination chemistry of ferric centers independent of and complementary to that from EXAFS data. Pre-edge information may aid the interpretation of EXAFS in unknown systems.

Using similar criteria, the coordination environments of Ti^{4+} in TiO_2–SiO_2 glasses have been deduced (*85, 105a*).

2. Energy Shifts and Coordination Charge

The position of an edge denotes the ionization threshold of the absorbing atom. The inflection in the initial absorption rise marks the energy value of the onset of allowed energy levels for the ejected inner electron (216). For a metal this represents the transition of an inner electron into the first empty level of the Fermi distribution (242) and in case of a compound the transition of an inner electron to the first available unoccupied outer level of proper symmetry. Chemical shifts in the absorption-edge position due to chemical combination (reflecting the initial density of states) were first observed by Bergergren (27).

In order to rationalize the shift effect properly, it is necessary to consider that the characteristic X-ray spectrum of an element is only in first approximation an atomic property since both the outer and inner levels in compounds are affected by chemical combination. In some cases it is sufficient to attribute chemical shifts in X-ray absorption spectra only to the changes in the position of the outer level, which directly participates in chemical bonding. The effect of chemical combination on the inner level is often barely perceptible, as in the case of compounds of nontransition elements, but becomes more significant for transition-element compounds (5). In the latter, the shift of the inner levels is governed mainly by the number of d electrons (5). It is not surprising that the absorption edges show greater dependence of the wavelength upon the physical and chemical state of the absorber for lower atomic numbers.

Chemical shifts also exist in X-ray photoelectron (106), X-ray fluorescence (21), and Mössbauer (123) spectroscopy. The common theme in all these phenomena is that inner electronic or nuclear energy levels are measurably affected by chemical changes in the valence electron distribution.

The study of chemical effects in X-ray spectra provides valuable information regarding electronic structure and bonding in chemical compounds (55, 182, 196). A variety of factors, like effective nuclear charge (245), oxidation state (154), coordination number, type of chemical bonding, electronegativity (136, 137), hybridization, relaxation energy, screening effect, inner level shifts, etc. have been considered with various degrees of success to account for K or L absorption-edge shifts in X-ray spectra of metals upon chemical compound formation (for bibliography and review, c.f. Refs 182 and 269). Notably, the energy positions of various absorption features in the V K-edge XAES spectra of various vanadium oxides correlate with the formal valences of vanadium. The absorption threshold (the position of the first and second peaks in the derivative curves), the energy of the pre-edge peak, and the $1s \rightarrow 4p$ transition above the absorption edge all shift to higher energies with an increase in oxidation state (Fig. 8). The chemical shifts follow Kunzl's law

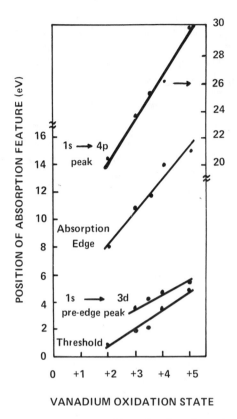

VANADIUM OXIDATION STATE

FIG. 8.   Oxidation state versus energy positions of various absorption features in the V *K*-edge XANES spectra of various vanadium oxides. From Wong *et al.* (*312*).

(*154, 182*). The positive shift in the threshold energy with valence increase [about 10 eV for S(*VI*) as compared to elemental sulfur] can be understood conceptually due to an increase in the attractive potential of the nucleus on the 1s core electron and a reduction in the repulsive core Coulomb interaction with all the other electrons in the compound. Numerous valency effects observed as shifts in the position of the main edge are reported by Srivastava and Nigam (*270*). However, the bond type is also of influence; in the case of covalent compounds and complexes the electronegativity of the bonding partner is of determining influence. Increasing covalent character of metal-ligand bonds leads to higher E_0 values (*101, 237a*). For strong covalent bonding the main edge position is more or less independent of the valency of the central metal ion and depends instead on the geometry of the complex ion (*197*).

More generally, the chemical shifts are due to a combination of valence, electronegativity of the bonding ligands, coordination number, and other structural features. These factors are expressed by Batsanov's (*19, 224, 225*) concept of coordination charge, η, as

$$\eta = Z - CN \qquad (6)$$

where Z is the formal valence of the central atom, C is the degree of covalency, which equals $1 - i$, i being the ionicity, and N is the coordination number. For a purely ionic material ($C = 0$, $\eta = Z$) the valence of a constituent atom equals its coordination charge.

The multiple-bond ionicity i is given by

$$i = 1 - \frac{Z}{N}\exp[-\tfrac{1}{4}(X_A - X_B)^2] \qquad (7)$$

and

$$I = 1 - \exp[-\tfrac{1}{4}(X_A - X_B)^2] \qquad (8)$$

is the single-bond ionicity. It follows that

$$\eta = ZI \qquad (9)$$

which is the charge appearing at the periphery of the atom as a result of chemical bonding to its ligands. The coordination charge concept states that as valence electrons are pulled away from the metal atom by the electronegativity X_B of the coordinating ligands, all other electrons of the central atom become more tightly bound in order to shield the unchanging nuclear charge. Hence, a K-shell transition must shift to higher energy with an increase in the product of valence Z and ionicity I. Batsanov's concept allows a much more systematic and general organization of the data than valence alone and has recently been used by Lytle *et al.* (*178*) to correlate the intensity of the L_{iii} white line in some Au, Pt, and Ir compounds and catalysts with coordination charge on the metal atoms, by Cramer *et al.* (*67*) in a study of the Mo K edge, and by Wong *et al.* (*312*) for a series of vanadium compounds.

Also, Suchet's method (*280*) for calculation of an effective charge q, which considers the effect of polarization and the sizes of the atomic radii for the cations and anions, is capable of interpreting the results of chemical shifts ΔE in X-ray absorption spectra in a satisfactory manner (*145, 145a, 248*). The effective ionic charge q is defined here as

$$q = n[1 - 0.01185Z/r' + Z'/r] \qquad (10)$$

where Z is the total number of electrons residing on the cation and r and n stand for the ionic radius and valency of the cation, respectively (*280, 282*). Primed symbols refer to the anion in the compound. In general, it is observed

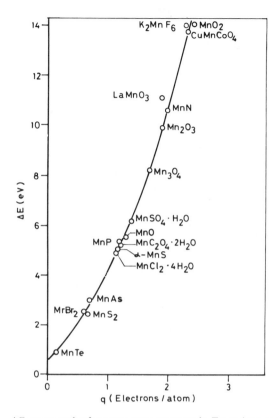

FIG. 9. ΔE versus q plot for manganese compounds. From Apte *et al.* (5).

that the absorption-edge shift to higher energy increases with higher effective nuclear charge on the absorbing ion. Figure 9 shows a plot between ΔE and such q values for manganese compounds (5), similar to that of Batsanov and Ovsyannikova (20).

Recently, Srivastava and Kumar (269) have devised another simple method for calculation of chemical shifts of X-ray absorption edges. This is illustrated in Fig. 10, where the absorption edge is indicated as a transition of a K or L core electron to the lowest part of the conduction band (E_c) above the Fermi level (E_F) and the energy difference between the edge for the compound and the metal by the chemical shift δE. The chemical shift corresponds to a shift in the bottom of the conduction bands of a metal and its compounds when they undergo chemical combination. It follows that

$$\delta E = E_c^S - E_c^M \tag{11}$$

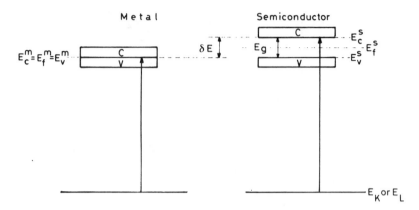

FIG. 10. Schematic energy diagram of metal (M) and intrinsic semiconductor (S). C, conduction band; V, valence band; E_c, bottom of the conduction band; E_v, top of valence band; E_f, Fermi level; $E_{K,L}$, K, L level; E_g, band gap. From Srivastava and Kumar (269).

where E_c^S and E_c^M are the energies at the bottom of the conduction bands of a semiconductor and metal, respectively. As in first approximation

$$\delta E = E_F^S - E_F^M \tag{12}$$

the chemical shift can be regarded as describing the change in the Fermi energy of a metal upon compound formation (269). The concept has been verified for a great variety of compounds.

As to the utility of the chemical shift concept, it is to be considered that on the basis of the empirical classification of K absorption spectra of transition-metal ions (256, 299, 300) and the application of ligand-field theory (64, 65), it is possible to determine the "net charge" on an atom, which in turn provides information about the coordination of the absorbing ion in a given complex. The observed relationships can be used to derive coordination environments in unknown systems. Structural suggestions, derived from absorption-edge data, can then further be verified by the EXAFS part of the absorption spectrum. As a consequence, absorption-edge measurements are useful in the characterization of the chemical states of metals in catalysts, metalloenzymes, etc. It would be desirable to have a theoretical substantiation of the observed empirical regularities.

The edge position is not always quantitatively informative. The edge features of 3d elements, which are pronounced in crystalline compounds, are blurred in the amorphous or highly dispersed state, probably because of mixing of various symmetries. This renders determination of the oxidation state by comparison with edge positions in crystals troublesome. The position of the pre-peak gives more reliable information when the final state is mainly determined by the absorbing atom. The pre-edge position is not reliable either in the case of strong covalency, and especially in complexes in extreme oxidation states.

3. Edge Widths

The variations in intensity in the different regions of the absorption edge, which give rise to a characteristic shape of the absorption-coefficient curve, may be correlated with many stereochemical features of a coordinated absorber. The X-ray intensity is given by the product of the transition probability and the density of states, and the latter depends upon the number, nature, and symmetry of the nearest neighbors. The features most relevant to the study of the stereochemistry of an absorber–ligand complex are (a) the shape of the edge, (b) low-energy absorption features (often indicating tetrahedral coordination, but with many deviations), and (c) splitting or broadening of the main peak.

The width of an absorption edge at half peak height of the derivative $d(\mu x)/dE$ can be used to determine the width of the valence band, which is a measure of the splitting of the degenerate energy levels and in turn is an indicator for the coordination symmetry around the absorber. For highly symmetric coordinations the edge width is very small, but it is quite large for highly distorted coordinations. Edge width is regarded as the energy difference between the edge position and the principal absorption maximum. Nigam and Srivastava (215) have established a semiempirical correlation between the edge width E_W of a metal in a complex and the number and nature of the surrounding atoms in terms of the overall metal–donor electronegativity difference, namely,

$$[E_W \sum (X_M - X_L)]^{1/2} = \text{constant} \tag{13}$$

(for a given metal in a given region), where the summation extends over all the atoms in the coordination sphere. In general, ionic solids give rise to sharply peaked edges, whereas in covalent compounds these are broadened (117). Provided other factors (metal ion vacancy, symmetry) remain the same, an increase in the overall covalent character of the metal–ligand bond, in general, results in a corresponding increase in the edge width and a decrease in the edge shift of the compound (152, 215). Correlation of the edge width with coordination stoichiometry is readily apparent in the significantly greater edge widths for tetrahedrally coordinated metal ions than for octahedral ones (141).

Figure 11 illustrates the general observation that second-row ligands induce broader, less resolved edges than do first-row ligands [as in the series

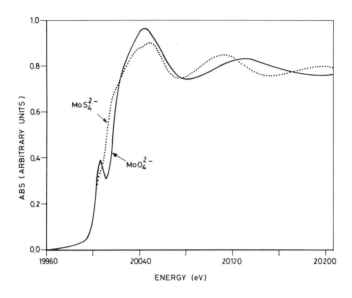

FIG. 11.   Edge and EXAFS spectra of molybdate (——) and thiomolybdate (⋯).

of vanadium halides, VF_3, VCl_3, and VBr_3 (296)]. Because of lifetime broadening, Mo K edges have a natural linewidth of about 6 eV (150). As a result, the sharp structure associated with the edges of lower Z elements is not observed. Nevertheless, it can be observed that the edges move progressively to higher energy as the Mo oxidation state is raised: Mo(II) ≪ Mo(III) < Mo(IV) < Mo(V).

4. Continuum Spectral Features

The smooth transition into the continuum is almost always convoluted with atomic transitions between the inner core state and outer bound-electron states which reflect themselves as bumps or shoulders on the edge itself. In certain cases, these transitions can provide electronic information about the absorbing metal ion, although the significance of the immediate "post-edge features" (in the 0–8 eV range) in relation to (electronic) structure is usually unclear. Maxima often correspond to temporary "trapping" of the electron by the potential due to the coordination cage. Multiple-scattered-wave SCF X_α calculations of the one-electron cross-section for K-shell photoabsorption in molecular complexes have been used to account for various fine-structural features (155). Shoulders or kinks on the rising absorption edge have been related to the beginning of the steplike continuum

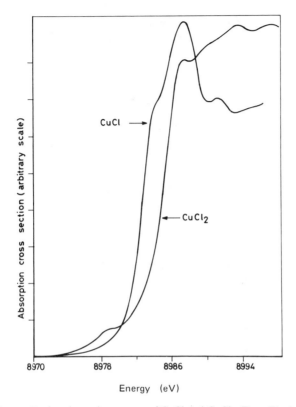

FIG. 12.   Copper K-edge absorption spectra of CuCl and $CuCl_2$. From Doniach *et al.* (*80*).

absorption, although other authors (*270*) prefer assignments to $1s$ to $(n + 1)s$ transitions. The major discontinuity may be considered as a dipole-allowed transition resulting from a "shape resonance" of the outgoing continuum electron in the molecular potential due to the ligands bound to the transition-metal atom and bears only a vague resemblance to the atomic states of isolated metal ions. The confining effect of the ionic cage depends both on the bond length and on the electronegativity or "hardness" of the confining cage. Typically, the strong post-edge resonance of Cu(I) and CuCl correlates with the closed nature of the tetrahedral cage (Fig. 12), whereas the weak resonance of $CuCl_2$ is ascribed to the open-geometry cluster of the square-planar coordination which allows a lower amplitude for "trapping."

Theoretical advancements will lead to improved analysis of the relationship of edge features to the chemical state of the absorber and to information about the binding site charge distribution and site symmetry.

E. Near-Edge X-Ray Absorption Spectroscopy
and Local Structure

1. X-Ray Absorption Near-Edge Structure

Figure 13 shows a pictorial view of the final-state radial wave functions relevant to core transitions in a molecule. Core transitions take place in an effective molecular potential seen by the excited photoelectron. Whereas for $E < E_0$, that is, below the continuum threshold (where E_0 is the energy of the core ionization potential), discrete transitions occur to the unoccupied valence states, photoelectrons excited in the photoionization process with

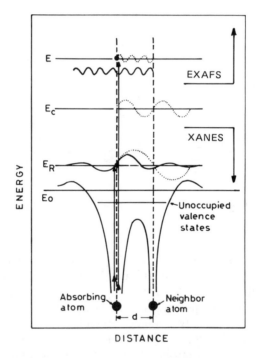

Fig. 13. Pictorial view of the final-state radial wave functions relevant for core transitions in a molecule. The core transitions take place in an effective molecular potential seen by the excited photoelectron. The final states in the continuum XANES region are quasi-bound multiple-scattering resonances (MSR), also called "shape resonances." Below the continuum threshold E_0 transitions to unoccupied valence states appear. E_0 is the energy of the core ionization potential (from ESCA). E_c is the energy where the wavelength of initially excited photoelectrons conforms to the interatomic distance. For $E < E_0$, discrete transitions to unoccupied valence states. $E_0 < E < E_c$, continuum XANES. For $E < E_c$, the EXAFS theory breaks down. The dotted curves show the wave functions of the initially excited photoelectron. From Bianconi (30).

low kinetic energy are strongly backscattered by the surrounding atoms. This process gives rise to the X-ray absorption near-edge structure (XANES), sometimes also indicated by the acronym NEXAFS (near-edge X-ray absorption fine structure).

XANES extends over an energy range of some tens of electron-volts above the absorption threshold E_0, where the wave number of the final-state electron (the internal photoelectron) is $k < 2\pi/R$, where R is the interatomic distance between the absorbing atom (the central atom) and the first coordination shell. This energy range is defined as that where the excited electron has enough kinetic energy to be in the continuum but its wavelength is larger than the interatomic distance between the central atom and its first neighbors. (According to a negative definition it is the least understood spectral region.)

In the XANES region the electron mean free paths are long; at the onset the mean free path is long enough (Fig. 14) that the excited electron will travel several unit cells and its wave function will resemble a stationary Bloch state. XANES appears as strong peaks before the threshold of the EXAFS oscillations. Although the EXAFS and XANES ranges join smoothly into one another, they are distinguished by an important theoretical criterion: in EXAFS the electron scattering is weak, modulations of the absorption cross section are $\sim 5\%$, and a single-scattering theory suffices to interpret data; in XANES electron scattering is much stronger, modulations can be large, and the interaction of the electron with the solid is no longer weak. In extreme

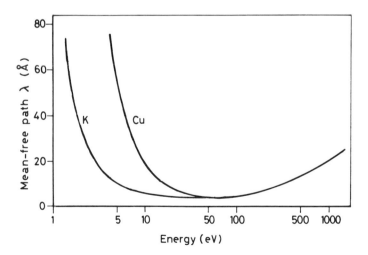

FIG. 14.   Energy-dependent mean-free path $\lambda_x(E)$ of the excited electron for K and Cu. The energy origin is at the Fermi level.

instances, it is more appropriate to regard the electron as being excited into a localized antibonding state of the system rather than to decribe it as an escaping plane wave. The final states in the continuum XANES region of the absorbing atom (covering broadly 8–50 eV above the edge in the $E_0 < E < E_c$ range, where E_c is the energy where the wavelength of the initially excited photoelectron conforms to the interatomic distance between the absorber and backscatterer) for simple molecules are quasi-bound states. The "shape resonances" arise from multiple scattering of the excited electron from the central atom (36). For the low electron kinetic energies ($E < E_c$), the single-scattering EXAFS theory breaks down. In the XANES region the excited electron is eventually backscattered many times within the system surrounding the central atom due to the high backscattering probability $A(k)$ at low k values (83). As a result, interpretation of near-edge structures is not straightforward and generally requires a full multiple-scattering treatment to model the data; computer codes are readily available (84). Figure 15 compares the scattering processes of the excited electron in the XANES and EXAFS regimes. Recently, the validity of the single-particle picture for XANES has been questioned (cf. Section III,B).

For weak scattering probes, in which only one scattering event is needed to describe the process, only the pair distribution function can be found and all

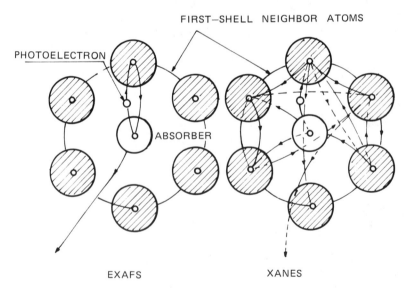

FIG. 15. Pictorial view of the scattering processes of the excited internal photoelectron determining the EXAFS oscillations (single-scattering regime) and the resonances in the XANES (multiple-scattering regime). From Bianconi (30).

directional information is lost. Thus XRD, neutron scattering, EXAFS, and HEED are all incapable of providing more than the pair distribution function, and many potential subtleties of the disordered state are inaccessible. The very property of weak scattering, which makes the experiment easily interpretable, removes the possibility of obtaining all structural details from the experiment. In fact, pair distribution functions obtained from a weak scattering, diffraction or EXAFS experiment give only the radius of the first coordination shell and provide no angular information about the arrangement of the atoms in that shell. For example, such events do not reveal the existence of well-defined molecular units (e.g., A_3B_3) in a disordered system. When such further information is desired, it is necessary to turn to the near-edge region to see the effects of higher correlations. Namely, in contrast to conventional single-scattering probes, XANES is sensitive to pairwise atomic correlations and to multi-atom correlations, including bond angles (*30*). This is entirely reasonable since strong scattering probes involve several orders of scattering. Radiation scattering *n* times contains information up to $2n$th-order correlations. Therefore XANES and EXAFS are complementary probes for local structural studies. The complementarity is obvious from the fact that EXAFS provides accurate information regarding the distance, nature, and number of ligands surrounding the absorber site but little geometrical information on the near neighbors because the electron scattering in the EXAFS region is spherically averaged. The limitation imposed on the extension of the EXAFS region at the low-energy side, due to the presence of the multiple-scattering process just beyond the edge (XANES region), puts a severe limit to the possibility of EXAFS to describe higher-shell near neighbors around the absorber in more detail.

In short, the main differences between the two spectral ranges may be summarized as follows.

	XANES	EXAFS
Scattering regime	Multiple	Single
Backscattering amplitude	Strong	Weak
Modulation of $\mu(E)$	Strong	Weak
Photoelectron kinetic energy	Low	High
Electron mean free path	Long	Short
Data analysis	Complex	Straightforward
Atomic correlations	Higher order	Pairwise
Information content	Bond data, site symmetry, C.N.	Bond lengths, ligand nature, C.N.

It is finally noticed that

(1)   the modulation in the photoabsorption cross section caused by the scattering from surrounding atoms is much larger in the XANES than in the EXAFS region;

(2)   XANES spectra are more sensitive to the scattering potential than EXAFS;

(3)   even though multiple-scattering effects are important in XANES, there are similarities with EXAFS, for example, backscattering from the atoms parallel to the X-ray polarization direction is enhanced;

(4)   broadening of the XANES structure (from 2 to 10 eV, depending on the material and on the energy above the Fermi level) is due to limitations of the lifetime of a core-hole state by many-body effects (principally Auger decay) and to decay processes of the excited electron.

A means of operationally separating the XANES and EXAFS regions is by examination of the derivative spectrum, taking into account that XANES is usually composed of strong and narrow peaks as opposed to the much broader and lower EXAFS oscillations.

2.   Coordination Geometry and Site Symmetry

As mentioned before, below about 8 eV beyond the edge the symmetry of the unoccupied electronic states and the effective atomic charge on the absorbing atom can be determined. This range is described by atomic or molecular effects if the *local character* of the final-state wave function is mainly determined by the atomic potential, or by the molecular potential due to the central atom and its neighbors. Strong "white lines" of mostly atomic character appear in the $L_{ii,iii}$ spectra of transition metals and rare earth compounds (*35, 307*). In these cases the atomic resonances near the edge are so strong that they obscure molecular effects.

On the other hand, the fine structure from about 8 to 50 eV above threshold is mainly structure dependent. Data analysis of the XANES region usually requires complex calculations and the area cannot usually be interpreted by simple fingerprinting techniques. Yet, some easily interpretable systems are found. In the case of molecules the invariance of near-edge resonance with physical state (free, surface chemisorbed, or solid) shows that these features are highly localized and only dependent on the intramolecular potential and geometry. In fact, XANES spectra of *molecules* like CF_4, and $GeCl_4$ are similar concerning the shape and intensity ratios of the peaks, despite the variations in the nature of the central and neighbor atoms. This indicates that the symmetry of a molecular cluster can be determined by comparison of XANES spectra. Also, the local site symmetry and coordina-

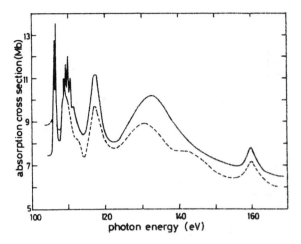

FIG. 16. $L_{ii,iii}$ edges of Si in tetrahedral SiF_4. ——, gas; ---, solid. From Sonntag (266a).

tion number in solids can sometimes be deduced by comparing to standard samples (191). Namely, as a result of the sensitivity to local structure, spectra of absorbers with similar environments but different long-range orders may be expected to be identical. Indeed, XANES spectra of *solids* with the absorbers in the closed tetrahedral coordination are similar to those of tetrahedral molecules in the gas phase (Fig. 16). This obviously implies that in these cases the spectral features are determined by multiple scattering inside the tight first coordination shell with short interatomic distances ("trapping" of the photoelectron). Figure 17 shows the similar XANES spectra for $[SiO_4]$ in bulk and in an SiO_2 overlayer; the differences in tetrahedral $[AlO_4]$ and octahedral $[AlO_6]$ units are shown in Fig. 18. In the case of more open, nontetrahedrally coordinated absorbers, XANES is strongly determined by the second and third neighbor shells (effect of long range on XANES).

F. COMPUTATIONAL SCHEMES OF LOW-ENERGY
ABSORPTION SPECTRA

The calculation of X-ray absorption coefficients has been an active field of theoretical physics almost since the birth of the quantum theory (275). Especially since the advent of modern computers, many calculations of X-ray absorption coefficients have appeared giving either a reasonable degree of accuracy over a wide energy range, including the strong, rapidly varying structure near an absorption edge (110), or detailed structure close to threshold (89, 127b). As to the former, simple approximate forms for the

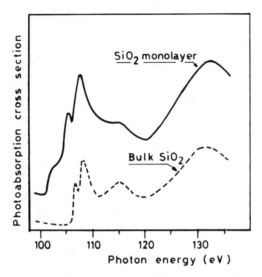

FIG. 17. L_{iii} XANES of Si in SiO_2 and surface XANES of a thin SiO_2 layer ($d = 5$ Å). From Bianconi (29).

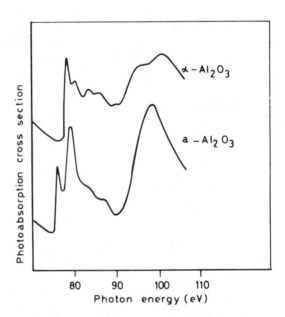

FIG. 18.   Characteristic $L_{ii,iii}$ XANES of tetrahedral $[AlO_4]$ units in amorphous Al_2O_3 and of octahedral $[AlO_6]$ units in crystalline α-Al_2O_3. From Bianconi (29).

atomic potential and initial-state wave functions can give X-ray absorption coefficients that agree well with experiment for virtually all atomic species. In this section we are interested in the computation of near-edge X-ray absorption fine structures.

On the whole, while emission spectra have been the subject of many calculations, generally less effort has been put into the calculation of absorption spectra. However, recently there have been a number of developments in the analytical aspects of the computation of X-ray edge absorption (*114, 115, 181, 181a, 220, 221, 226, 234, 287*). These schemes differ from a practical point of view. Comparison of calculated and experimental spectra provides a test of the theoretical and local component densities of states over a wide energy range. Theoretical and experimental results for absorption spectra are generally in good agreement. The fact that X-ray spectra provide a good check on calculated band structure over a large energy range is an important virtue of X-ray absorption measurements. Comparison of theoretical band-structure calculations with experiment sets the limits of validity of one-electron calculations for both ground- and excited-state properties of solids and allows evaluation of the contribution of many-body effects to the spectra. Absorption edges usually furnish a better vehicle for comparison to band-structure theories than do emission curves.

Absorption cross sections can be calculated in a most elementary way on the basis of the Fermi golden rule (*220, 221, 287*) by specification of the final states of the optical transition, calculation of the transition probability for electrons in the ground state, and summation of the probabilities over all possible final states. In the dipole approximation, the K spectrum will be the product of the p-like DOS and the probability of transition to the $1s$ state from the occupied band for the emission and to the unoccupied band for the absorption spectrum. The analysis leads to exact and general closed-form expressions for the total transition probability and integrated intensities of the optical absorption for both absolute and secondary absorption bands (never observed in X-ray absorption spectra). Expressions have been described which are valid beyond the threshold region. Calculated spectra are broadened by an energy-dependent Lorentzian function (including a spectrometer distortion function), a core, and lifetime broadening.

For the development of XANES theory it is necessary (i) to verify whether a one-particle Schrödinger equation with some potential obtained from local density theory describes the situation and (ii) to solve the Schrödinger equation for real systems. Recalling that LEED experiments are well descibed by such a simple Schrödinger equation in the 0–20-eV range (*232*), it is not unreasonable to expect the same to apply to XANES. However, the deep hole in the K shell resulting from X-ray excitation strongly perturbs the potential around the excited atom in a dynamic way and no rigorous treatment can be given for these more complex effects. Müller *et al.* (*205*) have established the important principle that the atomic potentials used in band-structure calculations can also be used to describe the electron excited by X-rays from the *K* shells; they report the first calculations based on a one-electron potential for the XANES spectrum of silver. The band-structure approach is computationally complex and only applicable to ordered solids, not to disordered matter, where XANES has its greatest impact. In order to overcome any assumptions about the long-range order, it is worthwhile considering an approach based on a finite cluster of atoms of dimensions comparable to the finite path length of excited electrons. For *small molecules* and small metal clusters a multiple-scattering theory based on a one-electron potential indeed describes many observed features (*76, 78, 79, 127, 155, 210*). On the other hand, in order to simulate a *solid*, large clusters (about 50–100 atoms) are needed, which have required further computational advancements (*84*). The various viewpoints are given below in more detail.

1. *Molecular-Orbital Approach to Edge Structures*

Molecular-orbital schemes have been used to describe qualitatively or semiquantitatively the effect on XAES of a variety of structural and electronic parameters (the formal oxidation state of the metal, the number of unpaired *d* electrons, the ligand-field symmetry and strength, the nature of ligands and metal–ligand bonding, the metal–ligand distances, etc.) (*28, 101, 256, 261, 270*). In the crude but simple one-electron picture the edge is described as transitions between the ground state and excited states originating from electronic configurations with a deep hole and the photoelectron in unoccupied antibonding orbitals.

Empirical correlations have been established between pre-edge band intensity and coordination charge (derived from MO valence-bond concepts) (*67, 224, 312*) or the composition of molecular antibonding orbitals, corresponding to bound states (*243*). Local orbital symmetry considerations explain simply the intensity of the *K* pre-edge of first-row transition-metal complexes (*28, 101, 256, 261, 270*) and the detailed edges in a variety of other organometallic complexes (*112, 258*).

Molecular-orbital approaches to edge structures differ for semiconducting and isolating molecular complexes. The latter and transition-metal complexes allow one to minimize solid-state effects and to obtain molecular energy levels at various degrees of approximation (semiempirical, X_α, ab initio). The various MO frameworks, namely, multiple-scattered wavefunction calculations (*76, 79, 127, 155*) and the many-body Hartree–Fock approach (*13*), describe states very close to threshold (bound levels) and continuum shape resonances.

Dehmer and Dill (76) have first applied a multiple-scattering method (MSM) to compute the discrete molecular states (132) and the low-energy part of the unbound photoionization continuum (78). In the calculation of the near-edge X-ray absorption structure of adsorbates and metal clusters by a one-electron theory in the self-consistent-field molecular potential scattered wave X_α (SCF-X_α) approximation (127), absorption cross sections for bound states and the continuum state are obtained separately. The oscillator strengths for the core-to-bound state transitions are derived using molecular orbitals in a SCF calculation on a cluster and are then converted to an absorption cross section following the method of Dehmer and Dill (76). Doniach et al. (79) have used the SCF-X_α method for interpretation of Cu K-edge spectra of low-symmetry molecular complexes. Although the method accounts for the principal spectral features seen for such complexes, the quantitative features of the calculated spectra are rather inadequate. It appears that ground-state properties are calculated rather well but excited-state energies are less satisfactory. The procedure tends to be fallacious in particular for highly polarizable complexes where a highly energy-dependent molecular potential may be needed. In fact, energy dependence of the effective molecular potential seen by the photoelectron in the threshold region is very important, as is well-known from band-structure calculations. Since just above threshold the escaping photoelectron moves quite slowly, the lack of many-electron screening is a considerable bias to the calculation of the threshold continuum cross section.

Despite these advances, a unique explanation of the whole XAES spectrum of molecular complexes is lacking, even employing *ab initio* MO calculations. Bair and Goddard (13) have carried out such *ab initio* SCF Hartree–Fock calculations with configuration interaction in order to elucidate the nature of the transitions involved in the K-edge X-ray absorption spectra of first-series transition-metal complexes. The method treats the electron–electron interactions in more generality than the X_α approach but suffers from the fact that the continuum states are modeled by localized orbitals. As it turns out, assignment of the $1s \to 3d$ and $1s \to 4p$ transitions in Cu K-edge spectra agrees with previous work, but the $1s \to$ "$4s$" transition is reassigned to an allowed $1s$-to-$4p$ transition plus shakedown. The $1s \to 3d$ transition, weak but not absent for centrosymmetric compounds, is a result of quadrupole coupling (112).

2. Band-Structure Approach

Within a single-particle treatment of the final electron state, XANES can be considered as due to long-range order (8), and consequently band-structure calculations may be applied. The band-structure approach [according to the conventional augmented-plane-wave (APW) and Korr-inga–Kohn–Rostoker (KKR) methods, the linear APW procedure (3), or other computational varieties (309)] is a cluster calculation with periodic boundary. This method of solving the equations assumes a perfectly regular lattice and is therefore unsuitable for general applications to disordered materials. In this formulation the modulation χ due to the scattering of the ejected electron on the surrounding atoms is proportional to the density of states N_l with angular momentum l determined by the orbital symmetry of the core state and the dipole selection rules ($l + 1$ orbital character); the final state is a Bloch state. An appealing feature of this scheme is that it includes multiple scattering to infinite order. The observed structures are explained in terms of hybridization and densities of states. In the KKR method of band-structure calculation, final states are found by matrix inversion and XANES can be calculated using this approach (84).

Describing the absorption cross section as

$$\sigma = \sigma_m(1 + \chi^{(1)} + \chi^{(2)} + \cdots) \qquad (14)$$

where $\chi^{(n)}$ represents the nth-order modulation by neighbors, the first term corresponds to EXAFS calculations (i.e., single scattering in a common approach to EXAFS and XANES). The so-called limited multiple-scattering approach allows assessment of the reliability of EXAFS calculations at lower energies, and, where EXAFS theory is unsuitable, higher order scattering contributions can be included. In case of face-centered cubic (fcc) Ni using a 7-shell cluster (135 atoms), single-scattering positions the absorption peaks correctly but amplitudes are made to fit only by including multiple-scattering contributions (100).

Müller et al. (205) have shown that the potential used in band theory, inserted into a single-particle Schrödinger equation, can describe the near-edge structure of the second-row d-band and rare-earth metals (204, 208). Some band-structure calculations extend in energy about 200 eV, that is, through the near-edge structure into the EXAFS region (Table I). This requires knowledge of conduction-band energies and wave functions to quite high energies. Deviations of the experiments from accurate single-particle calculations eventually point to many-body effects.

Band-structure calculations produce sharp thresholds. Since the initial and final states have finite lifetimes (the core hole decays by radiative or Auger electronic transitions from some occupied higher energy shells, while the excited electron loses energy by emitting plasmons or creating electron–hole pairs until it falls to the Fermi level), the single-particle results are convoluted with a Lorentzian function of the corresponding width. Consensus has yet to be reached on some of the reported issues regarding breadth and quantitative shape of the low-energy absorption edge. There are also ongoing debates on

TABLE I

Some Calculated Near-Edge Absorption Spectra

Element(s)	Edge	ΔE^a (eV)	Method^b	Reference
Cu	K	200	APW	*204*
Al	K	22	APW	*286a*
Ni	K	36	APW	*163*
Ni	$K, L_{\mathrm{ii,iii}}$	41	APW	*286*
Ni	K	30	KKR	*100*
Li	K	3	APW	*108*
Mg	$L_{\mathrm{ii,iii}}$	8	APW	*109*
Ca, Ni	L	17	APW	*192, 212*
Fe	K	50	KKR	*84*
Ti, Fe	K	10	APW	*226*
Ni, Cu, V, Fe	K	30	APW	*304*
Zr, Mo, Pd, Ag	K	60	APW	*205*
Ca, Ti, Cr, Co, Cu, Zn	K	75–200	APW	*206*
Ca, Ti, Cr, Co, Cu	$L_{\mathrm{ii,iii}}$	40	APW	*206*
Pd	K, L, M	200	APW	*208*
Pd	$M_{\mathrm{ii,iii}}, M_{\mathrm{iv,v}}$	250	APW	*206*
Sr, Zr, Nb, Ru, Rh, Pd	L_{iii}	40	APW	*206*
Sm, Gd–Lu	L	75	APW	*188*

^{a} Calculated energy range above Fermi level.
^{b} APW, self-consistent energy-band calculation by the augmented plane-wave method; KKR, Korringa–Kohn–Rostoker method for electronic band calculations in solids.

the relative importance of band-structure effects, Auger and phonon broadening, electron–hole scattering resonances, and the Nozières–De Dominicis many-body effect.

Table I summarizes the main results. Figure 19 shows the calculated Pd *K*-edge absorption coefficient in comparison to the experimental results (*208*), and Fig. 20 shows other band-structure results that reproduce all features of the single-particle spectrum (*204*).

3. *Scattering Formalism*

One basic assumption almost universally made in calculating XANES is the single-particle view, where it is assumed that a single-particle density of states can explain the various maxima and minima in the spectrum. This assumption has been strengthened by recent calculations (*83, 205, 206*) which appear to explain the positions of the measured features. The finite path length of the excited electrons limits analysis of the surroundings of the absorber to the electron mean-free path and precludes that XANES expresses

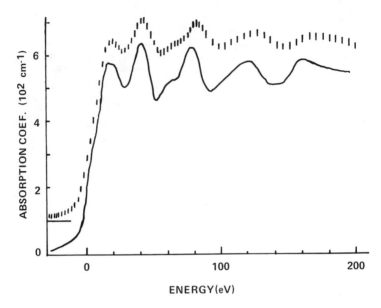

FIG. 19. Calculated K-edge absorption coefficient of palladium (continuous line) compared
with experimental results (discrete line). From Müller and Wilkins (208).

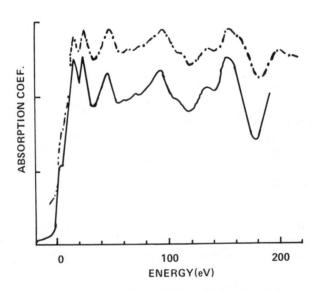

FIG. 20. Calculated K-edge absorption coefficient of copper (——) compared with experimental results (—·—·-). From Müller (204).

genuine long-range order. It is therefore reasonable to advance that X-ray absorption spectra can also be calculated according to a short-range scattering formalism. In a unifying view Schaich (252) has demonstrated that long-range- and short-range-order single-particle treatments of the final electron state essentially lead to the same result.

The simplest short-range approach is actually the EXAFS theory in which the effect of the central atom is included together with that of the neighboring atoms via single backscattering events; its validity is restricted to energies far above threshold, where the atomic cross sections become smaller. Breakdown of the single-scattering regime in the XANES region implies that analysis cannot proceed by simple Fourier transformation and must be achieved by comparison with more complex full multiple-scattering calculations. By including multiple-scattering events the short-range approach can be extended to low energies, and a variety of near-edge structures have successfully been computed. Durham et al. (83) have achieved the extension of the standard EXAFS formula beyond the high-energy range by including such MS events. The method, which is computationally involved, can distinguish between different models of local structure, much as LEED calculations enable surface structures to be determined (190, 232). This allows a more detailed description of the coordination geometry and site symmetry than with EXAFS alone. Higher shell effects in a fully multiple-scattering regime have been used successfully to match the K-edge structure of fcc Ag (83), fcc Cu, and α-Mn (105). Figure 21 shows the sequence of calculations for fcc clusters starting with one shell and running up to four shells. Clearly, the final calculation for a 55-atom Cu cluster yields good agreement with experiment.

Calculation of the multiple scattering by means of a real-space cluster approach is considerably more flexible than band-structure methods. Since this technique does not rely on crystal periodicity, it can readily be applied to interpret data for materials of arbitrary atomic arrangements. The sensitivity to higher order correlations has been shown. Fujikawa et al. (94, 96) favor short-range-order multiple-scattering XANES theory, in which atoms are not divided into shells but the scattered waves are classified into a direct term and a fully multiple-scattering term.

The nontrivial part in the computation of the photoabsorption cross section due to excitation of a core level is the evaluation of the final states ψ_f. In the scattering formalism one takes ψ_f to be a solution for the excited atom potential, which is improved by including single backscattering from neighboring atoms (6, 162), the spatial variation of the scattered wavelets (207), and multiple-scattering events (83). In scattering theory the observed structures are described in terms of matrix elements and interatomic distances (Table II). Müller (204) has provided the connection between the band-structure and scattering formalisms, which are two different ab initio methods

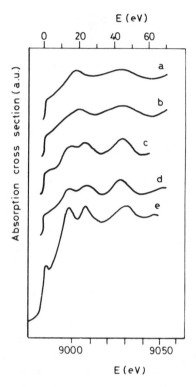

FIG. 21.   Multiple-scattering calculations for different cluster sizes of fcc Cu; a, 1 shell; b, 2 shells; c, 3 shells; d, 4 shells; e, experimental XANES spectrum. The shells consist of 12 atoms at 2.55 Å (first shell), 6 atoms at 3.61 Å (second shell), 24 atoms at 4.42 Å (third shell), and 12 atoms at 5.09 Å (fourth shell). From Greaves *et al.* (*105*).

TABLE II

Scattering versus Band-Structure Approach

	Scattering	Band structure
Final state ψ_f	Scattering state	Bloch state
Applicability	General	Periodic potential
Multiple scattering terms	Can be systematically improved	To infinite order
Description of spectral features	Interatomic distances Matrix elements	Energy bands, DOS Hybridization

of evaluating the final state. In terms of experimental and theoretical developments, XANES appears to have left its infancy.

IV. Near-Edge X-Ray Absorption Spectroscopy of Catalyst-Related Matter

A. CAPABILITIES AND RESTRICTIONS OF NEAR-EDGE X-RAY ABSORPTION SPECTROSCOPY AS AN ANALYTICAL TOOL

The capabilities of near-edge X-ray absorption spectroscopy that make it suited for structural studies are the following:

(1) applicability to any state of order;
(2) identification of element type;
(3) diagnostic value in monitoring chemical effects (symmetry of the unoccupied electronic states, effective atomic charge, d-band occupancy);
(4) determination of site symmetry around a particular species of atom in both periodic and nonperiodic arrangements; and
(5) measurements of bond lengths (in favorable cases only, e.g., in chemisorbed molecules) and angles.

It is a distinct advantage of XAES that the electronic and (some) structural properties (e.g., coordination geometry) can be determined in any aggregation state and morphology (up to atomic dispersion). XAES data may be obtained both for bulk samples and for the catalytic surface. It is also important to notice that the incident photon beam usually produces no damage to the material, as opposed to the action of several other probes. Moreover, the required information may be gathered under *in situ* reaction conditions. Above about 3 keV the measurements can be done in air. With modern X-ray photon sources data collection times are short. Among other advantages of XAES, it should be mentioned that absorption edges always exist and that the spectra are element specific (the inner-shell absorption steps occur at energies characteristic of Z). Obviously, near-edge spectra are also more free from interferences than extended X-ray absorption spectra: overlap of the short (~ 50 eV) data ranges with spectral features of contaminating absorbents is difficult.

X-Ray absorption data in combination with atomic theory and solid-state band-structure theory can yield detailed information about the ground-state electronic structure of solids on an energy scale on the order of meV. This holds particularly true for correlated narrow-band systems, such as the rare-earth and transition-metal compounds. In broad-band materials, such as the early $3d$ transition metals, band-structure effects are important, although XAS measurements indicate deviations from simple one-particle band-structure calculations (*250*).

Currently the feeling is that more structural information, for example, locations and site symmetries of the atoms in a particular shell, is contained in the XANES region of an edge absorption spectrum than in EXAFS. However, the large cluster size ($\sim 10^2$ atoms) required for XANES calculations is a considerable drawback.

Although XAES is a promising tool for the investigation of catalysts, amorphous materials, and complex systems, some limitations should be considered:

(1)   the restricted spectral range allows only few structural parameters to be determined (typically about three atoms);

(2)   XANES is at best a structural tool in combination with others, for example, in discriminating between hypotheses derived by other means;

(3)   XANES data analysis is considerably more complex than EXAFS analysis.

Besides theoretical limitations in the analysis of XANES, that is, the multiple-scattering treatment, other practical problems arise in the quantitative evaluation of XANES. These problems are in part the same as those influencing EXAFS analysis (cf. Ref. 18). It appears that their effects on the XANES absorption spectrum have not yet been considered with the same care as in case of EXAFS (*104*). For example, it is known that the thickness effect causes severe amplitude distortions, which are particularly significant for spectra having sharp maxima near the main edge (white line). Little systematic work has been carried out to assess its influence quantitatively. Small shifts in beam position and direction are especially disturbing in threshold studies since they can produce changes in the photon energy of 0.5 eV and more. The experimental setup can be arranged such as to locate relative edge positions with an accuracy of ± 0.1 eV (*163*). Actually no universally accepted procedures have been worked out for defining binding energy shifts indicative of changes in chemistry. The primary reason for lack of consensus in this area is the inability to fix the absorption edge at zero.

Difficult samples for investigation are those which present absorbing atomic species in a variety of configurations (heterogeneity). The presence of absorbers in different coordination geometries often leads to featureless XANES.

Several important catalyst-related materials are examined below on their X-ray absorption near-edge features. Specific details reported have been chosen in relation to the interest of the catalytic chemist.

B. METALS

The metal K-edge fine structure corresponds to the transition of a $1s$ electron of the absorbing atom to the unoccupied levels of p symmetry situated just beyond the Fermi level, and to any such hybrid levels in the conduction band which may have a p admixture (23, 24, 116, 199). The height of the absorption edge is related to the number of p electrons lacking. These transitions obey all selection rules, and in first-row transition elements the $4p$ orbitals are unoccupied; the $4p$–$5p$ distances are about 12 eV and the distance ratio $4p$–$5p$:$5p$–$6p$:$6p$–$7p$ ≈ 4:2:1. With the broadening of the higher levels, $5p$ and $6p$ absorption often overlap.

Early work on pure metals, solid-solution alloys, and intermetallic compounds has been reviewed by Azároff and Pease (9). Some of the very best X-ray absorption K-edge spectra for $3d$ transition metals were reported in 1939 by Beeman and Friedman (24), who applied band theory for their interpretation. Up to the early 1960s X-ray band spectra of metals were mainly explained in terms of a density of states multiplied by a transition probability.

According to Grunes (107) and others (181, 303), single-particle band-structure calculations (206) using the final-state potential with neglect of the core hole do well in predicting the observed X-ray absorption spectral features of metals past threshold (200 eV), which reflect the unfilled p density of states. The present insight is that the final-state rule (181, 303) for the calculation of X-ray absorption spectra of metals is satisfactory (72).

As mentioned before, removal of a core electron represents a strong perturbation, despite the fact that in a metal the core hole is more or less completely screened within the atomic sphere, which leads to a large effect on the core-electron binding energy. Many-body effects modify some absorption-edge features; strong effects on oscillator strengths and line shapes can be expected (173). In particular, neglect of many-body effects alters the lineshape at threshold. Citrin et al. (61) have carefully investigated the many-body effects for the K and $L_{ii,iii}$ absorption edges of Li, Na, Mg, and Al. By including processes such as phonon broadening and the MND many-body response of the conduction electrons in their calculated edges, they find that the K edges are rounded and the $L_{ii,iii}$ edges enhanced near threshold, in accordance with the experiment.

K-edge absorption spectra of $3d$ and $4d$ metals (205, 206) show similar XANES for metals with similar crystal structure. After recalibration of the energy scales, in order to remove the trivial kinetic energy dependence on lattice spacing, the observed peak positions are a signature of the environment of the absorber as determined by local geometry only. The amplitude of features due to specific bands are proportional to the hybridization strength or wave character (204).

The early $3d$ metals (Sc, Ti, V, and also Ca) are generally considered to be characterized by weak electron correlation in their ground states. Although band-structure theory seems appropriate to describe their excitation spectra, the $L_{ii,iii}$ near-edge structures show strong deviations from the prediction of single-particle theory. This is improved by taking into account core-hole-valence electron atomic-like interactions as well as the band structure (315). Other XAES spectra of metals are discussed in Section III,B,1.

The understanding of the near-edge absorption features of intermetallics is far from satisfactory. Band-structure effects are expected to play a role in determining the observed absorption characteristics, but calculations are not available for intermetallic compounds at sufficient energy above the Fermi level.

C. INSULATORS

Absorption spectra of insulators are not nearly so well understood as those of metals. Prototype materials of this category are the closed-shell solids (e.g., salts, oxides) with low polarizability. Because of the insulating nature of the oxides the valence charge is localized and cannot effectively screen the excited electron from the core hole, which is at variance to metals, in which screening is good. Therefore, many-body effects involving both the core hole and the excited electron are more important for oxides. The electron–hole final states for insulating materials are described by the Elliott exciton theory. The electron–hole interaction is stronger than any solid-state effect (i.e., the major relaxation effects are intra-atomic), so that in first-order approximation the situation in the solid is described by the atomic or ionic spectrum of the excited species. The atomic Coulomb-like limit applies only to the low-lying final states (3). Electron–hole effects influence the optical spectrum of insulators for many electron-volts above the band edge. In these materials the lifetime of the excited electron can be quite long, leading to rather detailed structure close to the edge. Cauchois and Mott (56) observed as early as 1949 a pre-edge white line for certain insulators. Generally, compounds having more ionic binding character show a sharp-peaked principal absorption maximum (270). For highly ionic absorbing atoms, the effect of the coordination atoms can be treated in terms of crystal-field arguments as a reasonable approximation for the qualitative understanding of the low-lying absorption peaks.

Systematic studies of ions have been carried out since the work of Beeman and Bearden (23), who investigated dissolved Ni^{2+}, Cu^{2+}, and Zn^{2+} ions. Splittings of the atomic $Cu^{2+}(1s)$ excitations are quite similar to those of the valence and Rydberg electron states of Zn^{2+}. In this sense, the past efforts

(57, 261) to analyze the absorption-edge features based on those $Z = N + 1$ atomic spectra cannot be faulted. This analogy is often used as a starting point (33, 261) and supposes that the $1s \rightarrow nl$ excited-state spectrum of atom $Z = N$ (nuclear charge) corresponds to the valence electron $\rightarrow nl$ spectrum of atom $Z = N + 1$. However, the ligands have a significant influence and introduce additional transitions not possible for the atomic ion.

The picture for $3d$ transition-metal-oxide near-edge structures is not as auspicious as for the corresponding metals (107). Although symmetry-based MO theory has been widely invoked (26, 101, 261, 283), it is probably inadequate. In the energy-band view of the X-ray absorption spectra of solids, the $L_{ii, iii}$ resonances are described as transitions to empty states in the partially filled d band of a metal or to states in the empty conduction band of an insulator. As the result of formation of a core hole in the absorbing atom, the atomic potential is strongly perturbed and covalent mixing will occur between the orbitals of the perturbed atom and the orbitals of the nearest neighbor atoms. A more appropriate description of X-ray transitions in insulating crystals is therefore given by calculations on a cluster of atoms which includes the absorber and its nearest neighbors. Kutzler et al. (155) have recently reported X_{α}-SW molecular-orbital calculations of the absorption cross sections of both bound and continuum K-edge transitions for a number of transition-metal complexes. However, even such ab initio SCF calculations (13, 155) show mismatch for several oxides (FeO, NiO, CuO), indicating failure of the one-electron theory for insulators. This suggests the need to consider long-range order (i.e., more atoms than a simple cluster) to describe the final-state wave function for transition-metal oxides. While some of the near-edge peaks are indeed attributable to one-electron MO-type transitions whose energies are modified by the presence of the core hole, other features have no such simple origin. Clearly, many-electron effects due to an incompletely screened core hole appear to be sufficiently important to alter the simpler MO transition scheme. More recently it has been indicated that the $L_{ii, iii}$ absorption near-edge fine structure in insulating Ni compounds (e.g., NiO) can be understood by means of many-body calculations on the basis of a NiX_6 cluster calculation (298).

Knapp et al. (144) show that for oxides containing $3d$ elements in spinel, perovskite, rocksalt, or zircon-type structures, the K-edge XANES spectra are quite independent of $3d$ electron occupation but instead nicely correlate with the crystal structure type. Various studies of Ti K edges of titanium oxides and other titanium compounds have been reported (40, 158, 172, 177, 297).

Calculations for the oxide XANES are clearly less developed than for metals, where the one-electron theory is adequate.

D. MOLECULAR COMPLEXES

Absorption edges often show shoulders corresponding to $1s \to$ valence and $1s \to$ empty bound-state transitions. Early analysis of these features has been based on state splittings of atomic spectroscopy (261), but significant changes are to be expected in these states upon molecular formation. Namely, although inner-shell photoabsorption in molecules takes place deep inside the core of one of the constituent atoms, where the local field is dominated by the "atomic" field of the nearby nucleus, the view in atomic physics, which assumes the exclusion of high-l partial waves from atomic cores, must significantly be changed in molecular physics under the effect of the molecular field. The particular area of molecular physics dealing with X-ray absorption cross sections of atomic subshells in different molecular environments has essentially taken off only since 1966 (cf. Ref. 74). With the development of SR sources progress in this area is now accelerated.

A single-particle effect that adds features in the X-ray absorption spectrum of molecules not present in that of atoms is the shape resonance (74, 75). (In the case of solids this effect, caused by a modification of the density of states due to the presence of the other atoms in the molecule, is automatically accounted for in band calculations.) Localization of the excited electron inside the molecule in states resulting from an effective potential barrier located near the electronegative atoms in the molecule causes strong absorption bands in free molecules and near the inner-shell ionization limits of positive ions in ionic crystals (74). Consequently, molecular inner-shell spectra depart markedly from the corresponding atomic spectra. The type of structure of an inner-shell photoabsorption spectrum depends on the geometry of the molecule, the nature of its ligands, etc., and can sometimes be used to determine the structure of the molecule.

For K-edge spectra of transition-metal compounds there are generally weak but distinct absorption features just before the onset of the main absorption edge. These features have been attributed to electronic transitions from $1s$ to nd, ns (nondipole), and np (dipole-allowed) empty states (261). For L-edge absorption, the transitions are from $2s$ and $2p$ to some higher, empty p and s or d states, respectively (50). Recently, the linearly polarized characteristic of SR has been used to determine the orientation dependence of the K-edge spectra in single crystals (156, 291) and has enabled certain types of geometric information about the X-ray absorption species to be deduced.

In studies on the absorption-edge fine structure in organometallic complexes, Pauling's valence bond theory may be used in explaining the fine structure.

In the case of covalent bonding in transition-metal complexes, the qualitative correlation between the observed fine structures and the calculated transitions indicates that XAS spectra provide a useful tool for verifying stereochemical deductions (256). Detailed edge studies of Cu(I) and Cu(II) complexes have been described by Blumberg et al. (38) and Brown et al. (49), where a comparison of the relative transition intensities has given useful chemical information. For example, only Cu(II) can give rise to a $1s \rightarrow 3d$ transition, the $3d$ levels being filled in Cu(I). For Cu(I) complexes, the $1s \rightarrow 4s$ transition is more intense than it is for Cu(II) complexes, consistent with the proclivity of Cu(I) for trigonal and distorted tetrahedral geometries which allow s-p mixing.

Azároff and Pease (9) and Srivastava and Nigam (270) review the application of XAES to coordination chemistry. Earlier reviews have also appeared (17, 117, 195, 249).

V. Near-Edge X-Ray Absorption Spectroscopy and
Applied Catalysis

This section intends to illustrate the utility of XAES methods in addressing problems of homo- and heterogeneous catalytic interest.

XAES is finding increased use in studies of catalysts, disordered solids, and surface structures as an unrivaled probe for gaining simultaneous information about the electronic structure and coordination geometry of absorbers in whatever physical state of a sample. It is therefore applicable to the study of active elements in amorphous and (micro)crystalline materials, in dilute and ultradisperse states (especially when fluorescence detection can be used), in controlled atmosphere or in reaction conditions, without limits to the particle size. Indeed, size effects can explicitly be explored. The topics considered include homogeneous catalysts, bulk versus small metal particle structure, metal–support interaction, in situ and dynamic catalyst characterization, poisoning and demetallization, etc. As will be obvious, the information gained by XAES is most valid in conjunction with other techniques, notably EXAFS (cf. Ref. 18). This has led to a more complete structural and electronic characterization and a better understanding of the catalytic activity of a wide variety of industrial catalysts, among which highly dispersed mono- and multimetallic catalysts (in particular the platinum group catalysts), metallic clusters, (mixed) oxide and hydrodesulfurization catalysts, metal–zeolite catalysts, etc. XANES and EXAFS are ideal for the study of various stages of catalyst preparation (coprecipitation, impregnation, etc.) and of complexes in solution. The fact that the physical framework for low photoelectron energy is considerably heavier than for higher energy is possibly the reason why catalytic applications of the low-energy region are more limited than those of the EXAFS region. Near-threshold studies prevail over past-threshold (continuum) XANES applications.

In order to use X-ray absorption techniques rationally in catalysis studies, it is important to first present a foundation for such studies. Systematic study of a wide range of chemical structures containing a specific absorber greatly facilitates evaluation of the effect of valence, site symmetry, coordination geometry, ligand electronegativity, and bond distances on various edge features in the XAES spectra of a single constituent atom. Trends observed for various spectral variables in such compounds represent a useful database for an improved understanding of absorption-edge theory and/or for characterizing and identifying the structure and bonding of atoms in unknown materials, such as catalysts. Such foundations now exist for several catalytically important absorbers, such as Pt (*186*), S (*268*), V (*312*), Fe (*243*), Mo (*122*; see also Fig. 22), and others.

Some topics have been omitted from this review. This holds for the structure and function of metal sites in metalloproteins and metalloenzymes in relation to enzymatic catalysis, for which the reader is referred to Cramer and Hodgson (*68*) and Doniach *et al.* (*80*). Also, chemisorption studies and the structure of adsorbate-covered surfaces are not considered in this review, which deals with XAES in transmission, thus characterizing bulk material. It is noted that even in the case of chemisorbed *atoms* XANES data analysis requires physically the definition of clusters of considerable size. On the other hand, the analysis is simplified for adsorbed *molecules*. Very pronounced near-edge effects (usually obtained by electron-stimulated Auger measurements) are observed for low-Z-atom(C, N, O, F)-containing chemisorbed

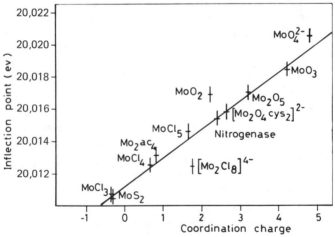

FIG. 22.   Correlation of Mo K absorption-edge energy with calculated coordination charge. From Cramer *et al.* (*67*).

molecules (*276, 278*), where the near-edge absorption fine structure is dominated by *intra*molecular scattering resonances with little influence from scattering contributions due to substrate surface atoms. This allows straightforward analysis. Consequently, the molecular orientation with respect to the substrate, bonding, and bond lengths of chemisorbed molecules can readily be derived from intramolecular scattering resonances in the near-edge range (*277*). Near-edge absorption at the C K edge can also be used to characterize carbonaceous layers (*308*).

A. Homogeneous Catalysts

Nickel one-component catalysts (Scheme 1) for linear olefin oligomerization have been studied by Peuckert *et al.* (*235*). The $\widehat{OP}$ nickel complexes are close models for the catalyst system in the Shell Higher Olefins Process

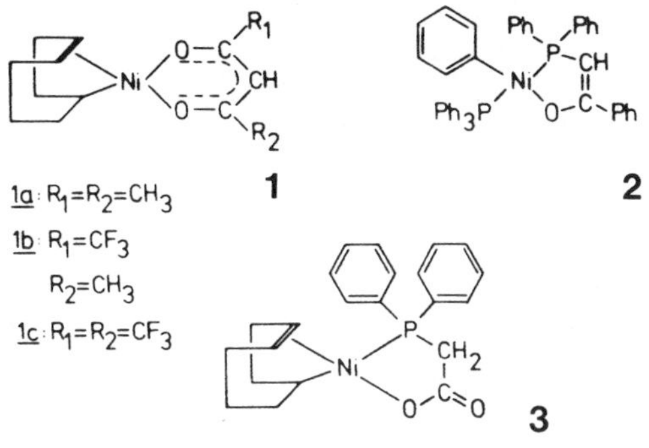

Scheme 1. From Peuckert *et al.* (*235*).

(SHOP). X-Ray near-edge absorption spectra (Fig. 23) show that the Ni K edge of $(C_8H_{12})_2Ni$ is typical of a tetrahedral d^{10} configuration, at variance to that of the $\widehat{OO}$ and $\widehat{OP}$ nickel complexes **Ic** and **3**, which are characteristic of a square-planar d^8 configuration. The measured photon energies (eV) of the (pre)edge range are as follows:

Ni	8331.6		8337.6	
(Cod)₂Ni	8332.8		8337.8	
3	8331.7	8333.9	8336.6	8339.0
Ic	8331.6	8333.4	8336.7	8339.0

XANES studies of ruthenium–polystyrene hydrogenation catalysts (*32*) show strong multiple-scattering resonance due to carbon atoms in the first coordination shell which exhibit high backscattering at low kinetic energy of the photoelectron, in accordance with EXAFS results.

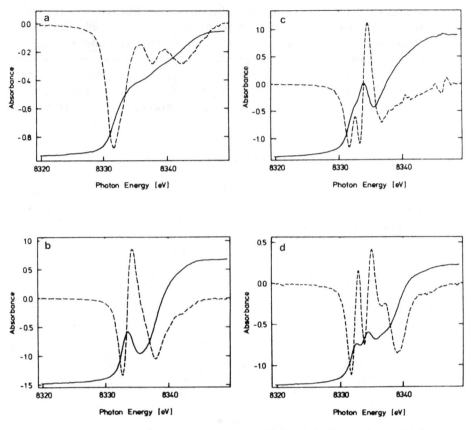

FIG. 23. K-edge X-ray absorption spectra (——) with first derivative (---) for (a) Ni foil, (b) $(C_8H_{12})_2Ni$, (c) $(\eta^3\text{-}C_8H_{13})Ni(CF_3COCHCOCF_3)$, and (d) $(\eta^3\text{-}C_8H_{13})Ni(PPh_2CH_2COO)$. From Peuckert *et al.* (*235*).

B.   CATALYSTS AND d-BAND CHARACTER

The catalysis literature contains a variety of examples of reactions which display periodic maxima in activity (*264*). In general, the catalytic activity can be correlated with the electronic configuration of the d orbitals (of the metallic bond in terms of Pauling's valence-bond theory) or with the strength of the metal–adsorbate bond. This is typically the case of C_2H_6 hydrogenolysis activities of Group VIII, VIIA, and IB metals, which have been correlated with the percentage d character of the metallic bond (*265*), a measure of the extent of participation of d orbitals in the bonding between atoms in a metal lattice (231). Correlation of metallic d-band character and catalytic activity (22, 44), derived from the similar trends for the percentage d character of *bulk metal* versus atomic number (Z) and catalytic activity of the *supported metal* versus Z, is one of the cornerstones of the electronic theory of catalysis. Yet, while the electron factor in catalysis has been recognized for many years, our understanding of the actual role remains primitive because of the difficulty in defining the d-electron character experimentally, especially in solids. Further progress thus rests upon our ability to determine this d-electron configuration in catalytic systems. It is now well understood that the white-line area of the L edges is related to the electronic configuration of the element, and quantitative atomic parameters can be measured.

Lewis (168) was the first to realize the power of XAS to follow up changes in the chemical state of a catalyst in the study of the reduction of a Pt catalyst in hydrogen.

Following up the advances in the understanding of the X-ray absorption white lines (50, 189, 203), L_{iii} X-ray absorption-edge spectra have been used to characterize the chemical state of the absorber by measurement of parameters directly related to the d state density on the catalytic atoms. In particular, the white lines of Pt metal have been a favored subject for quantitative evaluation. An increase in the relative area of the white line corresponds to a decrease in population of the $6s$, and, to a larger extent, of the $5d$ states in the valence band. There obviously exists a close correlation between the white-line area and edge shift since both reflect a decrease in electron population of the valence band. Not surprisingly, therefore, the coordination charge has also been correlated with the white-line area.

Difficulties in measuring white-line areas should not be underestimated and have given rise to controversies (69, 169), with immediate effects on the reliability of the results. Wei and Lytle (178, 307) have proposed a method for the evaluation of the electronic structure of transition metals which involves an analytical representation of the edge structure according to Breit–Wigner–Fano. They have shown that for platinum, iridium, and gold, the difference between the area of the L_{iii}-edge resonance in the pure metal and its compounds can be related to the amount of charge that is transferred to the ligands, calculated using the Pauling electronegativity scale. They used this relationship to discuss changes in the electronic structure of metal catalysts caused by the support and by chemisorption. Some workers (241) have followed these practices for other catalysts.

In view of their potential importance as a means of monitoring changes in the electronic structure of catalysts, well-founded procedures for quantitative analysis of X-ray absorption-edge data are clearly highly desirable. In the past the absence of such techniques has led to clearly contradictory results for

platinum incorporated in Y zeolite using different schemes of data evaluation
(*69, 99, 168*).

Horsley (*126, 127a*) has put the relation proposed by Lytle *et al.* on a more
reliable quantitative basis by using accurate calculations of the unoccupied *d*
orbital state using the X_α-SW method rather than semiempirical estimates.
Using simple deconvolution of the absorption-edge resonance, a relationship
has been established between the *total* vacant *d* orbital states for a series of
platinum compounds and the (corrected) *combined* areas of the L_{ii} and L_{iii}
absorption resonances, rather than the area of the L_{iii} resonance alone, as
earlier proposed. This is justified since the L_{ii} threshold resonance line arises
from transitions from the $2p_{1/2}$ core level to empty *d* states of $d_{3/2}$ character,
and the L_{iii} resonance line arises from transitions from the $2p_{3/2}$ core level to
empty *d* states of $d_{5/2}$ character.

Following up the earlier work (*99, 126, 167, 168, 174, 178*), Mansour *et al.*
(*183*) have tackled the problem of quantitative determination of the *fractional*
change in the number of *d*-band vacancies for a sample relative to a reference
material (platinum) rather than defining the total number of unoccupied *d*
states. The technique separates *s* from *d* contributions to the edge peak area,
and a correlation between the area under each of the L_{ii} and L_{iii} X-ray
absorption edges and *d*-band vacancies is given for platinum-containing
materials. The technique is demonstrated for a 2% Pt/SiO_2 catalyst. The
fractional change in the number of *d*-band vacancies in this catalyst relative
to Pt is 0.14. Even lower values are detectable.

Mansour *et al.* (*184*) have observed that the area under the Pt L_{ii} and L_{iii}
absorption edges for a Pt/Al_2O_3 catalyst decrease progressively to a value
characteristic of small supported crystallites of platinum with increasing
reduction temperature. The observed progressively decreasing area is indicative of reduction of platinum to a metallic (zero-valent) state. It is not
immediately obvious that the relationship obtained for platinum compounds
may be extended to other noble metal compounds. As pointed out by
Mansour *et al.* (*183*), there is in general not a simple relationship between the
area under the white line and the unoccupied *d* states.

Similar correlations with *d*-band character have been found by means of
Auger (*238*) and EELS (*239*). Where the need arises to pinpoint the oxidation
state of a particular transition element in multicomponent or heterogeneous
systems (as is often the case in the study of heterogeneous catalysts), a
technique such as EELS, which can focus on ultramicroscopic quantities and
still yield the number of *d*-electron states, adds further power to the X-ray
absorption method.

It has also been pointed out that XAS is a powerful tool for the
determination of the 4*f* occupancy number, in particular in rare-earth mixed-
valent compounds (*151a*).

C. Supported Metal Catalysts

An important matter of concern in catalyst studies is the control and determination of valence states. It is obvious from the physics of near-edge X-ray absorption spectroscopy that this technique is highly suited for the purpose. For example, the electronic state of cerium in Pd/CeO_2 and coimpregnated $Pd/Ce/Al_2O_3$ catalysts was examined by Ce L_{iii}-edge studies (164). On coimpregnated catalysts the valence of cerium was remarkably low (about $3+$, varying with dilution and Pd concentration). A correlation was established between the valence state of cerium and the selectivity for hydrogenolysis of methylcyclopentane (a good test reaction for electron transfer between support or the rare-earth and the transition-metal atoms). Partial reduction takes place during the catalytic experiment when CeO_2 is used as a support.

Materials' properties (e.g., miscibility, electronic and magnetic properties, etc.) also depend on particle size and/or geometry. In the ultrafinely divided state a metal exhibits properties different from those of the bulk state, a situation often exploited in catalysis.

Basic questions concerning metallic clusters relate to the variation of the electronic structure and binding energy with cluster size and geometry, the most stable cluster geometry and the possible significance of this geometry to the surface (and possibly catalytic) properties of small metal particles. It is also of interest to investigate in what fashion the binding energy per atom approaches the bulk metal value with increasing cluster size. Near-edge studies in conjunction with EXAFS are particularly indicated in answering some of these questions for single and bimetallic supported catalysts, leading to information on the micromorphology of the metal clusters, including bond distances, coordination number, thermal and static disorder, type of near neighbors, and interaction with the support (175). Tables III and VII show the great interest to clarify the physical, structural, and electronic phenomena associated with the size and morphology of supported noble metal particles. It is noted that much of the information about the electronic structure of catalyst particles cannot easily be obtained by other means. In this respect it should be considered that AES with an 8 to 10 Å Auger electron escape depth is a surface-sensitive technique for bulk metals but probes most of the "bulk" of microsurface particles due to their small size. Metal atoms in such clusters are almost all on the surface, giving these particles nearly atomic electronic properties. XAS provides detailed information about this state.

1. Size Effect

The adsorptive and catalytic properties of metals can be modified by altering their electronic structure (255), for example, through reduction of

TABLE III

X-Ray Absorption Studies of Monometallic Supported Metal Catalysts

Catalysts	Composition	Method	Edge	Reference	Year
Co/C	—	Threshold	Co K	*217*	1960
Co/SiO$_2$	0.8-6.2%	Threshold	Co K	*140*	1959
Ni/SiO$_2$	6.9%, 30 Å	Threshold	Ni K	*166*	1962
Pt/SiO$_2$	0.7%, <15 Å	EXAFS, edge	Pt L_{iii}	*260*	1983
Pt/SiO$_2$	2.0%	Edge	Pt $L_{ii,iii}$	*183*	1984
Co/SiO$_2$·Al$_2$O$_3$	1.0-3.2%	Threshold	Co K	*141*	1962
Co/γ-Al$_2$O$_3$	0.8-7.5%	Threshold	Co K	*140*	1959
Ru/γ-Al$_2$O$_3$	4-10%	Threshold	Ru L_{iii}	*151*	1983
Re/Al$_2$O$_3$	—	Threshold	Re L_{iii}	*313*	1977
Re/Al$_2$O$_3$	0.5%	Threshold	Re L_{iii}	*223*	1983
Pt/η-Al$_2$O$_3$	1.8%, 14-26 Å	Threshold	Pt L_{iii}	*167*	1963
Pt/Al$_2$O$_3$	—	Threshold	Pt $L_{ii,iii}$	*184*	1984
Pt/Al$_2$O$_3$	—	EXAFS, edge	Pt L_{iii}	*260*	1984
Au/η-Al$_2$O$_3$	8%, $D = 1$%, 2000 Å	EXAFS, threshold	Au L_{iii}	*18a* / *63*	1976 / 1979
Au/MgO	0.2-5%, $D = 45$-9% 20-100 Å	EXAFS, threshold	Au L_{iii}	*18a*	1976
Pt/TiO$_2$	1.7-3.1%, 15 Å	EXAFS, edge	Pt L_{iii}	*260*	1983
Pd/CeO$_2$	10%	Threshold	Ce L_{iii}	*164*	1985

particle size to about 10 Å (size effect). The intrinsic size effect leads to splitting of the band structure. Size and support effects are often superimposed. Size effects can best be revealed by measuring differences in edge areas for metal clusters of various size supported on the same carrier. Significant variations in bond energy are observed only for the tiniest particles.

Lewis (*166*) first noticed small differences between the Ni K-absorption edge of a metal (Ni) and metal particles (30 Å) dispersed on silica. Similar effects were observed by the same author on Pt/η-Al$_2$O$_3$ crystallites of size smaller than 35 Å (*167*), indicating differences in the electronic state as compared to bulk metal and closer association to atom-like energy states, in accordance with earlier suggestions by Mott (*203*). The perturbations of the X-ray absorption edge increase with decreasing particle size. The effect, consisting of lowered absorption coefficients for exciting an electron into the continuum, is understood if one considers that the electron excited to the continuum must leave its parent atom. For smaller crystals the final position of the excited electron will be more and more likely at the metal crystal surface. Since the density of unfilled electron energy states, $N(E)$, must be smaller for these positions, the average absorption coefficient decreases. In principle, XAS offers a method for measuring crystallite size. This is particularly important for the small particle sizes (below 30 Å) for which few good methods are available.

A case of a presumably pure size effect has been described by Apai *et al.* (*4*), who have observed shifts of the onset of the *K*-absorption edge toward higher binding energy as the Cu and Ni cluster sizes on carbon-supported catalysts decrease (Fig. 24), in agreement with photoemission results. The changes are more pronounced for Ni than for Cu as a result of differences in the *d*-orbital configuration. Some critical notes have recently been expressed by Vogel (*301b*). XANES spectra of small evaporated Au clusters (11 to 60 Å) (*14a*) and of small Cr clusters in neon (*200*) have also been published.

Recently the X_α-SW method has been applied to the Pt $L_{ii, iii}$ edge of small clusters (*127*). A Pt_{13} cubo-octahedral cluster calculation closely reproduces the $L_{ii, iii}$ edge structure of Pt bulk metal, whereas the calculated near-edge region in Pt_{10} and Pt_7 clusters shows progressive loss of structure as the coordination of the excited Pt atom is reduced. The near-edge structure of a Pt_{10} cluster shows substantial broadening of the white line when three

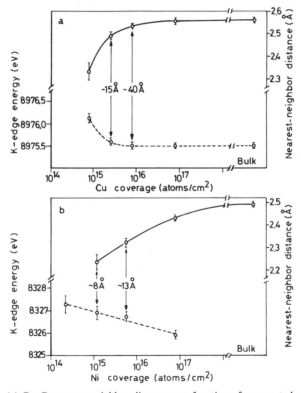

FIG. 24.   (a) Cu–Cu nearest-neighbor distance as a function of evaporated coverage (———). Relative change in the onset of the *K*-edge energy as a function of evaporated coverage (- - -). (b) Same as (a) for Ni. The lowest Ni coverage is indicative of atomic dispersion (no Ni–Ni nearest neighbors are detectable). The bulk Ni spectrum was recorded under different experimental conditions, thus precluding a comparison of *K*-edge energy. From Apai *et al.* (*4*).

oxygen atoms (representing support ions) are added. The experimentally observed narrowing of the white line in supported Pt catalysts as temperature is increased is therefore attributed to the loss of interaction between the Pt clusters and the support anions.

2. Metal-Support Interaction

The electronic structure of a metal particle may be affected by perturbations in its environment (gases, carrier, neighboring supported material) and is larger for smaller particles. In the case of supported catalysts the latter effect is termed the "support effect." Edge spectroscopy is a sensitive tool for detecting charge transfer between metal and support or adsorbates. "Pure" metal-support effects are difficult to observe, since these are frequently perturbed by size and electronic effects.

Metal-support interaction can be revealed by differences in XAS chemical shifts for metal particles of the same size but on different supports. As the particle size decreases the perturbation of the electronic structure by the particle environment (support effect) becomes larger. Such problems have recently also been investigated by means of XPS (129) for a series of Pt and Rh catalysts supported on γ-Al$_2$O$_3$ and TiO$_2$.

The effect of the support on the behavior of supported metals through electronic interactions, which becomes especially extremely important at high metal dispersions, is of course avoidable only in case of single crystal substrates.

With conventional supported catalysts the charge transferred is relatively small (11); only with very small aggregates or just a few metal atoms may the situation be rather different (12, 213). It has been speculated that by using a proper support a metal can be "changed" into another (e.g., Au into Pt, Pt into Ir, etc., as far as catalytic behavior is concerned). Metal-support interaction in its most dramatic form occurs in the so-called SMSI state (290), observed for Group VIII metals supported on reducible metal oxides (TiO$_2$, V$_2$O$_5$, MnO$_2$, Nb$_2$O$_5$, Ta$_2$O$_5$, etc.) after high-temperature reduction (289). The effect of this interaction, recently extended also to Group IB metals (60, 77), is to radically alter the chemisorptive and catalytic properties from those normally associated with supported metals.

In spite of numerous investigations of metals in the SMSI state (cf. bibliography of Ref. 153), the exact nature of the phenomenon is still controversial (131), although there exists a fairly compelling relation to support reducibility (14, 71, 128, 153, 288, 290). Changes in chemisorptive, catalytic, and structural properties in the SMSI state strongly suggest an electronic interaction at the metal-oxide interface with (whole or partial) electron transfer between a subjacent cation and a supported metal. Since the SMSI state apparently encompasses both a structural and an electronic component (290), XAS is a good means for physicochemical characterization, especially so in conjunction with other techniques (such as EPR and TPR) (260). XAS results lend little support to most explanations for SMSI, such as alloy formation, poisoning of the metal particles, charge transfer, encapsulation, and interaction between metal and titanium suboxides. According to Short *et al.* (260), highly dispersed impregnated and ion-exchanged Pt/TiO_2 catalysts are similar, with Pt–Pt distances smaller than in bulk Pt. The distance decreases with higher H_2 reduction temperature (Table IV). No EXAFS evidence is found for any Pt–Ti or Pt–O interaction that might explain the decrease in H_2 and CO chemisorption after high-temperature reduction, contrary to earlier suggestions (290), ruling out incorporation of Pt particles in the support, coverage of Pt particles with reduced TiO_x, or an epitaxial relationship between Pt and TiO_2. From a structural point of view, Pt–support interactions appear to be only of secondary importance. The first-shell coordination number (Table IV) and the absence of higher shells also shows that the effect is not due to crystallite size growth.

Edge studies indicate that the L_{iii}-edge position of all Pt/TiO_2 samples shift to higher energy at increasing H_2 reduction temperatures, which is suggestive of a positive charge on the absorber and consistent with considerations based on edge peak areas. These studies show clearly that the support (SiO_2 vs. TiO_2) is having a marked effect on the electron density of the metal. This holds, for example, for a non-SMSI state (after reduction with H_2 at 200°C) where SiO_2 is influencing Pt more than is TiO_2. Differences in unoccupied *d*-band character between Pt/TiO_2 and Pt/SiO_2 are significantly less than between these catalysts and Pt, and the catalysts appear to have more unoccupied *d* states than bulk Pt, indicating a transfer of charge to TiO_2.

TABLE IV

EXAFS Parameters of Pt/TiO_2

Sample	Reduction temperature (°C)	Reduction time (h)	Fitting technique		
			R (Å)	N_1	$\Delta\sigma^2 \times 10^3$ (Å^2)
3.1 wt % Pt/TiO_2	200	1.0	2.74	6.0	4.3
3.1 wt % Pt/TiO_2	300	1.5	2.70	5.3	5.9
3.1 wt % Pt/TiO_2	425	1.0	2.69	5.9	5.0
1.7 wt % Pt/TiO_2	200	2.0	2.73	5.0	10.7
1.7 wt % Pt/TiO_2	425	1.0	2.69	6.4	7.1
Pt foil					
EXAFS transform			2.77	12	1.7
Estimated errors			±0.03	±1	±0.5
Bulk Pt			2.77[a]	12[a]	

[a] Crystallographic values.

Studies by Short *et al.* (*260*) of the intensity of the white line of Pt on TiO_2 indicate that the charge transfer from metal to support is essentially quite small, in accordance with XPS data. In fact, no Siegbahn chemical shift of the metal in the SMSI state has been observed (*128a*), nor are there XPS indications for electron transfer from titania to the metal (*58, 257, 290*).

Using L_{ii} and L_{iii} X-ray absorption-edge spectroscopy, Mansour *et al.* (*184, 185*) have determined the electronic properties in terms of the change in the number of unfilled d states of Pt/Al_2O_3 and Pt/SiO_2 as a function of the reduction conditions and of the support relative to that of the bulk metal. SiO_2 induces the least perturbation in the electronic properties of the supported Pt, and thus impregnated $Pt(NH_3)_4(OH)_2/SiO_2$ is readily and fully reduced by H_2 at 200°C and Pt undergoes no further measurable changes in electronic properties with higher temperature reduction. Pt/Al_2O_3 still exhibits a significant amount of electron deficiency after 200°C H_2 reduction; this is largely eliminated by H_2 reduction at 450°C. Observed differences are related to the degree of interaction between the metal and the support and to changes of the electronic band structure which may be due to very small particles. The observed small, but significant, electron transfer from Pt to TiO_2 is already at the maximum level after H_2 reduction at 200°C. The change in d-electron vacancies for Pt/TiO_2 observed upon H_2 reduction at 425°C as compared to 200°C is much smaller (< 0.02 electrons/Pt atom) than the difference between Pt/TiO_2 and Pt/SiO_2 after H_2 reduction of both at 200°C (Table V). The SMSI effect is thus also *not* due to marked changes in the extent of electron transfer from TiO_2 to Pt, as previously suggested (*290*), but must be ascribed to more subtle changes in the electronic structure of the Pt atoms upon high-temperature reduction. A model for SMSI emerges from XAS studies in which Pt–Pt bonding has primary importance (*260*). Platinum clusters are expected to be raft-like, interacting in a nonperiodic way with the reduced support by means of occasional $Pt-Ti^{3+}$ and Pt–O bonds. Recently the existence of very thin "raft-like" nickel clusters on TiO_2 has also been claimed (*209*). Electronic interaction occurs between the band structure of TiO_2 and the developing band structure of the small Pt clusters. Recent analysis of the L-edge resonance lines in supported Ir catalysts indicates that the Ir particles have about 0.25 more d vacancies per atom than the bulk metal due to the presence of cationic Ir species anchored to the support that have survived the applied reduction treatment (*127a*).

D.   BIMETALLIC CATALYSTS

Analysis of the edge features (shape, position, etc.) gives useful indications of the electronic modifications undergone by the metal as a result of size effects, interaction with substrates, or the influence of alloying. For example,

TABLE V

X-Ray Absorption-Edge Parameters of Pt/TiO₂

Sample	Reduction temperature (°C)	Reduction time (h)	L_{iii}-edge shift (eV)	Normalized L_{iii}-edge area	Normalized L_{ii}-edge area	$f_d{}^a$	Number of unfilled d states per atom[b]
3.1 wt % Pt/TiO₂	200	1	+0.9	1.00	1.10	0.15	0.34
3.1 wt % Pt/TiO₂	300	1	+1.2	1.00	1.09	0.13	0.34
3.1 wt % Pt/TiO₂	425	1	+1.2	1.00	1.08	0.10	0.33
1.7 wt % Pt/TiO₂	200	2	+1.0	0.99	1.10	0.12	0.34
1.7 wt % Pt/TiO₂	425	1	+1.1	0.99	1.09	0.10	0.33
Pt foil	—	—	0.0	1.00	1.00	0.00	0.30
0.7 wt % Pt/SiO₂	450	2				0.23	0.37
Estimated errors			±0.5	±0.08	±0.08	±0.02	±0.006

[a] Fractional change in number of unfilled d states.
[b] Calculated assuming 0.30 unfilled d states per Pt atom in bulk Pt (50).

alloying of Pt with Ag causes a gradual occupation of the empty states and a commensurate decline in intensity of the white line (267).

It is of fundamental interest to establish whether the identity of metals in alloys is preserved. This requires sensitive tools to detect the changes in the electronic structure of the alloy components, such as XPS, XAS, and IR spectroscopy (of adsorbed species). The electronic structure of bulk alloys can adequately be studied by means of X-ray band spectra (threshold region), as recently reviewed by Azároff and Pease (10). Whereas alloy theory explains paramagnetism, conductivity, etc., in terms of the collective electron behavior, attempts to compare X-ray spectra to common densities-of-states curves of an alloy or to other collective electron models are dangerous (10). Since changes near the edge produced by alloying are usually small, any additional variations in the course of the experiment may present serious problems. Sensitivity to alloying differs from element to element. In metals with nearly filled bands, for example, the d band in nickel, any changes in the unfilled states are considerable and absorption spectra involving transitions to such states provide quite sensitive measures of alloying effects.

Early work on metal alloy catalysts was dominated by efforts to relate the catalytic activity of a metal to its electron band structure, but the existence of electronic structure effects has recently been questioned (236). Indeed, the electronic factor in multimetallic catalysis, based on the premise that the catalytic activity of a metal is determined by the electronic structure of the crystal as a whole, has now been overriden by the view that catalysis is determined by localized properties of surface sites, that is, chemical bonding effects similar to the ligand effects of organometallic chemistry. In this view, different types of atoms in the surface of an alloy largely retain their chemical identities, although their bonding properties may be modified. Features such as the percentage d character alone are not adequate for characterizing the catalytic activity of transition metals (263). An important effect of alloying is to dilute the active (adsorption) centers. More recently, other aspects have commanded attention, namely, the possibility of influencing the selectivity of chemical transformations on metal surfaces and of preparing metal alloys in a highly dispersed state.

A great advantage of X-ray spectroscopy over other experimental probes of multimetallic systems is that the X-ray process takes place in the immediate locality of an atom, so that the electronic structure of each atomic constituent can be studied separately.

As shown in Table VI, relatively few bimetallic catalysts have been studied by XAES methods. Yermakov et al. (313) were the first to take up the study of the Re oxidation state in reduced impregnated Re/Al_2O_3 and $PtRe/Al_2O_3$ catalysts by means of XAES techniques. It was concluded from edge shifts (0.3 eV shift of the Re L_{iii} absorption maximum per unitary change in

TABLE VI

X-Ray Absorption Studies of Multimetallic Catalysts

Catalyst	Composition	Method	Edge	Reference	Year
$Pt-Re/Al_2O_3$		Threshold	Re L_{iii}	*313*	1977
$Pt-Re/Al_2O_3$	0.3–0.6% M, Pt/Re = 1	EXAFS, threshold	—	*142*	1981
$Pt-Re/\gamma-Al_2O_3$	0.6–0.9% M	EXAFS, threshold	Re L_{iii}	*259*	1981
$Pt-Re/Al_2O_3$	0.9%	Threshold	Re L_{iii}	*223*	1983
$Os-Cu/SiO_2$	1–2%, Os/Cu = 1	EXAFS, threshold	Cu K, Os L_{iii}	*266*	1981
$Ir-Ru/Al_2O_3$	20% M	Threshold	Ru L_{iii}	*151*	1983
$Pd/Ce/Al_2O_3$	1–10% Pd, 1–8% Ce	Threshold	Ce L_{iii}	*164*	1985

oxidation number) that Re/Al_2O_3 catalysts show different reducibilities with variation in the concentration of the supported metal and temperature of catalyst reduction. Rhenium reduction appears to be promoted by supported platinum, but in general a heterogeneous oxidation state of rhenium is indicated (*313*). As a consequence of the heterogeneity of metal states in these systems, differentiation between a bimetallic cluster model (*262*) and a model of platinum stabilization by low-valence surface ions of the second component (*314*) was not achieved. More recently, and partly in contrast with the aforementioned work, Short *et al.* (*259*) and Kelley *et al.* (*142*) have interpreted the increase in the L_{iii}-edge white-line area observed for $PtRe/Al_2O_3$ catalysts relative to Re metal as an indication that Re is chiefly in a 4+ valence (ionic) state in the catalysts rather than in a zero-valence (metallic) state. This result has been confirmed by XPS data (Re 4f) of a $PtRe/Al_2O_3$ catalyst which had been reduced at 485°C (*223*). The greater white-line area of supported materials for a given energy shift than the bulk standards are suggestive of MSI. White-line area changes are a more sensitive probe of chemical changes than XPS. A comparison of white-line area, absorption-edge position, and photoelectron binding energy changes provides a direct separation of relaxation- and valence-state effects.

Ermakov *et al.* (*86*) have also characterized both $PtRe/Al_2O_3$ and $PtRe/SiO_2$ catalysts, prepared from $Pt(\pi-C_4H_7)_2$ and $[Re(OC_2H_5)_3]_3$, by means of Re L_{iii} X-ray absorption-edge spectroscopy. When $PtRe/SiO_2$ model systems, prepared by fixation of Re(II) ions to SiO_2 followed by the application of $Pt(\pi-C_3H_5)_2$, are used, rhenium maintains a mean oxidation state close to 2 after reduction of the catalyst. These findings conform to formation of a metal cluster bound to low-valence rhenium ions fixed to the surface of the support interacting with platinum clusters. This shows that interaction of the particles of a supported metal with low-valence surface ions of a modifying element is possible. The apparently contrasting XAES data for the PtRe system are likely to denote differences in catalyst structure as a result of variations in catalyst preparation and handling.

As indicated before, the edge structure is sensitive to the chemical environment of the absorber. In the case of OsCu/SiO$_2$ catalysts, the magnitude of the Os L_{iii} absorption threshold, which gives information on the electronic transition from the $2p_{3/2}$ core level to the vacant d states, is diminished by the presence of copper. This suggests an electronic interaction between Cu and Os in which the former contributes electrons to unfilled d states of the latter (266).

Ru L_{iii}-edge analysis shows interaction of the surface metal atoms ($> 50\%$) with chemisorbed oxygen or oxygen of the support in Ru/γ-Al$_2$O$_3$ and IrRu/γ-Al$_2$O$_3$ catalysts without the formation of an oxide phase (151). Interaction with oxygen increases with decreasing percentage of supported ruthenium. The structure of metallic ruthenium is less perturbed by interaction with oxygen upon deposition on χ-Al$_2$O$_3$ than on γ-Al$_2$O$_3$ and in bimetallic catalysts.

Finally, reactions of neopentane over (Pt, Ni)/Al$_2$O$_3$ catalysts show a pronounced maximum in hydrogenolysis rate at about 10 mole% Pt and for isomerization at 84 mole% Pt. At the same compositions, X-ray Pt L_{iii} absorption edges show modifications of the electronic structure of the alloy, suggesting a relationship to catalytic properties.

E. REACTIVITY OF SUPPORTED METAL CATALYSTS

H. P. Lewis (165) was the first to recognize that for a description of the effect of chemisorption on a metal in terms of the band structure of the metal, the methods of X-ray emission (for studying bands filled with electrons) and absorption spectroscopy (for studying unfilled bands) are well suited (Table VII). Changes in the electronic structure are quantified in terms of height and area of the edge peak and its position. In an early paper Lewis applied XAES to observe the changes in the unfilled energy bands of 30-Å Ni/γ-Al$_2$O$_3$ caused by oxygen chemisorption. Oxidation leads to surface Ni–O bonds similar to those in bulk NiO. The same author has also examined the perturbations of the Ni K edge of 6.9% Ni/SiO$_2$ catalysts due to small crystal size and hydrogen chemisorption (166). In a further study of the effect of crystal size and gas adsorption on the L_{iii} X-ray absorption edge of Al$_2$O$_3$-supported platinum, Lewis (167) concludes that absorption-edge data can be used to determine when gases penetrate the platinum lattice instead of

TABLE VII

X-Ray Absorption Studies of the Reactivity of Supported Metal Catalysts

Catalyst	Composition	Reagents	Method	Edge	Reference	Year
Ni/γ-Al$_2$O$_3$	4.5%	O$_2$	Threshold	Ni K	165	1960
Ni/SiO$_2$	6.9%	H$_2$	Threshold	Ni K	166	1962
Pt/η-Al$_2$O$_3$	1.8% (14–26 Å)	H$_2$, O$_2$	Threshold	Ni K	167	1963
Pt, Ni, Ir/support	($D = 1.0$–0.1)	O$_2$, CO	EXAFS, edge	Pt L_{iii}, Ni K, Ir L_{iii}	134	1978
Ir/Al$_2$O$_3$	1%	O$_2$	Threshold	Ir L_{iii}	178	1979
Pt/Al$_2$O$_3$	1%	O$_2$	Threshold	Pt L_{iii}	178	1979
Pt/γ-Al$_2$O$_3$	1.06–5.08% (<10–26 Å)	O$_2$, H$_2$, CO	EXAFS, edge	Pt L_{iii}	97	1981
Pt/γ-Al$_2$O$_3$	1%	H$_2$	XANES	Pt L_{iii}	251	1984
Pt/Al$_2$O$_3$	1% (<10 Å)	H$_2$	Edge	Pt L_{iii}	184	1984
Pt/SiO$_2$	0.7% (12 Å)	H$_2$	Edge	Pt L_{iii}	184	1984
Pt/SiO$_2$	6.7% (18 Å)	O$_2$	EXAFS, edge	Pt L_{iii}	241	1980
Pt/NaY	(10–30 Å)	O$_2$, H$_2$	Edge	Pt L_{iii}	99	1979
Pt/NaHY	(10 Å)	O$_2$, H$_2$	Edge	Pt L_{iii}	99	1979
Pt/NaCeY	(10 Å)	O$_2$, H$_2$	Edge	Pt L_{iii}	99	1979
Pt/NaY	15% (12 Å)	O$_2$, H$_2$	EXAFS, edge	Pt L_{iii}	241	1980
Pt/C	—	O$_2$, electrolyte	EXAFS, edge	Pt L_{iii}	45	1983
Pt/C zeolite	(5–10 Å)	O$_2$, H$_2$, electrolyte	EXAFS, edge	Pt L_{iii}	45	1983
PtRh/γ-Al$_2$O$_3$	2% Pt, 1% Rh	H$_2$	XANES	Pt L_{iii}	251	1984

being simply chemisorbed on the surface. Transfer of hydrogen atoms to the platinum energy bands can be expected to diminish the density of unfilled (by electrons) energy states, $N(E_F)$, and consequently $\mu(E_F)$. The reverse holds for oxygen and chlorine adatoms. The X-ray absorption-edge data can thus be used to follow the direction of electron transfer between platinum and the gas molecule.

There is a strong controversy in the literature about the polarization of the hydrogen–platinum bond (Pt^-H^+ or Pt^+H^-). Mignolet's (198) finding that the surface potential of platinum is negative after strong hydrogen adsorption favors Pt^+H^- formation. Conductivity and photoemission studies of Suhrmann et al. (284) are better in line with positively charged hydrogen, in conformance with the X-ray absorption-edge result. It would appear that this question has not yet been definitively settled (69, 98, 169).

Chemisorption of oxygen on the 1% Pt/Al_2O_3 and Ir/Al_2O_3 catalysts enhances the concentration of vacant d-electron states associated with Pt and Ir clusters. Similar conclusions have been reached by Gallezot et al. (99). The effect of chemisorbed oxygen is directionally the same as that for PtO_2 and IrO_2 in comparison to Pt and Ir, though smaller in magnitude. This could be a consequence of the metal–oxygen chemisorption bonds being less ionic in character than the metal–oxygen bonds in the bulk oxides.

Fukushima et al. (97) have investigated electronic and size effects for supported Pt/γ-Al_2O_3 catalysts (< 10-26 Å particles) as a result of CO and O_2 adsorption. Edge spectroscopic data of small Pt clusters indicate that the clusters are electron deficient when covered with H_2 (97, 99) and especially with O_2 and CO (97). Both CO and O_2 adsorption increase the Pt L_{iii} edge peak height. Information about electron withdrawal from the Pt atoms, contained in the edge position, is not nearly as sensitive to small fractional electron transfer and is also not as accurate. From edge data it appears that low-temperature oxidation of < 10 Å particles corresponds to Pt^{2+}, suggesting a chemisorption stoichiometry of about PtO. CO chemisorption reduces the electron density on Pt atoms as much as does structurally disruptive O_2 chemisorption.

F. DYNAMICAL AND *In Situ* CATALYST STUDIES

Dynamical studies of catalytic systems using dispersive X-ray absorption spectroscopy have been reported for the H_2 reduction process of $RhCl_3$ and H_2PtCl_6 (co)impregnated γ-Al_2O_3 (251). Acquisition time per spectrum was about 20 sec. The reduction process was followed on the basis of the white-line area at the Pt L_{iii} edge. Similarly, oxidation experiments have been carried out. The measurements show that dynamical changes of the catalyst systems undergoing redox reactions can be followed on the time scale of minutes, from which kinematic parameters can be obtained. Further technical improvements which will reduce the time scale of change to about 10 msec are expected. Other *in situ* XAES studies are reported in Sections V,H and V,J.

G. Sulfur Poisoning and Hydrodemetallization

As is well known, sulfur may poison catalysts during processing (*70*). A better understanding of the nature and distribution of sulfur-bearing feedstocks and organic sulfur-containing functional groups and their concomitant chemistry is thus desirable. Near-edge spectral features at the S K edge (at 2472.0 eV) are diagnostic for sulfur in minerals and specific organic moieties and have recently been used to probe the chemical and structural environments of sulfur in coal (*268*). Pyritic and organic sulfur-bearing functional groups (thiophenic) in coal were discriminated on the basis of well-differentiated pre-edge features. Applications in catalysis studies can readily be envisaged.

Since most industrial processes in oil production are critically affected by the variable trace amounts of vanadium and nickel, hydrodemetallization processes are required for upgrading heavy crude oils. For a better understanding of the mechanisms involved in the latter processes, it is highly desirable to know the metal environment in these heavy fractions. Vanadium K-edge XANES spectra of several asphaltenic fractions extracted from various crude oils have been investigated (*102*). The observation in the XAES spectra of a sharp pre-peak associated with dipole-allowed $1s \rightarrow E$ transitions (*103, 312*) is consistent with the presence of vanadylpetroporphyrins in these fractions. The present extensive knowledge of the vanadium near-edge absorption features greatly facilitates XAES studies of vanadium-containing catalysts and feedstocks.

H. Oxide Catalysts

Present knowledge of the electronic, surface, and catalytic properties of non-metals is substantially less advanced than that of metals. This is partly due to the fact that various physicochemical techniques are less rapidly applicable to non-metal catalysts. Relatively few metal oxide catalysts have been investigated by XAES techniques (Table VIII and Ref. *18*).

Several of the early oxide studies have already been mentioned in the introduction. The copper–alumina oxidation catalyst, which finds applications for the synthesis of glyoxal from glycol and as the principal component of base-metal formulations for automobile exhaust emission control, has

TABLE VIII

X-Ray Absorption Studies of Oxide Catalysts

Catalyst	Composition	Method	Edge	Reference	Year
Nickel oxides	—	Threshold	Ni K	119	1956
Supported nickel oxides	—	Threshold	Ni K	139	1956
Co/Al_2O_3	$0.8-7.5\%$	Threshold	Co K	140	1959
Co/SiO_2, $SiO_2 \cdot Al_2O_3$, Al_2O_3	—	Threshold	Co K	141	1962
$NaMnO_4/C$, SiO_2	—	Edge, near-edge	Mn K	217	1960
Co/C	—	Near-edge	Co K	217	1960
$(Co, Mo)O/Al_2O_3$	5% Co, 10% Mo	Near-edge	Co K	217	1960
$Co/Mo/Al_2O_3$	Various	Edge	Mo K	59	1984
CoO/Al_2O_3	—	Near-edge	Co K	217	1960
$Co/\gamma-Al_2O_3$	3%	EXAFS, edge	Co K	174	1976
$Mo/\gamma-Al_2O_3$	4%	EXAFS, edge	Mo K	174	1976
$(Co, Mo, P)/\gamma-Al_2O_3$	4.3% Co, 8.6% Mo, 2.9% P	EXAFS, edge	Co K, Mo K	174	1976
$Cu/\gamma-Al_2O_3$	—	Threshold	Cu K	310	1969
Cu/Al_2O_3	$1.8-16\%$	EXAFS, edge	Cu K	93	1978
$Cu/ZnO/Al_2O_3$	Various	EXAFS, near-edge	Cu K, Zn K	$301,301a$	1980/1985
$Zr\,M(PO_4)_2 \cdot nH_2O$ ($M = Cu, Co, Ni$)	—	EXAFS, edge	Cu K, Ni K, Co K	$1,2$	1981/1983
Hydrous titania	—	Near-edge	Ti K	217	1960
V_2O_5/TiO_2	Monolayer	EXAFS, XANES	Ti K, V K	$147-149$	1983
$V_2O_5/\gamma-Al_2O_3$	Monolayer	EXAFS, XANES	V K	147	1983
TiO_2, Cr_2O_3, FeO	—	XANES	Ti K, Cr K, Fe K	107	1983
Fe_2O_3, NiO, CuO	—	XANES	Fe K, Ni K, Cu K	107	1983
Adams' catalyst	—	EXAFS, edge	Pt $L_{ii,iii}$	186	1984
PdO/SiO_2, $\gamma-Al_2O_3$	$0.1-1\%$ Pd	XANES	Pd $L_{i,iii}$	$73a$	1983
MnO_2-SiO_2	55% Mn	EXAFS, edge	Mn K	51	1984

been studied by Wolberg and Roth (*310*), who used Cu *K*-absorption edge shifts (Table IX) to distinguish between CuO and an aluminate phase on Al_2O_3 supports. It was found that copper (from 3.3 to 10.3 wt%) on high-surface-area alumina (300 m^2/g) was typified by an aluminate, whereas CuO was predominant at a high (10.3 %) metal level on a lower-surface-area support (72 m^2/g). The aluminate phase was surprisingly detected even on material calcined as low as 300°C. Evidence was further found for magnetically dilute, isolated cupric ions (EPR). In general, mixtures of these phases are present in the catalysts. This calls for the need of a revision of the interpretation of catalytic oxidation activity data of copper oxide supported on alumina by Mooi and Selwood (*201*), who interpreted the data entirely in terms of the relative state of dispersion of the catalytic material on the support surface.

The interaction of cupric ions with alumina supports has subsequently been studied more extensively as a function of the support surface area, metal loading, and calcination temperature (*93, 279*) by means of EXAFS and X-ray absorption-edge shifts, in conjunction with XRD, EPR, XPS, and optical reflectance spectroscopy. These techniques, each sensitive to certain structural and electronic aspects, allow a unified picture of the phases present and the cation site location. Four Cu^{2+} ion sites are distinguished in the catalysts. In low concentrations (typically below about 4 wt.% Cu/100 m^2/g support surface area) Cu^{2+} ions enter the defect spinel lattice of the Al_2O_3 support. The well-dispersed "surface copper aluminate" has Cu^{2+} ions predominantly occupying tetragonally (Jahn–Teller) distorted octahedral sites, although

TABLE IX

X-Ray K-Absorption Edge of Copper on Alumina[a]

Copper (wt %)	Surface area of support (m^2/g)	Calcination temperature (°C)	State	ΔE (eV; ±0.45)
3.35	301	500	Ox.	7.94
3.35	301	500	Red.	1.72
3.35	301	500	Reox.	4.07
3.35	301	500	H_2O and reox.	7.94
10.30	72	500	Ox.	4.50
10.30	72	500	Red.	0.00
10.30	72	500	Reox.	3.86
10.30	72	900	Ox	7.73
$CuAl_2O_4$				7.94
CuO				3.86
Cu				0.00

[a] After Wolberg and Roth (*310*).

some tetrahedrally coordinated sites are also present. Heating the catalysts to 600°C results in an increase in the relative proportion of tetrahedral to octahedral Cu^{2+} ions. The "aluminate" phase is not necessarily a thin X-ray transparent domain analogous to bulk material, but the cation distribution of the surface spinel, which already forms at 300°C (310), can differ from that of bulk $CuAl_2O_4$. At higher copper concentrations, the adsorption sites in the support are saturated and the copper ions no longer enter the spinel. A discrete (micro)crystalline phase of CuO forms on the surface. In the EXAFS there is a distinctive structure in the 3–6-Å region of the radial distribution function (rdf) which uniquely indicates the presence of CuO. Above 600°C bulk copper aluminate (containing 60% tetrahedral and 40% octahedral sites) is formed.

X-Ray absorption spectra of the Cu K edge of (Cu, Cr)O/γ-Al$_2$O$_3$ auto exhaust catalysts are typical of Cu^{2+} bonded to oxygen, with little difference between the fresh and exhaust-cycled material (176). The Cr edge features are completely different in the two samples, and XAS and EPR data concur in indicating that the valence state is Cr^{5+} (chromyl bonding) in the fresh catalyst and Cr^{6+} in the octahedral environment in exhaust-cycled material. No low-energy absorption maximum specific of Cr^{6+} in a tetrahedral oxygen environment (chromate) was observed. As previously noted (237), it is suggested that Cr^{5+} is the active catalytic site, although no edge-shift data were reported to sustain the conclusions. The data are consistent with Cu and Cr atoms forming a spinel-like structure on (and/or within) the defect random γ-Al$_2$O$_3$ spinel support. By comparing Cu and Cr coordination numbers in the exhaust-cycled catalyst, it is evident that Cu occupies both octahedral and tetrahedral lattice sites. This is consistent with chemical evidence that Cr favors an octahedral site over Cu. The observed coordination numbers for Cu and Cr suggest surface site occupation and/or defects in the immediately surrounding lattice.

Fresh Cu/ZnO/Al$_2$O$_3$ low-temperature CO-shift catalysts have been reported (301, 301a) to contain CuO and ZnO in a state of considerable disorder. Edge-shift data do not lend support for the presence of ZnAl$_2$O$_4$ or CuAl$_2$O$_4$ spinels in the fresh catalyst. Reduced (ex situ) catalysts contain a fine copper dispersion in contact with ZnO (in good agreement with the higher stability of ZnO toward reduction), but no Cu$_2$O. It is useful to state the limits of this analysis. The simultaneous presence of various structural forms of an absorber in a complex system cannot readily be recognized. Detection limits probably exclude the observation of spinels, Cu(I)/ZnO solutes, or other low-concentration phases. Thus the XANES method is not expected to answer questions concerning the presence of a small fraction of dissolved Cu(I) species in a zinc oxide phase, which has been advanced as being the active phase (124, 194).

X-Ray absorption studies have also been applied to the determination of the electronic structure of Fe and Cu in spinels (*113*) and to clear up the relations in a dispersed manganese dioxide oxidation catalyst (cf. Section V,I) (*51*).

The $CoMo/Al_2O_3$ catalyst system has been studied by a variety of techniques in attempts to relate structural features of the system to catalytic performance. It is a remarkable property of this catalyst that despite high metal loading (about 15–20 wt.% based on oxides), poor crystallinity is developed. Consequently, traditional methods provide little information about the local structure of Mo and Co in these industrial catalysts and about structural changes with the preparative method. It is obvious that X-ray absorption studies are appropriate means for their characterization.

In early work, Keeling noticed that the shape and position of the Co K-absorption edges in supported Co/Al_2O_3 catalysts varies with concentration (0.8–7.5 wt.% Co). The edge corresponds to that of Co_3O_4 at 7.5% Co loading: the shift to lower energy with decreasing concentration was attributed to an increased proportion of divalent cobalt (*25*). The resulting model of the Co/Al_2O_3 system postulates two cobalt-containing states, namely, a dispersed phase predominant at low concentrations (<1.5%), consisting of Co^{2+} associated with the Al_2O_3 support, and a nondispersed Co_3O_4 phase predominant at higher concentrations.

For calcined industrial $CoMo/Al_2O_3$ catalysts for HDS applications (see Section V,J), many different structures have been proposed (*170, 294*). According to Lytle *et al.*, each cation is in nearly tetrahedral coordination and spinel-like sites are preferred. The shift in energy of the absorption edge was used to determine the valence (Co^{2+} and Mo^{6+}) and type of bonding. Qualitative interpretation of the Co and Mo K-edge spectra leads to the conclusion that Co occupies the same site in Al_2O_3 whether or not Mo is present. However, the environment of Mo is changed in the presence of cobalt.

Recent studies of this catalyst suggest a four- or six-coordinated Mo^{6+} ion, depending on loading and preparation technique (*91, 171, 187, 193, 240, 305*). Calcined $CoMo/\gamma$-Al_2O_3 catalysts (from various commercial sources: American Cyanamid HDS-2A, UOP, and Alpha Ventron) exhibit similar characteristics, attributed to two main Mo–O peaks (at 1.73 and 2.4 Å) in the Fourier transform (*227*). Chiu *et al.* (*59*) have resolved the near-edge Mo K-edge absorption features of unsulfided $CoMo/Al_2O_3$ precatalysts into three overlapping Gaussians, one of which has been ascribed (nominally) to the $1s \rightarrow 5p$ transition on the basis of the pseudo-atomic model of Shulman *et al.* (*261*). The results confirm EXAFS data showing that when the Mo loading increases, the *apparent* coordination number about molybdenum decreases; this is indicative of major distortions of the MoO_6 octahedra.

The changes in composition and structure upon H_2 reduction of silica-supported Co_3O_4 catalysts of different particle size (50 and 1000 Å) have been followed in detail by *in situ* real-time XANES (*54*). Fast-scan spectra, taken at 2.5-min intervals, could be interpreted in terms of their Co_3O_4, CoO, and Co contents. The rapid Co_3O_4-to-CoO transition occurs in the 250–275°C range for all catalyst samples, whereas the CoO-to-Co process shows a strong temperature dependence on particle size. Stronger interaction of the small particles with the support stabilizes the oxide phase. Sulfidation with H_2S/H_2 behaves similarly (forming Co_9S_8).

Pt $L_{ii,iii}$ edges of Adams' catalyst (*52a*) and a series of platinum compounds (viz., α-PtO_2, β-PtO_2, $Na_2Pt(OH)_6$, $Li_{0.6}Pt_3O_4$) were examined both by near-edge and EXAFS studies. Both edges show a strong absorption peak which is characteristic of Pt–O bonding, but with significant differences in height and shape of the peaks even in compounds with the same formal valence. Adams' catalyst seems to be mainly composed of a disordered form of α-PtO_2.

Large differences in Pd $L_{i,iii}$ XANES between SiO_2 and γ-Al_2O_3 supported PdO catalysts and PdO crystal are evidence for structural disorder of the first neighbor shell in the former and for PdO–substrate interaction (*73a*).

Vanadium Oxide-Based Catalysts

Vanadium exhibits a wide range of oxidation states ($-1, 0, \ldots, +5$) and coordination geometries (octahedral, tetrahedral, square pyramid, trigonal bipyramidal, dodecahedral) with various ligands in its compounds. This richness in chemical structures gives rise to a number of outstanding absorption features in the vicinity of the V K edge and L edge that are useful in the systematic study and understanding of the effect of bonding and coordination symmetry on the observed XAES spectra.

Wong *et al.* (*312*) have systematically investigated high-resolution V K-edge absorption spectra of a selected variety of vanadium compounds (oxides, salts, intermetallics, organovanadium compounds, minerals) in terms of valence, site symmetry, coordination geometry, ligand type, and bond distances. The pre-edge absorption in some vanadium compounds (among which the oxides VO, V_2O_3, V_4O_7, V_2O_4 and V_2O_5) has been analyzed semiquantitatively within a molecular-orbital framework and a simple coordination-charge concept (*312*). The XAES spectra have been used to elucidate V sites in coal (*191*) and crude oil (*311*).

Comparison of the XANES spectrum of the monolayer V_2O_5/TiO_2 (anatase) catalyst with those of vanadium compounds of different coordination types (Scheme 2) confirms the postulated two-shell structure derived from EXAFS, consisting of molecular species with two strongly double-bond terminal oxygens and two single-bond oxygen ligands bound to the support

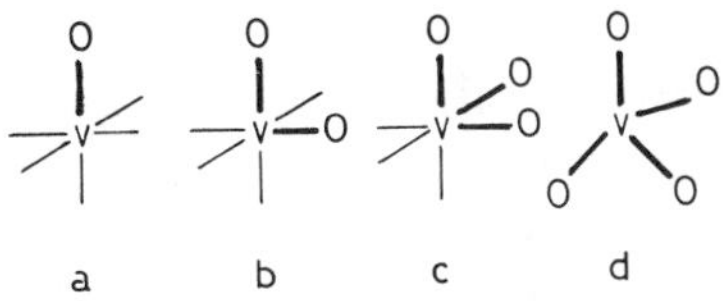

Scheme 2.

surface (*149*). The favored model (type **b**, Scheme 2) is remarkably close to that of $Mn_{1-x}V_{2-2x}Mo_{2x}O_6$ ($x = 0.36$), with the same coordination, and differs from V_2O_5 (type **a**) or VO_4^{3-} in solution (type **d**) (Fig. 25). The reduced amplitude of the double-peaked structure compared to the standard is ascribed to phase incoherence from the differing paths taken by the photoelectron. In turn, this may be on account of disorder in both bond length and angle and/or phase perturbations due to the varying electronic structure of the atoms in the material. In general, the simpler the fine structure of the absorption spectrum the greater is the degree of disorder around the absorber.

Comparison of the XANES spectra of the vanadium layer on TiO_2 prepared by two different routes (by alkoxide precipitation and via $VOCl_3$) shows close analogy. This indicates that the simple precipitation technique without chemical reaction between the vanadium reagent and the surface leads to a disordered monolayer on TiO_2. Similar data have also been reported for the V_2O_5/γ-Al_2O_3 system (*147*).

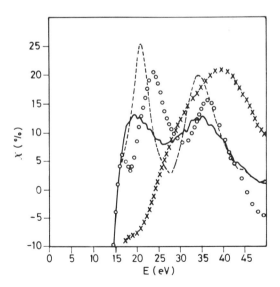

FIG. 25. V K-edge XANES spectra of a V_2O_5-TiO_2 catalyst (———), $Mn_{1-x}V_{2-2x}Mo_{2x}O_6$ ($x = 0.36$) (---), V_2O_5 (○), and VO_4^{3-} in solution (×). From Kozlowski *et al.* (*147*).

Stizza et al. (73, 274) have investigated amorphous vanadium phosphates, which are also of interest in relation to a XAS study of the butane–maleic anhydride (V, P)O catalysts (99a). From the V K edge useful information is obtained about the distortions in the vanadium coordination sphere [molecular cage effect on the pre-edge intensity (312)] and on the vanadium oxidation state. Notably, V^{4+} is silent to most spectroscopic methods. A mixed V^{4+}–V^{5+} valence state can be measured from the energy shift of the sharp core exciton at the absorption threshold of the 1s level of vanadium due to $1s \rightarrow 3d$ derived molecular orbitals localized within the first coordination shell of vanadium ions.

I. ALUMINOSILICATES AND ZEOLITES

Bianconi et al. have devoted XANES studies to the determination of the local structure of amorphous Al_2O_3 (15, 29) and SiO_2 (29, 31). In a review dedicated to problems in mineralogy and geochemistry, Waychunas and Brown (306) have considered crystalline and amorphous aluminosilicates.

Activated, deactivated, and thermally degraded dispersed manganese "dioxide" oxidation catalysts (MnO_2–SiO_2) for the complete vapor-phase oxidation of hydrocarbon and other vapors in pollution control applications have been characterized by XAS techniques by Brown et al. (51). In this catalytic system, consisting of manganese ions (~ 55 wt.%) dispersed through a glassy silica matrix, bulk and surface composition are identical. Table X shows a correlation between edge positions and first-shell Mn–O radii for reference materials from which the oxidation state of manganese in differently treated catalyst samples can be derived. Both the effective charge and oxidation number (281) vary linearly with the observed edge shift. It follows from Table X that in the activated form of the MnO_2–SiO_2 catalyst (heated in O_2 for 3 h at 350°C) the manganese ion centers are in a higher oxidation state than in normal dioxide, that is, close to Mn^V rather than Mn^{IV}, as indicated by a 2 eV greater Mn K-edge shift with respect to MnO_2. The valency state explains the markedly enhanced oxidative power of the catalyst. In the reversibly reduced form of the catalyst (treatment with H_2 for 5 h at 350°C) the oxidation state is between Mn^{II} and Mn^{III}. The edge shift found for the destroyed system (heated at 700°C in dry air for 3 h) is consistent with an intermediate $Mn^{III–IV}$ state.

As mentioned in the introduction, pioneering work on the use of XAS for catalysis studies has been carried out by Lewis, notably on platinum-supported Y-type zeolites (168). From examination of the X-ray Pt L_{iii} absorption edge of 0.5% Pt/CaY (with 10–60 Å metal particles), it is concluded that in the unreduced catalyst (prepared by substituting

TABLE X

X-Ray Absorption-Edge Positions, Edge Shift, First-Shell Radii (Corrected), and Related Data for MnO_2–SiO_2 *Catalyst Sample and Reference Materials*

Sample	Edge position[a] (eV)	Edge shift ΔE (eV)	First-shell radii (Å)	Oxidation number	Effective charge q[b]
Activated	6553.1	15.1	1.89	5.0[c]	3.46[c]
Reduced	6546.6	8.6	1.95	2.8[c]	1.83[c]
Destroyed	6548.3	10.3	1.91	3.4[c]	2.36[c]
α-Mn	6538.0	0	2.83	0	0
MnO	6543.6	5.6	2.22	2	1.35
Mn_2O_3	6548.5	10.5	1.92	3	2.03
MnO_2	6551.3	13.3	1.94	4	2.67
$KMnO_4$	6557.1	19.1	1.70	7	4.72

[a] Taken as the first major maximum in the derivative of the K-edge spectrum with ΔE measured with respect to the corresponding feature of α-Mn metal (6538.0 eV).

[b] Suchet (*281*); q is in units of electrons per atom.

[c] Derived by interpolation from reference data.

$Pt(NH_3)_4^{2+}$ for Ca^{2+}, followed by calcination) platinum carries a net positive charge (Table XI). In the reduced and outgassed sample platinum is converted to a zero-valent state (despite the small size of the platinum and the presence of divalent calcium ions). On average this platinum is smaller in size (10 Å) than that found in conventional reforming catalysts. The effect of exposing Pt/CaY to H_2 at 100 and 300°C is to convert platinum to a negative state. More recently, Gallezot *et al.* (*99*) have instead concluded from edge

TABLE XI

X-Ray Absorption Studies of Transition Aluminas, Silico-Aluminas, and Zeolites

Compound	Method	Edge	Reference	Year
Pt/CaY (0.5%; 10–60 Å)	Threshold	Pt L_{iii}	*168*	1968
Pt/Y (15%; 12 Å)	EXAFS, threshold	Pt L_{iii}	*241*	1980
Au/Y (15 Å)	EXAFS, threshold	Au L_{iii}	*46*	1984
$Pt_{8.7}Na_{37.2}H_{1.4}Y$	Threshold	Pt L_{iii}	*99*	1979
$Pt_{8.1}Na_{22.6}H_{17.4}Y$				
$Pt_{7.1}Na_{8.5}H_{33.3}Y$				
$Pt_5Ca_9Na_{13}H_6Y$				
$Pt_{7.6}Na_{30.9}H_{9.9}Y$	Threshold	Pt L_{iii}	*295*	1981
$Pt_{3.7}Ce_{12.6}Na_{12.4}Y$				
PtMoY				

work that hydrogen-covered platinum in Y zeolites is electron deficient (cf. also Refs. *69* and *169*). This controversy stands in connection with the data-handling procedures. Only lately has more detailed work established appropriate ways of analysis of X-ray absorption-edge data for the derivation of chemical information like the local charge density on the metal (*183*). In general, it appears to be well appreciated by infrared and XPS work (cf. Ref. *99*) that extremely small clusters stabilized in zeolite cages are electron deficient, but whether this is simply a consequence of their small size or of an interaction with the zeolite structure seems uncertain.

Boudart and Meitzner (*46*) have examined Au/Y zeolite samples. The properties of the Au L_{iii} edge are well known (*99, 126, 178*) and Au(III), Au(I), and Au can easily be distinguished on the basis of their edge resonances. Ion exchange between Y zeolite and $[Au(en)_2]Cl_3$ (en = ethylenediamine) followed by H_2 reduction [via a Au(I) intermediate] leads to 1-nm Au clusters in the supercages. Growth to larger (3-4 nm) crystallites is accompanied by removal of the metal to the external surface of the zeolite crystals. This indicates high mobility [the mp of bulk gold—1063°C—is suppressed to 327°C in the ultrafine state (*244*)] and lack of chemical interaction between gold and the zeolite.

J.   Sulfide Catalysts

As reviewed by Bart and Vlaic (*18*), several near-edge absorption studies of sulphided Mo/η-Al_2O_3, Mo/γ-Al_2O_3, CoMo/η-Al_2O_3, and CoMo/γ-Al_2O_3 catalysts have been carried out (*62, 246*). Noteworthy in particular is that Castner *et al.* (*54*) have used detailed rapid scanning (2 min) Co and Mo K near-edge spectroscopy for *in situ* H_2 and 2% H_2S/H_2 treatment of Co_3O_4 and supported CoMo/Al_2O_3 catalysts in order to determine compositional and kinetic changes. The experimental procedure consisted of reduction of Co_3O_4 to Co at 327°C, followed by sulfiding with 2% H_2S/H_2 at 302°C. On the basis of the concentration dependence of the Co species with time, it appears that the conversion of metallic cobalt and Co sulfide is a first-order process. In similar reaction conditions the reduction and sulfiding processes do not go to completion for supported catalysts. Also, Parham and Merrill (*228*) have dedicated a combined *in situ* XANES-EXAFS study to the popular sulfided CoMo/γ-Al_2O_3 catalysts in relation to activity in the thiophene HDS reaction.

Even without going through the rather involved multiple-scattering calculations, XANES spectra can readily be used as a fingerprint to ascertain the presence or absence of specific phases. This is in fact the way in which the technique has mainly found application so far in catalysis. By comparison of

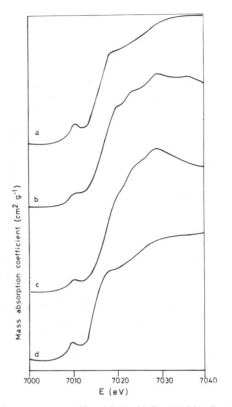

FIG. 26. Co K-edge spectra: a, (Co, Mo)/γ-Al$_2$O$_3$ (11.2% Co, 8.6% Mo); b, (Co, Mo)/γ-Al$_2$O$_3$ (2.2% Co, 7.1% Mo); c, 2.7% Co/γ-Al$_2$O$_3$; and d, Co$_9$S$_8$. From Sánchez et al. (246).

Co K-edge spectra of a series of HDS catalysts, prepared from (NH$_4$)$_6$Mo$_7$O$_{24}$·4H$_2$O and Co(NO$_3$)$_2$·6H$_2$O with H$_2$ reduction at 230°C and H$_2$S treatment at 340°C, it can readily be seen (Fig. 26) that the supported Co phase depends on the Co concentration (246). In the (11.2% Co, 8.6% Mo)/γ-Al$_2$O$_3$ catalyst, Co$_9$S$_8$ is likely to be present to a considerable extent, as confirmed by Co K-edge EXAFS analysis.

VI. Conclusions and Prospects

Near-edge X-ray absorption spectroscopy and techniques related to it by an extension of the photoelectron energy range or a modification in the experimental detection mode (EXAFS, surface XAS via Auger emission and photon-stimulated desorption, fluorescence and polarized XAS) have all matured in the last decade. The full potential of these techniques has barely been recognized.

The contributions of near-edge X-ray absorption spectroscopy to catalysis and surface science have grown considerably in the past few years, essentially since the development of synchrotron radiation sources. On a low interpretative level the application of threshold fine-structure effects to catalyst diagnostics has rapidly taken off as atoms in different structural or bonding conditions result in fingerprints that can readily be used for analysis of their structural and chemical state. Consequently, the near-threshold features have found considerable application in catalyst studies for the purpose of the definition of valence state, calculation of d-band character, coordination charge, etc., but also structural information has been deduced (site symmetry, coordination number). However, in order to make full use of the techniques the interpretation of next neighbor and higher lying shells should be improved by taking into account multiple-scattering effects and inelastic loss phenomena. Only then can a higher level of understanding be attained which makes full use of the information content in the post-edge region up to the onset of the EXAFS modulations. The current limitations to the applicability of near-edge techniques are not only theoretical but also experimental or rather practical. The prospects of accessibility to laboratory XAS spectrometers, linked to high-speed rotating anodes, is expected to reinforce the technique.

As pointed out, near-edge spectroscopy is at its best as an adjunct to other physicochemical characterization methods, such as other XAS techniques, and in combination with XRD, STEM, AES, thermal desorption, and catalytic activity measurements, can expect to bring new insights into the nature of basic catalytic reaction processes. *In situ* X-ray absorption near-edge experiments have the considerable advantage over others that they can be applied to *real* catalytic problems, sensing changes in the bonding electron distribution and coordination of the environment of the absorber and resulting in more closely matching models of structure and reaction processes. The combination of electronic and structural information in one technique is quite unique.

The recent accomplishments of near-edge X-ray absorption spectroscopy in catalysis studies are already quite impressive, in particular if one considers the limited availability of suitable X-ray spectrometers. Developments of catalytic interest have concerned the Shell Higher Olefin process, size effects, metal–support interaction, mono- and bimetallic catalysts (in particular the $PtRe/Al_2O_3$ system), the reactivity of supported metal catalysts, dynamical and *in situ* catalyst studies, and a variety of oxide and sulfide catalysts. Other catalytic problems are now coming within easy experimental reach, such as the study of sulfur poisoning and the nature of coking.

Whereas physicists have greatly advanced X-ray absorption theory, catalytic chemists can contribute to improving the general understanding by providing unique experimental data and samples, such as metal clusters of well-defined size in order to verify theoretical models on the effect of short- and long-range order in XANES. It is important to know the minimum size of a cluster to essentially reproduce a bulk density of states. Also, further efforts are needed to study very dilute species (0.01 at. %) in catalysts, using wiggler lines and fluorescence detection.

ACKNOWLEDGMENTS

The author wishes to thank the Weizmann Institute of Science for a Meyerhoff Visiting Professorship and expresses his gratitude to the Department of Structural Chemistry for kind hospitality.

REFERENCES

1. Alagna, L., Prosperi, T., and Tomlinson, A. A. G., in "EXAFS and Near-Edge Structure" (A. Bianconi, L. Incoccia, and S. Stipcich, eds.), p. 303. Springer-Verlag, Berlin and New York, 1983.
2. Alagna, L., Tomlinson, A. A. G., Ferragina, C., and La Ginestra, A., Proc. Congr. Chim. Inorg., 14th, 1981 p. 224 (1981).
3. Andersen, O. K., Phys. Rev. B: Solid State [3] 12, 3060 (1975).
4. Apai, G., Hamilton, J. F., Stöhr, J., and Thompson, A., Phys. Rev. Lett. 43, 165 (1979).
5. Apte, M. Y., Mandé, C., and Suchet, J. P., J. Chim. Phys. 79, 325 (1982).
6. Ashley, C. A., and Doniach, S., Phys. Rev. B: Solid State [3] 11, 1279 (1975).
7. Azároff, L. V., ed., "X-Ray Spectroscopy." McGraw-Hill, New York, 1974.
8. Azároff, L. V., Rev. Mod. Phys. 35, 1012 (1963).
9. Azároff, L. V., and Pease, D. M., in "X-Ray Spectroscopy" (L. V. Azároff, ed.), p. 284. McGraw-Hill, New York, 1974.
10. Azároff, L. V., and Pease, D. M., in "Advances in X-Ray Spectroscopy" (C. Bonnelle and C. Mandé, eds.), Chapter 2. Pergamon, Oxford, 1982.
11. Baddour, R. F., and Deibert, M., J. Phys. Chem. 70, 2173 (1966).
12. Baetzold, R. C., Surf. Sci. 36, 123 (1972).
13. Bair, R. A., and Goddard, W. A., Phys. Rev. B: Condens. Matter [3] 22, 2767 (1980).
14. Baker, R. T. K., Prestridge, E. B., and Garten, R. L., J. Catal. 56, 390 (1979); 59, 293 (1979).
14a. Balerna, A., Bernieri, E., Picozzi, P., Reale, A., Santucci, S., Burattini, E., and Mobilio, S., Phys. Rev. B: Condens. Matter [3] 31, 5058 (1985).
15. Balzarotti, A., Bianconi, A., Burattini, E., Grandolfo, M., Habel, R., and Piacentini, M., Phys. Status Solidi B 63, 77 (1974).
16. Balzarotti, A., De Crescenzi, M., and Incoccia, L., Phys. Rev. B: Condens. Matter [3] 25, 6349 (1982).
17. Barinskii, R. L., J. Struct. Chem. (English Transl.) 8, 805 (1967).
18. Bart, J. C. J., and Vlaic, G., Adv. Catal., in press (1986).
18a. Bassi, I. W., Lytle, F. W., and Parravano, G., J. Catal. 42, 139 (1976).
19. Batsanov, S. S., and Ovsyannikova, I. A., Khim. Svyaz Poluprovodn. Termodin. Akad. Nauk Beloruss. p. 93 (1966); C.A. 67, 37755j (1967).
20. Batsanov, S. S., and Ovsyannikova, I. A., in "Chemical Bonds in Semiconductors and Thermodynamics" (N. N. Sirota, ed.), p. 65. Consultant's Bureau, New York, 1968.

21. Baur, K. A., and Raff, U., *Helv. Phys. Acta* **45**, 765 (1972).
22. Beeck, O., *Discuss. Faraday Soc.* **8**, 118 (1950).
23. Beeman, W. W., and Bearden, J. A., *Phys. Rev.* **61**, 455 (1942).
24. Beeman, W. W., and Friedman, R. M., *Phys. Rev.* **56**, 392 (1939)
25. Beeman, W. W., and Bearden, J. A., *Phys. Rev.* **61**, 455 (1942).
26. Belli, M., Scafati, A., Bianconi, A., Mobilio, S., Palladino, L., Reale, A., and Burattini, E., *Solid State Commun.* **35**, 355 (1980).
27. Bergengren, J., *Z. Phys.* **3**, 247 (1920).
28. Best, P. E., *J. Chem. Phys.* **44**, 3248 (1966).
29. Bianconi, A., *Surf. Sci.* **89**, 41 (1979).
30. Bianconi, A., *in* "EXAFS for Inorganic Systems" (C. D. Garner and S. S. Hasnain, eds.), Daresbury Lab. Rep. DL/SCI/R17, p. 13. Daresbury (U.K.), 1981.
31. Bianconi, A., and Bauer, R. S., *Surf. Sci.* **99**, 76 (1980).
32. Bianconi, A., Dell'Ariccia, M., Giovannelli, A., Burattini, E., Cavallo, N., Patteri, P., Pancini, E., Carlini, C., Ciardelli, F., Papeschi, D., Pertici, P., Vitulli, G., Dalba, G., Fornasini, F., Mobilio, S., and Palladino, L., *Chem. Phys. Lett.* **90**, 257 (1982).
33. Bianconi, A., Doniach, S., and Lublin, D., *Chem. Phys. Lett.* **59**, 121 (1978).
34. Bianconi, A., Incoccia, L., and Stipcich, S., eds., "EXAFS and Near Edge Structure." Springer-Verlag, Berlin and New York, 1983.
35. Bianconi, A., Modesti, S., Campagna, M., Fischer, K., and Stizza, S., *J. Phys. C* **14**, 4737 (1981).
36. Bianconi, A., Petersen, H., Brown, F. C., and Bachrach, R. Z., *Phys. Rev. A* **17**, 1907 (1978).
37. Blair, R. A., and Goddart, W. A., *Phys. Rev. B: Condens. Matter* [3] **22**, 2767 (1980).
38. Blumberg, W. E., Peisach, J., and Powers, L., unpublished results (1980).
39. Boehm, G., Faessler, A., and Rittmayer, G., *Naturwissenschaften* **41**, 187 (1954); *Z. Naturforsch. B* **9B**, 509 (1954).
40. Böke, K., *Z. Phys. Chem* [N.S.] **10**, 45, 59, 326 (1957).
41. Boland, J. J., Crane, S. E., and Baldeschwieler, J. D., *J. Chem. Phys.* **77**, 142 (1982).
42. Bonham, R. A., *in* "Electron Spectroscopy, Theory, Techniques, and Applications" (C. R. Bonnelle and A. D. Baker, eds.), Vol. 3. Academic Press, New York, 1979.
43. Bonnelle, C., and Mandé, C., eds., "Advances in X-Ray Spectroscopy." Pergamon, Oxford, 1982.
44. Boudart, M., *J. Am. Chem. Soc.* **72**, 1040 (1950).
45. Boudart, M., Dalla Betta, R. A., and Ng, C. F., SSRL (Stanford Synchr. Radiation Lab.) Proposal No. 598Mp, p. VII-40 (1983).
46. Boudart, M., and Meitzner, G., in Ref. *125*, p. 217.
47. Brown, F. C., *in* "Synchrotron Radiation Research" (H. Winick and S. Doniach, eds.), p. 61. Plenum, New York, 1980.
48. Brown, F. C., *Solid State Phys.* **29**, 1 (1974).
48a. Brown, G. S., and Doniach, S., *in* "Synchrotron Radiation Research" (H. Winick and S. Doniach, eds.), p. 353. Plenum, New York, 1980.
49. Brown, J. M., Powers, L., Kincaid, B., Larrabee, J. A., and Spiro, T. G., *J. Am. Chem. Soc.* **102**, 4210 (1980).
50. Brown, M., Peierls, R. E., and Stern, E. A., *Phys. Rev. B: Solid State* [3] **15**, 738 (1977).
51. Brown, N. M. D., McNonagle, J. B., and Greaves, G. N., *J. Chem. Soc. Faraday Trans. 1* **80**, 589 (1984).
52. Bunker, G., and Stern, E. A., *Phys. Rev. Lett.* **52**, 1990 (1984).
52a. Cahen, D., and Ibers, J. A., *J. Catal.* **31**, 369 (1973).
53. Carlson, T. A., Krause, M. O., and Moddeman, W. E., *J. Phys. (Orsay, Fr)* **32**, C4-76 (1971).

54. Castner, D. G., Watson, P. R., and Dimpfl, W. L., SSRL (Stanford Synchr. Radiation Lab.) Act. Rep. Proposal No. 628M, P. VII-57 (1983); see also *ACS Symp. Ser.* **288**, 144 (1985).

55. Cauchois, Y., "Les spectres de rayons X et la structure électronique de la matière." Gauthier-Villars, Paris, 1948.

56. Cauchois, Y., and Mott, M. F., *Philos. Mag.* [7] **40**, 1260 (1949).

57. Chan, S. I., Hu, V. W., and Gamble, R. C., *J. Mol. Struct.* **45**, 239 (1978).

58. Chien, S. H., Shelimov, B. N., Resasco, D. E., Lee, E. H., and Haller, G. L., *J. Catal.* **77**, 301 (1982).

59. Chiu, N.-S., Bauer, S. H., and Johnson, M. F. L., *J. Catal.* **89**, 226 (1984).

60. Chung, Y.-W., Xiong, G., and Kao, C.-C., *J. Catal.* **85**, 237 (1984).

61. Citrin, P. H., Wertheim, G. K., and Schlüter, M., *Phys. Rev. B: Condens. Matter* [3] **20**, 3067 (1979).

62. Clausen, B. S., Topsøe, H., Candia, R., Villadsen, J., Lengeler, B., Als-Nielsen, J., and Christensen, F., *J. Phys. Chem.* **85**, 3868 (1981).

63. Cocco, G., Enzo, S., Fagherazzi, G., Schiffini, L., Bassi, I. W., Vlaic, G., Galvagno, S., and Parravano, G., *J. Phys. Chem.* **83**, 2527 (1979).

64. Cotton, F. A., and Ballhausen, C. J., *J. Chem. Phys.* **25**, 617 (1956).

65. Cotton, F. A., and Hanson, H. P., *J. Chem. Phys.* **25**, 619 (1956).

66. Cotton, F. A., and Hanson, H. P., *J. Chem. Phys.* **28**, 83 (1958).

67. Cramer, S. P., Eccles, T. K., Kutzler, F., Hodgson, K. O., and Mortenson, L. E., *J. Am. Chem. Soc.* **98**, 1287 (1976).

68. Cramer, S. P., and Hodgson, K. O., *Prog. Inorg. Chem.* **25**, 1 (1979).

69. Dalla Betta, R. A., Boudart, M., Gallezot, P., and Weber, R. S., *J. Catal.* **69**, 514 (1981).

70. Dalla Betta, R. A., Piken, A. G., and Shelef, M., *J. Catal.* **40**, 173 (1975).

71. Dautzenberg, F. M., and Wolters, H. B. M., *J. Catal.* **51**, 26 (1978).

72. Davis, L. C., and Feldkamp, L. A., *Phys. Rev. B: Condens. Matter* [3] **23**, 4269 (1981).

73. Davoli, I., Stizza, S., Benfatto, M., Gzowskii, O., Murawski, L., and Bianconi, A., *in* "EXAFS and Near Edge Structure" (A. Bianconi, L. Incoccia, and S. Stipcich, eds.), p. 162. Springer-Verlag, Berlin and New York, 1983.

73a. Davoli, I., Stizza, S., Bianconi, A., Benfatto, M., Furlani, C., and Sessa, V., *Solid State Commun.* **48**, 475 (1983).

74. Dehmer, J. L., *J. Chem. Phys.* **56**, 4496 (1972).

75. Dehmer, J. L., and Dill, D., *Phys. Rev. Lett.* **35**, 213 (1975).

76. Dehmer, J. L., and Dill, D., *J. Chem. Phys.* **65**, 5327 (1976).

77. Delk, F. S., II, and Vävere, A., *J. Catal.* **85**, 380 (1984).

78. Dill, D., and Dehmer, J. L., *J. Chem. Phys.* **61**, 692 (1974).

79. Doniach, S., Berding, M., Smith, T., and Hodgson, K. O., in Ref. *125*, p. 33.

80. Doniach, S., Eisenberger, P., and Hodgson, K. O., *in* "Synchrotron Radiation Research" (H. Winick and S. Doniach, eds.), p. 425. Plenum, New York, 1980.

81. Dow, J. D., *Comments Solid State Phys.* **6**, 71 (1975).

82. Dow, J. D., Robinson, J. E., and Carver, T. R., *Phys. Rev. Lett.* **31**, 759 (1973).

83. Durham, P. J., Pendry, J. B., and Hodges, C. H., *Solid State Commun.* **38**, 159 (1981).

84. Durham, P. J., Pendry, J. B., and Hodges, C. H., *Comput. Phys. Commun.* **25**, 193 (1982).

85. Emili, M., Incoccia, L., Mobilio, S., Guglielmi, M., and Fagherazzi, G., in Ref. *125*, p. 317.

86. Ermakov, Yu., I., Kuznetsov, B. N., Ovsyannikova, I. A., Ryndin, Yu. A., and Startsev, A. N., *in* "Heterogeneous Catalysis" (D. Shopov, A. Andreev, A. Palazov, and I. Petrov, eds.), Part 1, p. 163. Publ. House of the Bulgarian Acad. of Sciences, Sofia, 1979.

87. Fabian, D. J., ed., "Soft X-Ray Band Spectra and the Electronic Structure of Metals and Materials." Academic Press, New York, 1969.

88. Falicov, L. M., Hanke, W., and Maple, M. P., eds., "Valence Fluctuations in Solids." North-Holland Publ., Amsterdam, 1981.

89. Fano, U., and Cooper, J. W., *Rev. Mod. Phys.* **40**, 441 (1968).

90. Fischer, D. W., and Baun, W. L., *J. Appl. Phys.* **39**, 4757 (1968).

91. Fransen, T., Van der Meer, O., and Mars, P., *J. Catal.* **42**, 79 (1976).

92. Fricke, H., *Phys. Rev.* **16**, 202 (1920).

93. Friedman, R. M., Freeman, J. J., and Lytle, F. W., *J. Catal.* **55**, 10 (1978).

94. Fujikawa, T., *J. Phys. Soc. Jpn.* **50**, 1321 (1981).

95. Fujikawa, T., *J. Phys. Soc. Jpn.* **51**, 2619 (1982).

96. Fujikawa, T., Matsuura, T., and Kuroda, H., *J. Phys. Soc. Jpn.*, **52**, 905 (1983).

97. Fukushima, T., Katzer, J. R., Sayers, D. T., and Cook, J., *Proc. Int. Catal. Congr., 7th, 1980* p. 79 (1981).

98. Gallezot, P., *Catal. Rev.—Sci. Eng.* **20**(1), 121 (1979).

99. Gallezot, P., Weber, R., Dalla Betta, R. A., and Boudart, M., *Z. Naturforsch. A* **34A**, 40 (1979).

99a. Garbassi, F., Bart, J. C. J., Tassinari, R., Vlaic, G., and Lagarde, P., *J. Catal.* **98**, 317 (1986).

100. Geere, R. G., *Proc. Int. Conf. Prog. X-Ray Stud. Synchrotron Radiat.,* Strasbourg (1985).

101. Glen, G. L., and Dodd, C. G., *J. Appl. Phys.* **39**, 5372 (1968).

102. Goulon, J., Esselin, C., Friant, P., Berthe, C., Muller, J. F., Poncet, J. L., Guilard, R., Escalier, J. C., and Neff, B., *Proc. Int. Conf. X-Ray Stud. Synchrotron Radiat.,* Strasbourg (1985).

103. Goulon, J., Retournard, A., Friant, P., Goulon-Ginet, C., Berthe, C., Müller, J. F., Poncet, J. L., Guilard, R., Escalier, J. C., and Neff, B., *J. Chem. Soc., Dalton Trans.* p. 1095 (1984).

104. Goulon, J., Goulon-Ginet, C., Cortes, R., and Dubois, J. M., *J. Phys.* **43**, 539 (1982).

105. Greaves, G. N., Durham, P. J., Diakun, G., and Quinn, P., *Nature (London)* **294**, 139 (1981).

105a. Greegor, R. B., Lytle, F. W., Sandström, R. D., Wong, J., and Schultz, P., *J. Non-Cryst. Solids* **55**, 27 (1983).

106. Grim, S., and Matienzo, L. J., *Inorg. Chem.* **14**, 1014 (1975).

107. Grunes, L. A., *Phys. Rev. B: Condens. Matter* [3] **27**, 2111 (1983).

108. Gupta, R. P., Freeman, A. J., and Dow, J. D., *Phys. Lett. A* **59A**, 226 (1976).

109. Gupta, R. P., and Freeman, A. J., *Phys. Rev. Lett.* **36**, 1194 (1976).

110. Gurman, S. J., *J. Phys. C* **16**, 2987 (1983).

111. Gurman, S. J., and Pettifer, R. F., *Philos. Mag. B* **40**, 345 (1979).

112. Hahn, J. E., Scott, R. A., Hodgson, K. O., Doniach, S., Desjardins, S. R., and Solomon, E. I., *Chem. Phys. Lett.* **88**, 595 (1982).

113. Hannoyer, B., Dürr, J., Calas, G., Petiau, J., and Lenglet, M., *in* "Solid State Chemistry 1982" (R. Metselaar, H. J. M. Heyligers and J. Schoonman, eds.), p. 551. Elsevier, Amsterdam, 1982.

114. Hänsch, W., and Ekardt, W., *Phys. Rev. B: Condens. Matter* [3] **24**, 5497 (1981).

115. Hänsch, W., and Minnhagen, P., *Phys. Rev. B: Condens. Matter* [3] **26**, 2772 (1982).

116. Hanson, H. P., and Beeman, W. W., *Phys. Rev.* **76**, 118 (1949).

117. Hanson, H. P., *Dev. Appl. Spectrosc.* **2**, 254 (1963).

118. Hanson, H. P., and Knight, J. R., *Phys. Rev.* **102**, 632 (1956).

119. Hanson, H. P., and Milligan, W. O., *J. Phys. Chem.* **60**, 1144 (1956).

120. Hedin, L., *J. Phys. (Orsay Fr.)* **39**, C4-103 (1978).

121. Hedin, L., and Lundqvist, S., *Solid State Phys.* **23**, 1 (1969).

122. Hedman, B., Penner-Hahn, J. E., and Hodgson, K. O., in Ref. *125*, p. 64.

123. Herber, R., *Adv. Chem. Ser.* **68**, 8 (1967).

124. Herman, R. G., Klier, K., Simmons, G. W., Finn, B. P., and Bulko, J. B., *J. Catal.* **56**, 407 (1979).
125. Hodgson, K. O., Hedman, B., and Penner-Hahn, J. E., eds., "EXAFS and Near Edge Structure III." Springer-Verlag, Berlin and New York, 1984.
126. Horsley, J. A., *J. Chem. Phys.* **76**, 1451 (1982).
127. Horsley, J. A., *Proc. Int. Conf. Prog. X-Ray Stud. Synchrotron Radiat.*, Strasberg (1985)
127a. Horsley, J. A., and Lytle, F. W., *ACS Symp. Ser.* **298**, 10 (1986).
127b. Hubbell, J. H., and Viegele, W. J., *NBS Tech. Note (U.S.)* **901** (1976).
128. Huizinga, T., and Prins, R., *J. Phys. Chem.* **85**, 2156 (1981).
128a. Huizinga, T., and Prins, R., "Studies in Surface Science and Catalysis," Vol. II, p. 11. Elsevier, Amsterdam, 1982.
129. Huizinga, T., Van't Blik, H. F. J., Vis, J. C., and Prins, R., *Surf. Sci.* **135**, 580 (1983).
130. Hulbert, S. L., Bunker, B. A., Brown, F. C., and Pianetta, P., *Phys. Rev., B: Condens. Matter* **B30**, 2120 (1984).
131. Imelik, B., Naccache, C., Condurier, G., Praliaud, H., Meriaudeau, P., Gallezot, P., Martin, G. A., and Vedrine, J. C., eds., "Metal–Support and Metal-Additive Effects in Catalysis." Elsevier, Amsterdam, 1982.
132. Johnson, K. H., *Adv. Quantum Chem.* **7**, 143 (1973).
133. Jørgensen, C. K., *in* "Advances in X-Ray Spectroscopy" (C. Bonnelle and C. Mandé, eds.), p. 225. Pergamon, Oxford, 1982.
134. Katzer, J. R., and Sayers, D. E., *Symp. Catal. Mater., Mater. Res. Soc., 1978* Paper C-9 (1978).
135. Kawata, S., *Philos. Mag. B* **49**, 185 (1984).
136. Kawata, S., and Maeda, K., *J. Phys. F* **3**, 167 (1973),
137. Kawata, S., and Maeda, K., *J. Phys. C* **11**, 2391 (1978).
138. Kawata, S., and Maeda, K., *Philos. Mag. B* **50**, 731 (1984).
139. Keeling, R. O., *Bull. Am. Phys. Soc.* [2] **1**, 138 (1956).
140. Keeling, R. O., *J. Chem. Phys.* **31**, 279 (1959).
141. Keeling, R. O., *Dev. Appl. Spectrosc.* **2**, 263 (1962).
142. Kelley, M. J., Freed, R. L., Swartzfager, D. G., Katzer, J. R., and Short, D. R., *Prepr. Div. Pet. Chem., Am. Chem. Soc.* **26**, 407 (1981).
143. Kim, K. S., *J. Electron Spectrosc. Relat. Phenom.* **3**, 217 (1974).
144. Knapp, G. S., Veal, B. W., Pan, H. K., and Klippert, T., *Solid State Commun.* **44**, 1343 (1982).
145. Kondawar, V. K., and Mandé, C., *X-Ray Spectrom.* **5**, 2 (1976).
145a. Kondawar, V. K., and Mandé, C., *J. Phys. C* **9**, 1351 (1976).
146. Kossel, W., *Z. Phys.* **1**, 119 (1920); **2**, 470 (1920).
147. Kozlowski, R., Pettifer, R. F., and Thomas, J. M., *J. Phys. Chem.* **87**, 5176 (1983).
148. Kozlowski, R., Pettifer, R. F., and Thomas, J. M., *J. Chem. Soc., Chem. Commun.* p. 438 (1983).
149. Kozlowski, R., Pettifer, R. F., and Thomas, J. M., *in* "EXAFS and Near Edge Structure" (A. Bianconi, L. Incoccia, and S. Stipcich, eds.), p. 313. Springer-Verlag, Berlin and New York, 1983.
150. Krause, M. O., and Oliver, J. H., *J. Phys. Chem. Ref. Data* **8**, 329 (1979).
151. Kravtsova, E. A., Erenburg, S. B., Mazalov, L. N., Kochubei, D. I., Babushok, O. P., and Zamaraev, K. I., *Kinet. Katal.* **24**, 969 (1983).
151a. Krill, G., *Proc. Int. Conf. Prog. X-Ray Stud. Synchrotron Radiat.*, Strasbourg (1985).
152. Krishna, V., Prasad, J., and Nigam, H. L., *Inorg. Chim. Acta* **20**, 193 (1976).
153. Kunimori, K., Matsui, S., and Uchijima, T., *J. Catal.* **85**, 253 (1984).
154. Kunzl, V., *Collect. Czech. Commun.* **4**, 213 (1932).

292 JAN C. J. BART

155. Kutzler, F. W., Natoli, C. R., Misemer, D. K., Doniach, S., and Hodgson, K.O., J. Chem. Phys. **73**, 3274 (1980).
156. Kutzler, F. W., Scott, R. A., Berg, J. M., Hodgson, K. O., Doniach, S., Cramer, S. P., and Chang, C. H., J. Am. Chem. Soc. **103**, 6083 (1981).
157. "Landolt-Börnstein Tables" 6th ed., Vol. I, Part 4, p. 769 (by A. Faessler). Springer-Verlag, Berlin and New York, 1955.
158. Launois, H., Petiau, J., Bondot, P., Loupias, G., and Gautier, F., LURE (Lab. Utilisation Rayonnement Electromagn., Orsay, Fr.) Act. Rep. (1982).
159. Leapman, R. D., and Grunes, L. A., Phys. Rev. Lett. **45**, 397 (1980).
160. Leapman, R. D., Grunes, L. A., and Fejes, P. L., Phys. Rev. B: Condens. Matter **26**, 614 (1982).
161. Lee, P. A., Citrin, P. H., Eisenberger, P., and Kincaid, B. M., Rev. Mod. Phys. **53**, 769 (1981).
162. Lee, P. A., and Pendry, J. B., Phys. Rev. B: Solid State [3] **11**, 2795 (1975).
163. Lengeler, B., and Zeller, R., Solid State Commun. **51**, 889 (1984).
164. Le Normand, F., Hilaire, L., Bernhardt, P., Krill, G. and Maire, G., LURE (Lab. Utilisation Rayonnement Electromagn., Orsay, Fr.) Act. Rep. 1983-1985, p. 141 (1985).
165. Lewis, P. H., J. Phys. Chem. **64**, 1103 (1960).
166. Lewis, P. H., J. Phys. Chem. **66**, 105 (1962).
167. Lewis, P. H., J. Phys. Chem. **67**, 2151 (1963).
168. Lewis, P. H., J. Catal. **11**, 162 (1968).
169. Lewis, P. H., J. Catal. **69**, 511 (1981).
170. Lipsch, J. M. J. G., Ph.D. Thesis, Eindhoven Institute of Technology (1968).
171. Lipsch, J. M. J. G., and Schuit, G. C. A., J. Catal. **15**, 174 (1969).
172. Loeffler, B. M., Burns, R. G., Tossell, J. A., Vaughan, D. J., and Johnson, K. K., Geochim. Cosmochim. Acta, Suppl. **5** (3), 3007 (1974).
173. Longe, P., in "Advances in X-Ray Spectroscopy" (C. Bonnelle and C. Mandé, eds.), Chapter 15. Pergamon, Oxford, 1982.
174. Lytle, F. W., J. Catal. **43**, 376 (1976).
175. Lytle, F. W., Greegor, R. B., Sinfelt, J., and Via, G. H., J. Mol. Catal. **20**, 389 (1983).
176. Lytle, F. W., Sayers, D. E., and Moore, E. B., J. Appl. Phys. Lett. **24**, 45 (1974).
177. Lytle, F. W., Wei, P. S. P., and Bassi, I. W., SSRL (Stanford Synchr. Radiation Lab.) User's Meet., Oct. 27-28 (1977).
178. Lytle, F. W., Wei, P. S. P., Greegor, R. B., Via, G. H., and Sinfelt, J. H., J. Chem. Phys. **70**, 4849 (1979).
179. Mahan, G. D., Phys. Rev. **163**, 612 (1967).
180. Mahan, G. D., Solid State Phys. **29**, 75 (1974).
181. Mahan, G. D., Phys. Rev. B: Condens. Matter [3] **21**, 1421 (1980).
181a. Mahan, G. D., Phys. Rev. B: Condens. Matter [3] **25**, 5021 (1982).
182. Mandé, C. M., and Sapre, B., in Ref. 43.
183. Mansour, A. N., Cook, J. W., and Sayers, D. E., J. Phys. Chem. **88**, 2330 (1984).
184. Mansour, A. N., Cook, J. W., Sayers, D. E., Emrich, R. J., and Katzer, J. R., J. Catal. **89**, 462 (1984).
185. Mansour, A. N., Cook, J. W., Sayers, D. E., and Stern, E. A., J. Chem. Phys. (in press).
186. Mansour, A. N., Sayers, D. E., Cook, J. W., Short, D. R., Shannon, R. D., and Katzer, J. R., J. Phys. Chem. **88**, 1778 (1984).
187. Massoth, F. E., Adv. Catal. **27**, 265 (1978).
188. Materlik, G., Müller, J. E., and Wilkins, J. W., Phys. Rev. Lett. **50**, 267 (1983).
189. Mattheis, L. F., and Dietz, R. E., Phys. Rev. B: Condens. Matter [3] **22**, 1663 (1980).
190. May, J. W., Adv. Catal. **21**, 151 (1970).

191. Maylotte, D. H., Wong, J., Peters, R. L. St., Lytle, F. W., and Greegor, R. B., *Science* **214**, 554 (1981).
192. McCaffrey, J. W., and Papaconstantopoulos, D. A., *Solid State Commun.* **14**, 1055 (1974).
193. Medema, J., Van Stam, C., De Beer, V. H. J., Konings, A. J. A., and Koningsberger, D. C., *J. Catal.* **53**, 386 (1978).
194. Mehta, S., Simmons, G. W., Klier, K., and Herman, R. G., *J. Catal.* **57**, 339 (1979).
195. Meisel, A., *Phys. Status Solidi* **10**, 365 (1965).
196. Meisel, A., Roentgenspektren Chem. Bindung. Akad. Verkagsges. Geest u. Portig, Leipzig (1977).
197. Meisel, A., and Keilacker, H., *Z. Phys. Chem.* **247**, 32 (1971).
198. Mignolet, J. C. P., *J. Chim. Phys.* **54**, 19 (1957).
199. Mitchell, G., and Beeman, W. W., *J. Chem. Phys.* **20**, 1298 (1952).
200 Montano, P. A., Purdum, H., Shenoy, G. K., Morrison, T. I., and Schulze, W., *Surf. Sci.* **156**, 228 (1985).
201. Mooi, J., and Selwood, P. W., *J. Am. Chem. Soc.* **74**, 2461 (1952).
202. Moruzzi, V. L., Janek, J. F., and Williams, A. R., "Calculated Electronic Properties of Metals." Pergamon, Oxford, 1978.
203. Mott, N. F., *Proc. R. Soc. London* **62**, 416 (1949).
204. Müller, J. E., in Ref. *125*, p. 7.
205. Müller, J. E., Jepsen, O., Anderson, O. K., and Wilkins, J. W., *Phys. Rev. Lett.* **40**, 720 (1978).
206. Müller, J. E., Jepsen, O., and Wilkins, J. W., *Solid State Commun.* **42**, 365 (1982).
207. Müller, J. E., and Schaich, W. L., *Phys. Rev. B: Condens. Matter* [3] **27**, 6489 (1983).
208. Müller, J. E., and Wilkins, J. W., *Phys. Rev. B: Condens. Matter* **29**, 4331 (1984).
209. Mustard, D. G., and Bartholomew, C. H., *J. Catal.* **67**, 186 (1981).
210. Natoli, C. R., Misemer, D. K., Doniach, S., and Kutzler, F. W., *Phys. Rev. A* **22**, 1104 (1980).
211. Neddermeyer, H., *Phys. Rev. B: Solid State* [3] **13**, 2411 (1976).
212. Negel, D. J., Papaconstantopoulos, D. A., McCaffrey, J. W., and Criss, J. W., *in* "X-Ray Spectra and Electronic Structure of Matter" (A. Faessler and G. Wiech, eds.). Academic Press, New York, 1973.
213. Nicolau, C. S., and Thom, H. G., *Z. Anorg. Allg. Chem.* **303**, 133 (1960).
214. Nigam, H. L., and Srivastava, U. C., *Inorg. Chim. Acta* **5**, 338 (1971).
215. Nigam, H. L., and Srivastava, U. C., *J. Chem. Soc., Chem. Commun.* p. 761 (1971).
216. Nordfors, B., *Ark. Fys.* **18**, 37 (1960).
217. Nordstrand, R. A. Van, *Adv. Catal.* **12**, 149 (1960).
218. Norman, D., Stöhr, J., Jaeger, R., Durham, P. J., and Pendry, J. B., *Phys. Rev., Lett.* **51**, 2052 (1983).
219. Nozières, P., and De Dominicis, C. T., *Phys. Rev.* **178**, 1097 (1969).
220. Ohtaka, K., and Tanabe, Y., *Phys. Rev. B: Condens. Matter* [3] **28**, 6833 (1983).
221. Ohtaka, K., and Tanabe, Y., *Phys. Rev. B: Condens. Matter* [3] **30**, 4235 (1984).
222. Oliveira, L. N., and Wilkins, J. W., *Phys. Rev. B: Condens. Matter* [3] **24**, 4863 (1981).
223. Onuferko, J. H., Short, D. R., and Kelley, M. J., *Vacuum* **33**, 856 (1983).
224. Ovsyannikova, I. A., Batsanov, S. S., Nasonova, L. I., Batsanova, L. R., and Nekrasova, E. A., *Bull. Acad. Sci. USSR, Phys. Ser. (Engl. Transl.)* **31**, 936 (1967).
225. Ovsyannikova, I. A., and Nasonova, L. I., *Zh. Strukt. Khim.* **11**, 548 (1970).
226. Papaconstantopoulos, D. A., *Phys. Rev. Lett.* **31**, 1050 (1973).
227. Parham, T. G., and Merrill, R. P., *J. Catal.* **85**, 295 (1984).
228. Parham, T. G., and Merrill, R. P., *Proc. Int. Congr. Catal., 8th, 1984* p. II-421 (1984).
229. Parratt, L. G., *Phys. Rev.* **56**, 295 (1939).

230. Parratt, L. G., *Rev. Mod. Phys.* **31**, 616 (1959).
231. Pauling, L., *Proc. R. Soc. London, Ser. A* **196**, 343 (1949).
232. Pendry, J. B., "Low Energy Electron Diffraction." Academic Press, London, 1974.
233. Pendry, J. B., *Comments Solid State Phys.* **10**, 219 (1983).
234. Penn, D. R., Girvin, S. M., and Mahan, G. D., *Phys. Rev. B: Condens. Matter* [3] **24**, 6971 (1981).
235. Peuckert, M., Keim, W., Storp, S., and Weber, R. S., *J. Mol. Catal.* **20**, 115 (1983).
236. Ponec, V., *Proc. Int. Vac. Congr., 9th* [*Int. Conf. Solid Surf., 5th*], *1983* p. 25 (1983).
237. Poole, C. P., and MacIver, D. S., *Adv. Catal.* **17**, 223 (1967).
237a. Prasad, J., Krishna, V., and Nigam, H. L., *J. Chem. Soc., Dalton Trans.* p. 2413 (1976).
237b. Prins, R., and Koningsberger, D., eds., "X-Ray Absorption Spectroscopy and Applications of EXAFS, SEXAFS and XANES." Wiley, New York, 1985.
238. Rao, C. N. R., Sarma, D. D., and Hegde, M. S., *Proc. R. Soc. London, Ser. A* **370**, 269 (1980).
239. Rao, C. N. R., Thomas, J. M., Williams, B. G., and Sparrow, T. G., *J. Phys. Chem.* **88**, 5769 (1984).
240. Ratnasamy, P., and Sivasanker, S., *Cat. Rev.—Sci. Eng.* **22**, 401 (1980).
241. Renouprez, A., Fouilloux, P., and Moraweck, B., *in* "Growth and Properties of Metal Clusters" (J. Bourdon, ed.), p. 421. Elsevier, Amsterdam, 1980.
242. Richtmyer, F. K., Barnes, S. W., and Ramberg, R., *Phys. Rev.* **46**, 843 (1934).
243. Roe, A. L., Schneider, D. J., Mayer, R. J., Pyrz, J. W., Widom, J., and Que, L., *J. Am. Chem. Soc.* **106**, 1676 (1984).
244. Ross, J., and Andres, R. P., *Surf. Sci.* **106**, 11 (1981).
245. Salem, S. I., Chang, C. N., and Nash, T. J., *Phys. Rev. B: Condens. Matter* [3], **18**, 5168 (1978).
246. Sánchez, J., Dalla Betta, R. A., and Boudart, M., SSRL (Stanford Synchr. Radiation Lab.) Act. Rep. Proposal No. 696M, p. VII–93 (1983).
247. Sanner, V. H., Ph.D. Thesis, Uppsala University, Uppsala, Sweden (1941).
248. Sapre, V. P., and Mandé, C., *J. Phys. C* **5**, 793 (1972).
249. Sawada, M., Tsutsumi, K., Shiraiwa, T., Ishimura, T., and Obashi, M., *Annu. Rep. Sci. Works, Fac. Sci., Osaka Univ.* **7**, 1 (1969).
250. Sawatzky, G. A., *Proc. Int. Conf. Prog. X-Ray Stud. Synchrotron Radiat., Strasbourg* (1985).
251. Sayers, D. E., Bazin, D., Dexpert, H., Jucha, A., Dartyge, E., Fontaine, A., and Lagarde, P., *in* Ref. *125*, p. 209.
252. Schaich, W. L., *Phys. Rev. B: Solid State* [3] **8**, 4028 (1973).
253. Schaich, W. L., *Phys. Rev. B: Condens. Matter* [3] **29**, 6513 (1984).
254. Schaich, W. L., *in* Ref. *125*, p. 2.
255. Schwab, G.-M., Block, J., and Schultze, D., *Angew. Chem.* **71**, 101 (1959); *Naturwissenschaften* **44**, 22 (1957); **46**, 13 (1959).
256. Seka, W., and Hanson, H. P., *J. Chem. Phys.* **50**, 344 (1969).
257. Sexton, B. A., Hughes, A. E., and Foger, K., *J. Catal.* **77**, 85 (1982).
258. Sham, T. K., *J. Am. Chem. Soc.* **105**, 2269 (1983).
259. Short, D. R., Khalid, S. M., Katzer, J. R., and Kelley, M. J., *J. Catal.* **72**, 288 (1981).
260. Short, D. R., Mansour, A. N., Cook, J. W., Jr., Sayers, D. E., and Katzer, J. R., *J. Catal.* **82**, 299 (1983).
261. Shulman, R. G., Yafet, Y., Eisenberger, P., and Blumberg, W. E., *Proc. Natl. Acad. Sci. U.S.A.* **73**, 1384 (1976).
262. Sinfelt, J. H., *J. Catal.* **29**, 308 (1973).
263. Sinfelt, J. H., *Adv. Catal.* **23**, 91 (1973).
264. Sinfelt, J. H., *Prog. Solid State Chem.* **10** (2) 55 (1975).

265. Sinfelt, J. H., *Science* **195**, 641 (1977).
266. Sinfelt, J. H., Via, G. H., Lytle, F. W., and Greegor, R. B., *J. Chem. Phys.* **75**, 5527 (1981).
266a. Sonntag, B., *J. Phys. (Orsay, Fr.)* **39**, C4-9 (1978).
267. Sorokina, M. F., and Nemonov, S. A., *Bull. Acad. Sci. USSR, Phys. Ser. (Engl. Transl.)* **31**, 1039 (1967).
268. Spiro, C. L., Wong, J., Lytle, F. W., Greegor, R. B., Maylotte, D. H., and Lamson, S. H., *Science* **226**, 48 (1984).
269. Srivastava, K. S., and Kumar, V., *J. Phys. Chem. Solids* **42**, 275 (1981).
270. Srivastava, U. C., and Nigam, H. L., *Coord. Chem. Rev.* **9**, 275 (1973).
271. Stern, E. A., *Phys. Rev. B: Solid State* [3] **10**, 3027 (1974)
272. Stern, E. A., *AIP Conf. Proc.* **64** (1980).
273. Stern, E. A., *Phys. Rev. Lett.* **49**, 1353 (1982).
274. Stizza, S., Davoli, I., Tomellini, M., Marcelli, A., Bianconi, A., Gzowski, A., and Murawski, L., in Ref. *125*, p. 331.
275. Stobbe, M., *Ann. Phys. (Leipzig)* [5] **7**, 661 (1930).
276. Stöhr, J., Baberschke, K., Jaeger, R., Treichler, R., and Brennan, S., *Phys. Rev. Lett.* **47**, 381 (1981).
277. Stöhr, J., Gland, J. L., Eberhardt, W., Outka, D., Madix, R. J., Sette, F., Koestner, R. J., and Doebler, U., *Phys. Rev. Lett.* **51**, 2414 (1983).
278. Stöhr, J., and Jaeger, R., *Phys. Rev. B: Condens. Matter* [3] **26**, 4111 (1982).
279. Stults, B. R., and Friedman, R. M., *Symp. Catal. Mater., Mater. Res. Soc., 1978*, Paper C-5 (1978).
280. Suchet, J. P., *J. Phys. Chem. Solids* **21**, 156 (1961).
281. Suchet, J. P., *Phys. Status Solidi* **2**, 167 (1962).
282. Suchet, J. P., *J. Electrochem. Soc.* **124**, 30C (1977).
283. Sugiura, C., and Nakai, S., *Jpn. J. Appl. Phys.* **17**, Suppl. 17-2, 190 (1978).
284. Suhrmann, R., Wedler, G., and Gentsch, H., *Z. Phys. Chem.* [N.S.] **17**, 350 (1958).
285. Swarts, C. A., Dow, J. D., and Flynn, C. P., *Phys. Rev. Lett.* **43**, 158 (1979).
286. Szmulowicz, F., and Pease, D. M., *Phys. Rev. B: Solid State* [3] **17**, 3341 (1978).
286a. Szmulowicz, F., and Segall, B., *Phys. Rev. B: Condens. Matter* [3] **21**, 5628 (1980).
287. Tanabe, Y., and Ohtaka, K., *Phys. Rev. B: Condens. Matter.* [3] **29**, 1653 (1984).
288. Tauster, S. J., and Fung, S. C., *J. Catal.* **55**, 29 (1978).
289. Tauster, S. J., Fung, S. C., and Garten, R. L., *J. Am. Chem. Soc.* **100**, 170 (1978).
290. Tauster, S. J., Fung, S. C., Baker, R. T. K., and Horsley, J. A., *Science* **211**, 1121 (1981).
291. Templeton, D. H., and Templeton, L. K., *Acta Crystallogr., Sect. A* **A36**, 237 (1980).
292. Teo, B. K., in "EXAFS and Near-Edge Structure" (A. Bianconi, L. Incoccia, and S. Stripcich, eds.), p. 11. Springer-Verlag, Berlin and New York, 1983.
293. Teo, B. K., *J. Am. Chem. Soc.* **103**, 3990 (1981).
294. Topsøe, H., in "Surface Properties and Catalysis by Non Metals" (J. P. Bonnelle, B. Delmon, and E. Derouane, eds.), p. 329. Reidel Publ., Dordrecht, Netherlands, 1983.
295. Tri, T. M., Massardier, J., Gallezot, P., and Imelik, B., *Proc. Int. Congr. Catal., 7th, 1980* Part A, p. 226 (1981).
296. Tullius, T. D., Ph.D. Thesis, Stanford University, Stanford, California (1978).
297. Vainshtein, E., Shurakowski, E. A., and Stari, I. B., *Zh. Neorg. Khim.* **4**, 245 (1959).
298. Van der Laan, G., *Proc. Int. Conf. Prog. X-Ray Stud. Synchrotron Radiat., Strasbourg* (1985).
299. Van Nordstrand, R. A., *Adv. Catal.* **12**, 149 (1960).
300. Van Nordstrand, R. A., in "Handbook of X-Rays" (E. F. Kaelble, ed.), Chapter 43. McGraw-Hill, New York, 1965.
301. Vlaic, G., Bart, J. C. J., Cavigiolo, W., and Mobilio, J., *Chem. Phys. Lett.* **76**, 453 (1980).

301a. Vlaic G., Bart, J. C. J., Cavigiolo, W., Pianzola, B., and Mobilio, S., *J. Catal.* **96**, 314 (1985).

301b. Vogel W., *Surf. Sci.* **156**, 420 (1985).

302. von Barth, U., and Grossmann, G., *Solid State Commun.* **32**, 645 (1979).

303. von Barth, U., and Grossmann, G., *Phys. Scr.* **20**, No. 39 (1979).

304. Wakoh, S., and Kubo, Y., *Tech. Rep. Inst. Solid State Phys. (Univ.Tokyo)*, Ser. *A* No. 925 (1978) (unpublished).

305. Walton, R. A., *J. Catal.* **44**, 335 (1976).

306. Waychunas, G. A., and Brown, G. E., in Ref. *125*, p. 336.

307. Wei, P. S., and Lytle, F. W., *Phys. Rev. B: Condens. Matter* [3] **19**, 679 (1979).

308. Wendin, G., *Phys. Rev. Lett.* **53**, 724 (1984).

309. Williams, A. R., Kübler, J., and Gelatt, C. D., *Phys. Rev. B: Condens. Matter* [3] **19**, 6094 (1979).

310. Wolberg, A., and Roth, J. F., *J. Catal.* **15**, 250 (1969).

311. Wong, J., Lytle, F. W., Maylotte, D. H., and Via, G. H., *Fuel* (in press).

312. Wong, J., Lytle, F. W., Messmer, R. P., and Maylotte, D. H., *Phys. Rev. B: Condens. Matter* [3] **30**, 5596 (1984).

313. Yermakov, Yu. I., Kuznetsov, B. N., Ovsyannikova, I. A., Startsev, A. N., Erenburg, S. B., Sheromov, M. A., and Mironenko, L. A., *React. Kinet. Catal. Lett.* **7**, 309 (1977).

314. Yermakov, Yu. I., Kuznetsov, B. N., and Ryndin, Yu. A., *J. Catal.* **42**, 73 (1976).

315. Zaanen, J., Sawatzky, G. A., Fink, J., Speierl, W., and Fuggle, J. C., *Proc. Int. Conf. Prog. X-Ray Stud. Synchrotron Radiat., Strasbourg* (1985).

316. Zimkina, T. N., and Fomichev, V. A., *Dokl. Akad. Nauk SSSR* **169**, 1304 (1966).

Index